TROM PUBLISHING
PROUDLY PRESENTS

MARVIN N. CARR

POSITIVELY NEGATIVE

COPYRIGHT PAGE

Publisher: Trom Publications
5013 N. Stevens St.
Spokane, Washington 99205

Positively Negative:

The logo <u>Dragonfly</u> is the trade mark of Trom Publications.

Printed in the United States of America.

DEDICATION

This Novel Is Dedicated to

My Beloved Wife

Betty L. Carr

ACKNOWLEDGEMENTS

To Beverly F. Mahrt, Bev's Secretarial & Typesetting, Inc., Spokane, Washington for her patience and ability to transpose my hieroglyphics.

Reprinted by permission of The Putnam Publishing Group/Jeremy P. Tarcher, Inc. from *CROSS CURRENTS* by Robert O. Becker. Copyright© 1990 by Robert O. Becker.

Bayliner, U.S. Marine, Brunswick, and Olympic Center Boat.

Chilkoot Charlie's, Anchorage, Alaska

PROLOGUE I

Our planet earth spews magnetic lines of force from the North to the South Pole, not necessarily in that order. Our atmosphere extends approximately 250 miles above the earth.

The massive magnetic lines of force that envelop our world act as a protective shield from the cosmic rays. However, they can filter and enter where the magnetic field is induced. Our magnetic field also protects us from the powerful solar winds that flow around our earth.

The sun, with its thermonuclear furnace and massive explosions and other storms, dictates our weather conditions daily and seasonally. The sun discharges light and magnetic forces that pull and twist the earth's magnetic field. Sunspots are magnetic storms, one spot is as large as the size of our earth.

Our moon's magnetic field dictates the tides of our ocean and all bodies on earth. The Milky Way is the galaxy of which we are part and is made up of over four million stars. It is approximately 15 billion years old. Magnetic forces are magnets of our universe, and any major disruption causes turbulence that would envelop our earth.

Jupiter, the fifth in order from the sun, could be affected by the magnetic electrical field that is changing above earth, thus affecting the planet Uranus, which is the seventh planet from the sun. It could throw and hurl our sync.

Orion, 50,000 times brighter than our sun, is another example of how small we are in the spectrum of our galaxy. There are trillions of galaxies.

The continuous nuclear explosions on the sun are like billions of hydrogen bombs existing constantly. Collisions of atoms are in such disarray that polarity is changed from positive to negative. Protons and electrons, although not the same size, collide. The sun defuses elements in space that, in themselves, are magnetic, and they again collide in a non-random cycle. Upon colliding again and exploding, the mass reduces its up to ten-fold, but the magnetic field intensifies. Matter and antimatter explode, producing energy in many forms.

Many stars regenerate themselves by time and position. A small star is trapped when its magnetic forces are over and come from a larger star. If they do not collide, the smaller star usually transfers the mass of energy until its mass becomes overfed, at which time a spectacular nova explosion begins.

Death of stars which earth was at one time, are composed of only two components, hydrogen and helium, that continue to enhance life in our universe. New stars are approximately 90 percent hydrogen and 10 percent helium. As these gasses cool, the star gets brighter and condenses in size. Its matter is dispersed from its surface into space. It is blown into the universe. New stars show up regularly, are thermonuclear and are made up of loose gas.

As in the sun, the collapsing and reduction in size, its magnetic force increases. Its spent energy is released as light through space.

Later this new star will be known as a pulsating star. This is the changing of fuel and energy. The mass is reducing, but the nuclear fission enhances, as does the magnetic field. Some of these new stars make it. They wander through space as white stars. They have a grand gravity and magnetic field, but their cores are turning to iron, the last phase of nuclear fission-- their cores, still molten masses of hot compressed liquid iron.

Some stars have been exploding in a massive, uncontrollable fashion, that their mass is so highly condensed that a spoonful would weigh over a billion tons.

The Northern Lights (Aurora Borealis) is a projection of gases that have been expounded from the sun. The magnetic fields at the poles have less resistance to the sun's magnetic field; thus, the wavy, strange glow of seams of light are viewed in a haze of swept light of majestic beauty.

Our earth tilts is at 24.74 percent. Seventy-one percent of our earth's surface is made up of oceans. Our volcanic eruptions that spurt silicone and elements from the deep haunts of our earth, are hurled into space, but most of the matter positions itself to the outer layers of our atmosphere. This matter will rotate as a cushion, to our planet earth, as long as the earth survives. As with any major catastrophe, more than one, acting in odd sequences, can unite and become a compelling force.

It all started with one of our most capacious stars--the sun. Its light takes approximately one million years, at the speed of light, to reach our galaxy.

A star must have magnetic properties if it is to expand and emit energy. Mass is the key. Star and planet mass is known as character. Hydrogen is converted to helium, and three helium into carbon, oxygen and silicone. Core temperature is another key to mass. These deaths and births are a continuing mode in our universe.

Our jet streams have changed as our forests are being cut, and huge holes appear in the ozone. Different winds find new channels of currents over our earth.

Tornadoes, hurricanes and earthquakes will be more violent. Our whole fragile system, as we know it, will be perceived as having been terrorized. The spectrum of our universe could be dangerously altered. Incidents of the last few years have been devastating, some worse than have ever been recorded by man. This will be the norm. Catastrophic extremes will continue and grow stronger in might and magnitude.

PROLOGUE II

The weird phenomenon is that all creatures of this planet have within themselves an aurora of magnetic force, emanating from within. In a true sense, we are all magnets, and we do act upon the prevailing electrical magnetic forces, wherever they come from. This conduces and influences happiness in one slot of the spectrum, to depression and worse.

All of these facts are not surprising to our group of four who represent the S.A.W. group. The discontent, measured with massive weather changes, has soundly demonstrated and dictated the well being of life on our earth.

We are mathematically being induced with increasing magnetism that affects our cells, tissues and organs. The physiological character by which our minds reflect and respond to our environment, is the aggregate of social and cultural conditions that influence our very lives.

Other factors flow with everyday life and predicate our very actions. A cloudy day, defined by a cloud, is a visible mass of water or ice in the form of fog, mist, or haze, suspended in air. It is one of the masses of obscuring matter, in interstellar space, such as our sun, stars, planets and galaxies.

A cloudy day affects every human being and animal on earth. If two days pass and the shimmering sun is still alluding us, more people are affected by light annoyance, some more amplified than others. Three days of gray, with no sun, and tempers rise to another possible hostile level.

In fact, your body is an electric magnetic force. An electrocardiograph is a test for revealing the functional action of the heart. This is a test that measures the electrical potential across the heart muscle. No electricity is put in your body. Twelve electrodes are placed after applying ointment between the electrodes and the skin. These twelve leads give the doctor an excellent picture of the electrical impulses as they reflect your heart function. When your own electrolytes are out of sync, damage or death usually follows.

When electrical current is present, there is also a magnetic field in its place. A power line, motor, star, or planet, or an electrical lightning charge, all possess magnetic fields, <u>only the intensity changes</u>. This magnetic and electrical process is a marriage with us every day. The suns, moon, planets, novas and cellular dust all emit magnetic fields.

Consider that an atom is composed essentially of small, positively charged heavy nucleus, surrounded by a comparatively large arrangement of electrons.

This prelude to the thought that all matter on our earth, and around it, including the trillions of stars and galaxies, all are at play in our world.

Atoms are electrified particles, and thus present an electro-magnetic force. Understanding and predicting changes of these forces would relate to a wealthy enterprise.

Iron is the final end of fusion, the last on the nuclear table. Iron is the mate to magnetic forces--

the last marriage. These are the facts of life. The
unveiling of the magnetic nightmare, the terminus,
the vector, of the final act!

Chapter 1

Mr. Albert Dryfuss had just left the conference room of a little known corporation, called S.A.W. whose corporate management was comprised of four millionaires. This closely held private organization had just adjourned. He left with a smile and in his valise a certified check made out in his name for one million dollars.

Albert is six feet tall, with light brown hair, and faint strains of buff grey. He has been on this earth 61 years. His brown eyes seem to dance when he talks, matching his low, soft voice. There is a faint scar on the left side of his neck providing the light is fractured just right, exposing the smooth skin that had been sutured. He seldom wears glasses. A neat man, he always seems to wear the right clothes and always smells good. Most likely his finest attribute is in the unusual way he is perceived when in charge or command. He is never domineering, and has the gracious knack of making you feel good, even when you screwed up.

He shed his business suits for wool slacks, and a soft thick colorful velour sweater for this new assignment. He had been commissioned by S.A.W. for his genius, and the incorruptibility of electromagnetic forces. Daily increases and saturation promote intense bombarding throughout the spectrum. This corrupts and breeds physical and mental stress in every human being and all creatures--above, on, and in our world. He was allotted one year to produce results.

The S.A.W. Corporation, on given days, would wager millions, in the futures and options market at the New York Stock Exchange. Any advance knowledge on extreme weather conditions, or areas affected, would tilt the increase and the odds of their fortunes in a matter of hours. Its likeness was similar to the ability to access the winner of a horse race only in a different but preempting time zone.

Albert selected his home base carefully. He leased quarters on a quiet, dead-end street, named, unfortunately, Downer Street. The house appeared to have been built in the years of World War II. The house had small rooms--three bedrooms, a kitchen, and a bathroom. However, the substantial site was simply furnished, but lacked beauty. He had set up a checking account at the National Bank of Alaska, purchased sheets, towels, and all the ingredients for three, including himself. There was no phone, and that relieved him of having one disconnected.

There were three most important matters to which he had to attend. One was, although not prudential in order, that he had to know the arrival time of the ship which carried his precious cargo. He checked his billfold, and left his new residence, in a new white four wheel drive Scout, which he had purchased a week ago. Why he headed for the shoreline was of no consequence. A phone, a pay phone, was his objective. He called Bayliner in Seattle, Washington. He repeated the invoice number and said he would hold to ascertain the name of the ship and date of arrival.

The news was good. The freighter *Delta* would arrive within three days and aboard would be his unique 38 foot cruiser, a nifty craft with added features and some special equipment. He had commissioned them to inscribe, in flare gothic letters, "The Magnetic" on the transom only.

He had previously obtained, very discreetly, the phone numbers of the two knowledgeable colleagues he most wanted in his new venture, neither of whom he had met. He was aware of their astounding knowledge in each of their fields of electromagnetic forces. He decided to call Mr. Tom Spoones first. He knew his age--it was 36, and Spoones was tops in determining if magnetic volume is the factor that predicates human reaction. Miss Norma Moon, age unknown, was a master, in view of her published papers on the magnetic strength at the poles and deviations of magnetic moments. He had been fortunate enough to receive a photo of Miss Moon. She appeared to be a nice looking 5'6" or 5'7" brunette. Her eyebrows were masterly arched to enhance her femininity. Her file indicated that her husband had been shot in a hunting accident, and that she was still a widow.

He dialed and linked up with Mr. Spoones. He introduced himself and was surprised at how much Tom knew about him. Dryfuss queried, and then just plain asked if Spoones would be interested. His salary would be a guaranteed $85,000 for one year, room and board included. Albert also told Tom that he hoped to have another member, if possible, and the three of them would convene as soon as possible.

Tom was excited and said, "I don't suppose you can tell me what I will be working on?" Dryfuss replied, "No, but it's right down your alley. If you are aboard, you must be secret about it and tell no one."

Tom reacted, "I like it better all the time."

Albert said, "I must say again, don't even mention my name, or where you are going."

Tom asked, "Where is it I'm going?"

Albert answered, "As quickly as you can leave your present employment, without arousing curiosity, take a flight to Seattle, Washington. Then take a flight on Alaska Air to Anchorage, Alaska. Have you fare for the flights?"

"Yes," Tom replied.

"Good," said Albert. "Now do you have a paper and pen?" "Yes," was the reply.

Tom was given a number to call from Seattle, to let him know the time and flight number, then Albert said,

"By the way, that phone call from Seattle will be taped. Don't expect my voice."

Tom said, "It appears to me that you have everything under control. May I ask who our third partner is?"

"No," Albert said, "it's not that I don't want to tell you, but I have no confirmation of our third party. When can I expect you, Tom?"

He replied, "Three days, four at the most."

"Excellent," Albert replied, and added, "Pack some warmies."

Tom replied, "I will Albert. I may call you Albert?"

Albert said, "Of course, all on a first name basis. Have a great flight." Albert hung up, pleased by his first recruit.

He made his second call and was told that number was changed. He wrote down the automatic number and redialed, the sixth ring produced pleasant voice, "Yes, hello."

Albert replied, "Is this Miss Norma Moon?"

"Yes," she answered.

He said, "My name is Albert Dryfuss, I've read your accomplished works on moments of electromagnetic forces."

She said, "Are you Albert Dryfuss?"

"Yes," he said happily, and added, "I have a proposition to present to you if you are available. It is immediate and be prepared for one year of scientific work with me and Mr. Tom Spoones. You will be guaranteed one year's salary at $85,000, plus room and board. It will just be the three of us. I can't get into specifics, what our work will be, but it is secret. Are you interested?"

He waited for her answer, then said, "Are you still there?"

Her answer, "Of course I'm still here, but you whipped my sails."

He said, "May I call you Norma?"

"Yes," she replied.

He went on, "Norma, this has to be done very discreetly. No one is to know where we are, and especially what our project is. The pay is reasonable, the three of us would make a great team, and we meet all the requirements."

She asked, "What requirement is that?"

He replied, "I can't tell you that over the phone, but if you can get out of your present commitments, and join us as soon as possible, I will explain."

"Sure, she said, "Where do I go?"

He replied, "Pack some warmies and take the first flight to Seattle, Washington. Then arrange a flight on Alaska Air to Anchorage, Alaska. Do you need any money forwarded?"

She said, "You really are serious about this."

"Yes," he answered, and trailed, "Very serious indeed. When will you arrive?"

She responded, "In three to five days."

"Great," Albert responded. "Have you paper and pen"?

"No, I have paper and pencil."

He laughed. "Please write this number down. When you get your flight number in Seattle, call this number, and leave information of your date and flight number. It will be recorded. Did you get that?"

"Yes," she said, "I expect there is a reason for not booking my destination flight from here."

"Yes, Norma, your trail will be hard to pick up from Seattle to Anchorage."

She said, "That middle initial "S" in your name must stand for Sherlock."

"Yes," he replied, it truly does." He continued, "Welcome aboard, Norma. Have a great flight. I'll be at the airport to greet you."

"Okay, Albert, thank you. See you in three or four days."

"Christ," he said to himself, "this has been a hot damn day."

They both arrived within six hours of each other at the international airport. Tom had arrived first. Albert had a recent photo of Tom too, so there was no problem in recognition. Tom wore a classic genuine grin, as the two men grasped and shook hands. Both men appraised the other silently.   They engaged in some small talk until they arrived at the baggage carousel.

Tom said, "This is one state that I have always wanted to visit."

Albert countered, "This will be your home for at least a year.   I hope you brought some warmies."

"Some," he said, then continued, "I thought I would see what the natives were wearing. Oh, there's one of my bags."   Tom snatched it, and pointed at the matching bag. Albert grabbed it and they headed for the exit.

Tom asked, "Do you have a car?"

"You bet, a durable Scout. We will all have access," and they put the luggage inside. When they got in and closed the doors, Albert reached between the seats and brought out a second set of keys on a horseshoe key case.   "Here, Tom, I'll have a set made for our third party."

Tom thought to ask who that might be, but Albert probably would have told him if he wanted him to know at this stage. Tom was noting the lay of the land, and landmarks. . .   They braked in front of their new drab, unpretentious home.

Albert said as they unloaded the luggage, "I know it's not much, but Downer Street is a dead end, and rather private, and it does have three small bedrooms.

Each man handled the same bag to inside their new home. Albert led Tom to a hallway and opened the door on the left. "This is the same as the other bedroom, Tom. Get settled and browse about. This is your home. Enjoy."

"Thanks, Albert," Tom said.

Albert said, "I have some messages to check. Bathroom is at the end of the hall."

He left with the Scout. He checked in at the answering service which he had set up for incoming calls. Sure enough, Norma had called and related her flight number. He asked the thin looking woman if she would call the international airport and ask when the flight would arrive and if it was on time. She obliged willingly as she was being paid by the day, in addition to a small fee for each call she handled.

He thanked her and glanced at his watch. He had two hours to kill. He drove his Scout north on Arctic Street then on Northern Lights to "L" Street, to the waterfront, and surveyed the many areas. He had yet to find a slip for his boat. He parked and entered a shabby office located just off Knik Arm. He asked a blonde, rather stout lady who was at the desk if the owner was in.

She stated gruffly, "Mister, what makes you think I'm not the owner?"

He said, "I apologize, a nice attractive woman, in this not too attractive surroundings. I was looking to find a slip for my boat up."

She perked up, shifted her weight, and said, "You came to the right harbor, mister."

He said, "Do you mind if I inspect your floats?"

She winced and questionably said, "No go ahead, I have a few open."

He asked, "Any vacancies for a 38 foot boat?"

"Yes, I have two left," and looked at her chart. "Numbers 14 and 28."

He asked, "You do have diesel fuel and water?"

She put up her hand, "Mister, I've been an owner here for nine years. I can, and have supplied all kinds. My rules are simple: pay in advance and no dumping of any kind in my waters, even a cigar ash."

He said, "Those rules sound very environmental."

She snapped, "Are you a smart ass, or an inspector, or. . .?"

He did an about face, saying, "No, I'm sorry, I'll just check the docks." He went through the office door to the docks. There were two other entrances leading from the outside. All kinds of weathered signs forbidding fishing, to sex --this harbor had not been updated for many years, yet it still seemed to be stable. The birds and the sea had made their mark. He liked the locale, not in the very business portion of the city, yet not too long a drive. He noted that her docks had one big stall. The largest would tie to a 45 foot boat.

He got caught up in the excitement of imagining his boat gently riding in slip No. 28. It was the best of the two. Most of the slips were occupied, none of the crafts looked ritzy, but no junkers showed their colors. He checked the depth gauge and noticed the water looked clear for this type of portage. He returned to the shabby office with the unhappy owner.

"Well," she said, "I rent or lease, no less than week, and one more thing, mister." She slurred the mister, "although we just might do business, I have the final word, and if for any reason, you disobey the rules, you have just 48 hours to paddle your boat out of my waters."

He asked, "How much for slip No. 28?"

"Week or month," she spat.

He asked, "How much a month?"

She spoke out loudly, "If you comply and pass my tests, I'll let you have it for $160 a month."

He said, "I'll need it for at least a year."

A little spittle formed in the corner of her mouth. Her eyes blinked.

He said, "A fair price would be $1800, in advance, of course."

She snapped her pencil in two, tossed it aside and snatched another. She asked nicely, "What's the name of your boat?"

He replied, "The Magnetic."

She said fluttering, "You have a deal master, ah, I mean mister."

He asked, "Who should I make this check out to?"

She quickly answered, "To me, of course."

He waited, she thought of her reply and restated, "Make it out to Ruth Dander, this establishment is called "White Tern.""

He wrote out the check, and handed it to her.

She looked it over very carefully and said, "Mr. Dryfuss, when will you be tying up to my dock?"

"Hopefully in two days, three at the most, but I want my lease to start today, Mrs. Dander."

She said smartly, "It's Miss, not Mrs. If your check clears, there is only one item remaining to conclude your mooring in slip No. 28."

He patiently asked, "And what might that be, Miss Dander?"

She pushed her swivel chair back and stood up. "I will not tolerate my waters to become a sewer. No dumping of human waste, or any litter." She paused, then said, "If there is so much as a trace of oil, gas, or diesel, or a leakage from your engines, you're out of my waters." She sat and waited for his response.

He said, "I will happily comply with all of your regulations, and I commend you on the preservation of our waters. Good day, Miss Dander."

She said loudly, "Yeah, good doing business with you, Mr. Dryfuss."

He took one long look at the area, then headed for the airport to intercept Norma.

The flight was only seven minutes late. He watched the huge craft land and eagerly awaited the incoming passengers. Albert saw her fresh face, eyes sparkling under her beautiful arched eyebrows. He waved as he approached her and said, "Welcome to Alaska, Miss Moon."

She laughed and stated, "What's with this Miss Moon. Norma will do just fine, as it did on the phone."

"My apologies, Norma," he said and quickly added, "It was like pinching myself to make sure you were really here. How was the flight? Let me take your carry on."

She replied, "The flight was great, food, not too bad, and no, I can carry this."

He gently touched her elbow and said, "Let's get your luggage."

She said, "I'm really excited to be here and look forward to working with you."

He flushed, "Thank you, but I am the one who is honored to have you as a colleague."

She smiled and asked, "Is this a 'no smoking' area?"

"Yes," he said.

"Damn," she responded.

He stated, "Norma, it will be a just a few more minutes, and you will be able to recalibrate your sympathies."

She laughed and said, "Yes, my sympathetic nerve endings are indeed flying." She pointed at her bags coming out of the conveyor to the carrousel. He grimaced, as he hauled two large suitcases out of the rotating machine.

Norma was unaware of his twinge of pain and asked, "How far is it to our quarters?"

He replied, "Just a few miles. You can light up now."

"No," she said, trailing, "I'll wait until I get in the car." She looked about at the bustling new scenery, the machines, and the people.

He said, "This way. That white Scout is our transporter. We all will share the conveyance, and I'll have a set of keys made for you." He loaded her baggage in and slipped inside.

She immediately lit up, a long cigar-cigarette, and sighed as she exhaled. She exclaimed, "I've always wanted to visit Alaska."

Albert smiled and said, "You and Tom should make great pals. That's what he said. Tom Spoones is a dandy in his field, as you are. We'll make a great team," he added, as he drove. "Here we are," he said, as he pulled to the curb.

Tom was loading up the coffee pot when the front door opened and he saw his new co-worker. Albert did the introductions. They both smiled, storing memory phases of each other.   T o m said, "Here, let me help you with that," and took one suitcase from Albert.

Albert said, "This way, Norma," and pointed toward the hall. "To the right," he said, "this is your home away from home.

Norma said, "Please put them on the bed." Both men did as they were told.

Tom said, "I'll be right back," and slipped out, back to the kitchen.

He said, "I know it's kind of drab, but our abode is rather private, on a dead-end street. It's called Downer Street. The bathroom is at the end of the hall, only one. I'll let you unpack. If you need anything, just holler, Norma."

Tom just missed colliding with Albert at her door, but didn't spill the cup of steaming coffee he had in his hands.

"Mmmm," she cooed, "that looks inviting." She offered him a cigar-cigarette, but Tom declined, stating that he didn't smoke. She lit up and took the cup and saucer from him and blew through her sensuous lips to cool, then sipped. Her taste buds rejoiced as the flavor reacted to her responders.

"Tom, you make one hell of a cup of java."

Tom smiled and said, "I'll let you get settled. Welcome to Alaska, and I'm looking forward to working with you."

"Thank you, Tom, both of you have been grand." Two hours passed and Norma and Tom had placed their goods in their rooms to their liking. Albert was checking and noting in his small pocket-size notebook. He sighed and looked at his watch. He ran his fingers through his hair and set out to find his new friends. They both were talking at the kitchen table; however, they stopped when Albert approached them.

Albert asked, "Is anyone hungry?"

Norma looked at Tom. Tom thought it over, and said, "No, I'll dine at breakfast." Norma joined in, "I'm still full."

Albert said, "There are snacks in the refrigerator if you change your minds. I just would like to say that I am happy to have you here. I have the unpleasant task to ask for two volunteers for two very necessary duties--cooking and laundry. It would be advantageous for each of you to select one of the two."

Tom spoke first, "Count on me for laundry detail."

Norma was about to reject her role as a female stereotyped for cooking chores. They both waited. Norma said, "I'll try for now. Have you the makings for breakfast?

Albert replied, "Yes, there's probably enough for three days, nine meals. Oh, yes, is it all right with both of you to have breakfast at 7:30

or 8:00 a.m.? They both agreed. Albert said, "I would suggest we all set our alarm clocks for 7:00.

Norma said, questioningly, "How in the hubs of hell will that give me time to freshen up, and have prepared three breakfasts?"

Tom laughed, and said, "It doesn't."

Norma looked at Albert, but he wanted no part of it. She grinned, and thought to herself, "We'll see who the comic is in the morning." She got up and set her alarm for 7:00 a.m. Norma stood in the doorway and announced to Albert, "Why is it we all have alarm clocks but no TV?"

Albert stopped, about to address her question. "Norma, Tom I apologize. I just ran out of time. I will deal with that tomorrow, and thank you for reminding me. Have a good night."

Norma felt a twinge of regret about the clock and TV jingle. The bedrooms were quite scant of furniture, much less the room to house it. Her room held only a bed and a small oak four foot high dresser, with four drawers with a faint and hazy mirror on top that was attached with pins. The lighting fixtures were the same as about 45 years ago. Some time in the middle to late '40's. Norma looked into the mirror, stepped back, over to the right, then left. The mirror was tilted. She discovered if she bent her knees, her reflection was reasonably adequate. She reminded herself to ask how many beauty parlors there were here. She pulled down the green, wrinkled shade, and stood in the center of her little bedroom. She folded her arms, and traversed to take it in simultaneously, like a prisoner sentenced to a cell. She graciously smiled, and tested the bed. She sat down hard on it. She bounded, not the mattress.

"Lord," she said aloud. She peeled down the bed covers and sheets, and was heartened to see they were new. She felt the cotton. she would bet even money that Albert, buying them new, washed them with fabric softener. This nicety reminded her of her discourteous crack about the TV. She stripped off her clothes and put on a light blue set of pajamas, stretched, and climbed in the hard bed. She got up and turned off the overhead light. Another thing she wanted was a lamp and something to hold it up, by her bed. And ashtrays. She told herself to relax and get some rest. It worked.

The next morning, Albert heard banging on the bathroom door. He shouted through it, "I'm not finished yet, and you are slowing me down."

Norma and Tom shouted in unison, "How much longer." They looked at each other, surprised to have made identical remarks, then smiled at each other.

Norma said very seriously, "This isn't going to work. I have to pee."

Tom said, "Is it an emergency."

Norma crisply said, "Christ, you want me to go right here to prove it?"

Tom grimaced and boldly opened the bath-room door and shouted, "King's X, Norma's bladder is on the verge of non-containment. Better heed, or we'll have a mess to clean up."

Albert left his goods on the counter, and stepped out of the bathroom. Norma brushed him lightly in her haste as she closed the door.
Albert turned and asked Tom, "What the hell is this King's X thing?"

Tom, looking a little sheepish said, "Didn't you have King's X when you were a kid?" Albert replied, "Hello no, what is it, a code?"

Determinedly, Tom said, "King's X, is a sort of stand-off. It means that everything up to that point, King's X is of no consequence, conditions being equal. But King's X stops everything and becomes the most prudent stop, no matter what the situation is at that time."

Albert stated, "Congress never heard of that." He then raised his voice, and added, "Miss Moon, Norma, have you finished with your emergency? I would like to complete what I started."

She raised her voice and stated, "Seeing that I was designated, not by a show of hands, to be the cook, and if you want your breakfast on time, I'm going to complete my morning duties. And besides, possession is nine points of the law."

Albert and Tom looked foolishly at each other and both shrugged their shoulders. Albert broke the quandary. He straightened up and fallaciously stated, "I must say, this informative meeting of my staff has made great inroads. In a matter of hours, I was absconded from the bathroom, with a King's X, and duly informed, that if I wanted breakfast, I was to wait until the cook finished her affairs."

Tom absorbed it all.

Albert smiled and said, "Now, let that be a lesson to you, when you are in command."

Tom asked, "What lesson is that?"

Albert replied, "Don't fool around with Mother Nature."

The bathroom door opened and Norma, with dignity, said "Thank you gentlemen. Don't be late for breakfast. I wait for no man."

Tom raised his voice, so Norma could hear as she retreated, "You made your point."

"Amen," said Albert.

Norma located the coffee, the pot, and turned on the element. Then she returned to her bedroom once again and dressed in her smock for the day's chores. Albert was at the table in his chair as Norma checked on the coffee and found the cups and saucers.

Albert said, unknowingly, "Are you a morning person?"

She quickly replied, "Do men have babies?" Albert slouched a little and questioned her response.

Tom joined them in the kitchen and sat down.

Albert asked Tom, "Did you have a good night's rest?"

"You bet," he replied, then continued, "that's one adversity I have overcome. I can sleep anywhere, any time zone, jet lag, fine, or poor beds. It makes me no never mind. By the way, let me congratulate you on your selection of firmness in the mattress. It's a fine supporter. God, I'm famished."

Albert said, "Amen. I'm a bit hungry myself. Good to start the day with a wholesome meal!" He glanced at his watch, fourteen minutes to eight. Both men raised their noses as the fragrance from the coffee drifted and began to tease their sensors.

Tom asked Norma, "How was your night, Norma?"

She replied, "The flip-flop side of your buddy talk to our boss."

Albert thought, "God, I hate that word 'boss'."

Tom said, "Don't look at me. We're not all morning people," and when he heard what he said, he was sorry he got involved.

Norma stood above Albert, and used her manicured index finger to aim at him, "I'm not a morning person. When I have had my cigarette and coffee, then I can face the world."

Albert looked at his watch--five minutes to eight. Norma retrieved four pieces of bread and

put them in the old chrome commercial toaster, checked the setting, poured the coffee and asked,
"Are these going to be your official chairs for every meal?" That caught both of them off guard. Tom shrugged and stared at Albert. Albert realized she was waiting for an answer and said, "This is your kitchen, Norma. As long as you are cooking, place us as you will." She said, "Spoken as a true boss." Tom, please move over there." He did, and pulled his saucer with him. She placed the butter and milk on the table. The toaster chugged up noisily. She found a small plate and piled the four pieces on top of each other. She took the carving knife, and cut the toast diagonally. The men watched. She then cut it again diagonally and smiled. She set the toast where all could easily reach it. She looked around, not seeing what she wanted, opened the cabinets and retrieved a saucer and set it beside her cup. She sat down and smiled, "Eat up gentlemen." They looked at each other in disbelief. She said, "Go ahead," and reached in her smock pocket and pulled a pack of cigar-cigarettes out and lit the thin, long cylinder with her silver lighter. "Mmmm," she said, and added, "Isn't this a wonderful morning?" She picked up a piece of toast and dipped a corner in her coffee and musingly nibbled at its end with contemplation and smiled at her two breakfast companions.

Tom delicately picked up a tiny toast point and averted the eyes of Albert.   Albert looked down, across, took a sip of coffee and asked, "Norma is this it, coffee and toast?"

She quickly replied, with merriment in her eyes, "Yes, I'm a coffee, toast, and cigarette morning girl, not necessarily in that order." She paused and asked, "Anyone take sugar?"

Both simultaneously answered "No!" to which Norma graciously replied, "That's my men."

Norma had them bewildered, two great scientists. All it took was a touch of femininity. Albert noted the time--two minutes after eight. Norma had a satisfying, almost an addictive smile, and she blew another round of smoke. She perkily had another quarter of toast, and buttered it, like a painter, in simulated strokes. Four unbelieving eyes watched her every move as she again dunked and nipped the corner off the toast in complete contentment.

Albert selected a piece and buttered it, dunked, and, with two bites, it was gone. His stomach was waiting to be fulfilled.

Norma plucked another piece of toast, buttered it, and stated, "Not bad for the first day. Despite the interruptions, I beat the deadline by one minute." No one else was speaking. She paused, and then said, "Hey guys, lighten up, this is a gorgeous morning." She pulled and released a hue of smoke and waited.

Albert ran his napkin over his lips and said, "Norma, how much do you weigh?" She recrossed her legs and replied, "Gentlemen, don't ask a lady her weight or age." Tom sensed the fencing and was happy to be a silent audience. Albert picked up his coffee, sipped, then smiled at Norma to get her attention, and stated, "Norma, this is a fine cup of Java, a real good, get going juice in the morning. How much do you think I weigh?"

Norma sipped her coffee and answered, "Are you serious," she asked straight faced, "naked, or with your clothes and shoes?" Tom could not hold his snicker and followed it with a grin. He was enjoying this encounter. She blew her last cloud and ground her short stub.

Albert fanned the smoke and answered, "For scientific dimensions, let's say without clothes or shoes." Norma said, "stand up."

Tom cracked up and snickered, then laughed. Albert drank a bit of coffee, set his cup down, pushed back the chair, and stood up. Norma slowly eyed him up and down. Albert's face turned pink. He felt like she was seeing him standing in front of her nude. She put her index finger to her chin and presented herself as carefully analyzing her subject. Albert couldn't stand it any longer. He sat down, trying to hide his indignity. Finally, looking into Albert's eyes, she said, "Well, I could be quite accurate, as a scientist, if you were naked." Tom laughed again and glanced at Albert, only to receive a desperate glare--a man-to-man look for help. Tom turned his attention toward

Norma. Albert was about to speak; Norma held up her hand and stated,

"But, unscientifically, or would you rather?"

"No, no," Albert stated, and mumbled, "Get on with it."

She smiled, Tom laughed and she said, "Okay, your weight nude--one hundred and sixty pounds of flesh." Tom snorted, and even Albert broke into a wide grin.

Albert said, "Congratulations, you came within three pounds." I just wanted you to know that to maintain my frame I need, and want, three squares a day, especially at breakfast."

Norma knew she was being delicately chastised and stated, "I'm a toast and coffee girl. I believe you both were a bit hasty in prescribing duties. I'm very good at laundry. I never have liked cooking."

Tom eased himself out of his chair and asked, "Anyone for warm-ups?"

They both said, "Please," then smiled. Tom addressed both of them. "I am your chef."
Norma spoke first, "How about bacon and eggs, Albert."

Albert grinned and replied, "I like my eggs over easy, bacon crisp. Are you preparing pancakes?"

Before Tom could answer, Norma stated, "Tom, please baste two eggs. I also like my bacon crisp, and more toast and coffee."

Tom turned around, found an apron, put it on, gathered up pans and covers, and lightened the ingredients in the refrigerator. Albert spoke loud enough for both to hear. "By the way, if you are asked any questions, say that you are amateur geologists."

Norma asked, "What is our real objective?"

"Yes," said Tom, as he cracked the eggs, "I would like to know also."

Albert stated, "In theory and fact, both of you are unique in your fields of electromagnetic forces. We are going to evaluate and base weather and human factors across the galaxy to earth. All matter is comprised of magnetic forces as you know. We are going to chart the sun's solar magnetic flares, trade winds, and every change in earth's electromagnetic forces. The effect determines our world weather patterns humanity. If we succeed, we will know what's going to happen before any national or international weather stations would or could react.

"The atomic experimentation, and theoretical considerations hold that the atom is composed essentially of small positively charged comparatively large arrangements of electrons. Our bodies are made up of trillions of atoms and millions of responders. This, then, precludes the thought that all matter on our earth and surrounding it, including the trillions of stars and galaxies, all play in each and every day. Atoms are electrified particles and, thus hold an electromagnetic force in place. Understanding all phases would represent predictable changes, before they can be detected."

Albert sheepishly grinned and stated, "I didn't mean this to be a lecture, but the probabilities are immense."

Norma said, "That is interesting," and Tom stated that he never thought he would be a weather forecaster.

Albert said, "It goes far beyond that, but I can't emphasize enough that all information we collect must be absolutely confidential. We, that is I, will then forward our collective findings to Chicago (Morse, to Van Buren Street in code.")

Tom shouted over the hissing bacon, "Just call me James Bond."

Norma responded, "Hot damn, I do enjoy a mystery."

Albert stated, "So you shall. We will all enjoy unraveling the mysteries of the universe." Then he laughed and added, "The pay's not bad either."

Norma said, "What is our pay? Exactly how firm was that figure?"

Tom interrupted Norma by tapping her on the shoulder and saying, "Speaking of firm figures, here's your basted eggs and three slices of crisp bacon."

"Thank you, Tom," she said, then added, "Tom, will you please bring me a napkin?"

Tom handed out napkins and captured a wink from Albert. Tom retreated to his stove, filled a plate and sat it in front of Albert saying, "There you go, two over easy and four crisp bacon."

They both dove into their food. They grinned and chewed. Norma said, "Tom, more toast, please."

Albert said while chewing his bacon, "There's orange juice in the fridge, hint, hint." Tom delivered on time.

Tom filled his plate, two straight up and four pieces of bacon. He sat down, and rubbed his hands together. "I worked up an appetite." He picked up his fork and looked at both of his companions. They had just about finished eating, and Norma was attacking a smoke.

Tom took a big bite and said, "I could say, I hate to eat alone, but I won't." He continued to devour his breakfast eagerly.

Norma said firmly, "Are we all going to be treated as equals?"

Tom dipped his toast in the yellow, still crunching on a bit of bacon. He looked up to see the reaction on Albert's face in response to that loaded question. Albert asked, "What exactly do you mean?"

Norma said, "I hope you treat me as an equal, in all ways, including my salary."

"Oh, that," said Albert. "Absolutely."

Norma reached over with her left hand, and plucked the last piece of bacon from Tom's plate, and quickly nibbled at the end of it, mixing it with her smoke.

Tom said, in a flustered manner, "What the hell! That was my bacon you scrounged."
Norma blew a ball of smoke and demurely said, "Scrounged is not what I did. I had only three pieces of bacon, compared to both male portions of four."

Albert laughed, and Tom slowly grinned. Albert sipped his orange juice and watched Tom's body language.

Tom said, chewing lightly, "I just assumed."

Norma shut him off, interrupting and briskly stating, "That's the problem, too much assuming by males."

Tom had to laugh and then scooted back in his chair and said, "Okay, King's X."

Albert snorted, "Oh God, not that again!"

Tom continued, "In our meal preparations I will divide up the food into three equal portions, Norma. I thought you said you were a toast and coffee girl."

Norma replied, "And a cigarette girl, if you want to be exact." She paused, then said, "Only when I eat alone."

Albert laughed, patted his stomach and sincerely said to Tom, "That was a dandy breakfast."

Tom smiled and said, "You're welcome," then turned his head in Norma's direction as he picked up his coffee.

Norma congenially said, "I couldn't have cooked it better myself."

Tom grinned and sopped up the remaining food on the plate with half a piece of toast. He said, "I don't really know how to interpret that. I remember you saying you were not a good cook."

Albert said, "oh, oh."

Tom looked at him as Norma said, "There you go again. I said I never did like cooking, not that I wasn't a good cook." Then she added, "don't pull that King's X on me."

Albert was enjoying the mood and made a mental note that this gal was what he would call an exquisite person, a crackerjack. He had made two good choices. Albert decided to spring a bit of good news and said, "Very shortly we will have at our disposal a 38 foot cruiser, fully equipped, with instruments and some stores, and a spectacular built-in radio magnetic antenna. It will arrive, hopefully, on the freighter *Delta* tomorrow. The name of our working yacht is colorfully called "The Magnetic." All we will need is fuel, water, and food complements to make her seaworthy."

Norma was delighted and said, "Call me Tugboat Annie."

Tom turned to Norma and said, "Okay, Annie, I am 'Barnacle Bill the Sailor.'"

Albert said, "That must make me 'Captain Nemo.'" They all laughed and warmly appreciated each other's candor.

Norma said, "I'll say this just once, Captain Nemo. You referred to the craft as a she."

"Oh, oh," said Barnacle Bill. They laughed until tears twinkled in their eyes and rolled down their cheeks. Each of them wanted to freeze this happy moment.

Albert recovered and said, "The *Magnetic* has two heads, a fine feature."

Norma said, "Tom, I told you so."

Tom looked into her mischievous eyes and played her game and said, "You told me so?"

"Yes," she said laughing, "Two heads are better than one." They repeated their happy affair. Albert had to take his glasses off and clean the salty tears from his lenses.

Norma tapped another cigarette out, fired up another, and pleasurably inhaled and blew a billow of smoke that layered above them. Albert said that he was looking forward to navigating on the sea, and maybe, best of all, no more fiascos like the one which transpired this morning at the bathroom, and hopefully, looking at Tom, no more surprise K'Xing.

Tom inquired, "You said that if we are being asked or pumped for information, we are to be amateur geologists. Anything special in that field?"

Norma said, "I was going to ask the same question, but I didn't want to appear pushy."

Albert said, "Take your choice--gold, oil, and even diamonds, although that find is further away in upper Canada. If we click and compile our information correctly, I would venture to say that both of you, and I, of course, would not have to worry about the menial buck for the rest of our lives."

"Wow," said Tom.

Norma said, "Another wow from the opposite sex. Is this a grant of some kind?"

"No," Albert said, "but they have paid one year in advance, and I was assured, that if this is successful, there will be a bonus for all of us. We all have some knowledge to pass to each other, and by combining the latest techniques and teamwork, we can unravel the hidden mysteries of the heavens. Yes, and one more thing, some of the instruments we need are already on board our craft, our home away from home. If there is a special tool you need, please let me know and I will do my best to acquire it."

Norma spoke up, "There's one tool I could use."

Albert asked, "What might that be?" and Norma replied, "A bathtub."

Tom snickered and Albert smiled and said, "Norma, I think not. We can't have strangers in here, or visitors, we must be careful. However, there is a place you can bathe or spa in this fair city, Alaska Spa and Bath on Campbell Street. Remember, we won't be here very long before our boat will be our home."

Norma said, "You can't blame me for trying. I love to soak in warm water."

Tom said, "I'm happy we haven't got a tub. The john would be occupied, and I would have to King's X you while you were soaking."

Norma replied, "I soak alone, and, King's X or not, I wouldn't be about to get out of a warm tub to let anyone in."

Tom asked, "Are all females the same about bathing and soaking?"

"You bet we are. You males make us nervous. We have to relax, and that is the direct poop."

Albert said, "Good Lord, you are direct. I hope I don't come across to make you nervous. That is not my intention."

Tom cut in and said defensively, "I would hate to think that you would have to bathe just after a conversation with us."

"My God," Norma retaliated, "I'm talking about generalities. If that imaginary tub was that important, I would have mentioned it before committing myself to a year in Alaska with two male strangers."

Albert thought about that and said, "Yes, I can see that we take you for granted, and that it is entirely different from another person's perception."

Tom said, "Norma, no offense, but you make men out as ugly, the enemy."

Albert held up his hands and stated, "Oh no, Tom, or Norma, I'm not going to sort out the discussing of unsolvable mysteries of males versus females."

"See," Norma said, they waited, and she obliged them both. "Albert, you said male versus female, always domineering. Why didn't you say, female versus male?"

Albert said, "Whoa, whoa."

Norma quipped, "I'm not a horse."

Tom said, "Yes, Albert, she's not a female horse." That got a smile all around.

Albert said, "Let me say this. I chose each of your for your edification in your fields of electromagnetry. Example of my legitimate policy as to sex, color, etc. Go ahead, go ahead, ask one another about the fee I promised each of you. With your permission, Norma, Tom answer the question."

Tom screwed up his face and nervously said, "Eighty-five thousand, with room and board." Norma, Albert and Tom exchanged looks. Finally little smiles reflected in each other's eyes. Norma got up from her chair, and displaying her femininity, kissed Albert on the cheek, much to Albert's surprise and delight. Then she approached wide-eyed Tom and duplicated her welcoming gesture. Tom and Albert blushed but were grateful. Norma recaptured her chair and wore a placid smile. Only that wise, feminine gesture, could have cleared and levitated the air of sexuality. They all felt closer as friends and associates.

Chapter 2

Tom got up and said, "Let me complete my morning performance and clear away the dishes."

Norma got up and said, "I'll give you a hand, if you like."

Tom responded, "I like, thank you."

As they whisked the table of its contents, Albert spoke to both of them. "Have either of you read the paper of the zoologist from the University of Hawaii who isolated single-domain magnetic crystals from the sinus of a same bone in a yellow fin tuna and a chinook salmon?" Neither had heard of it. He went on, "It was shown by the magnetite synthesized, rather than geological processes. He found abundant nerve endings that entered the magnetic tissue, and the crystals were organized in chains, much like those in the simple cell magnetic bacteria. Each crystal, apparently, was fixed in place, but free to rotate slightly in response to external magnetic forces. This being found in birds, insects, and animals, including humans, the magnetic direction is enacted not by instinct, but by the very magnetism of our environment.

Norma said, as she was drying, "The flow of electrical currents in the brain actually does produce a magnetic field that can be analyzed several feet away from the head. I used a M.E.G. and it indicated a vector of direct current running between the front and back of the brain. I also noted that the cycles have identical-similar relations to the cycles of the day, daily tidal changes in the earth, and galaxy relationship."

She put her towel down and turned to where she faced Albert and said, "Albert, I'm speaking as a scientist, not a woman, but the earth's magnetic and lunar cycle of 28 days, is also the pattern of female menstruation."

Tom said, "I'll be damned." Tom then said, as he "chore boyed" the sticky pan, "The electro-magnetic field about us only has to be changed slightly in force, less than any magnetic force to move a compass needle. Yet our bodies, yours too, Norma, act upon it."

Albert said, "Yes, you both have a grasp of it. Our complete magnetic foundation is so fragile. The air we breathe is loaded with horrendous man-made electromagnetic fields. We are being totally saturated. Just 100 years ago only the solar, earth, light, and lightning were the true sources of electro-magnetic charges." Tom stopped, and faced Albert, and they both stood side by side, learning back-wards a little to support their buttocks with the counter.

Albert stated, "It is most probably that the massive releasing of unnatural energy sources, in fact, is affecting our minds and bodies. Every device contains energy, has electrical properties and is crashing into our brains and our bodies. Our satellites, more every year, solar phones, TV, radio, F.M. radar, fax, all transmissions and power stations all over the world."

Norma said, "Maybe we don't need a TV at that! These talks and discussions scare the pants off me, figuratively speaking, of course. But this is powerful information."

Albert said, "Yes, this field and its proven facts and context has yet to be published to the peoples of the world. Yet, it has been proven, without doubt, that this relentless spasm of micro and electromagnetic short and high frequency waves cause early cataracts. This and a host of destructive battering is constantly being ravaged on our bodies in forms of stress, vision changes, menstrual cycles, skin rashes, birth defects, headaches, nausea and fatigue. Can you imagine if the newspapers ran headlines to alert the population, there would be a catastrophe?"

Tom said, "We know about these dramatic powers, yet the machines we use to gather this data and the source can be devastating."

Norma added, "An earlier study on women conducted in Oakland, California, wasn't given high priority, if I remember right, and I'm sure I do."

Tom cut in, "I believe your memory," and Albert nodded in respect.

She continued, "A study was conducted representing 1583 pregnant women who used computers more than 24 hours. Researchers found that those women had twice the rate of miscarriages as their female counterparts who did the same work but did not use computers."   Albert stated. "Sometimes it's maddening, professionals that we are, that we can't get these messages to the public."

Tom turned back to his dish washing. All of them were contemplating the horrors that were going on--and unwary people were the victims.

Tom said, as they rejoined Albert at the table, "I really don't know if it is the worst, but surely ranks on the one to ten scale. In 1973, the recommendation of the Navy's SANGUINE study, which was sponsored by the Pubic Service Commission on findings of health hazards. In a five year study, the Navy recommended that a three Milligass field was safe under thousands of miles of high transmission lines, from coast to coast. The magnetic field at the edge of a 50 foot right-of-way below them was 100 Milligass. They didn't announce this, and the study was changed to read that 100 Milligass was not a hazard and, therefore, acceptable. The very people, our so-called watch dogs of the federal government, betrayed us once more."

Albert said, "I just want to thank both of you again for being so forthright and spontaneous in leaving your present world and joining in our new world of solving, or at least predicting Nature's wonders."

Tom turned to Albert and said, "I just want to clear up a nonchalant notion I have to ask, but, if I am intruding on foreign, or forbidden land, let me know." Norma lit up her cylinder and netted an aura of blue haze about them.   Albert   said, "Please ask, you have peaked my interest."

"I assume," Tom said, then hesitated and continued, "that this was a rather urgent assignment. I just have one little question."

Norma couldn't stand it and blurted out, "Oh, for God's sakes Tom, spit it out." She blew another wreath of haze out of the left side of her smile.

Tom looked foolish and said, "I shouldn't have even brought it up."

Norma turned in her chair, put her right index finger on Tom's nose, and stated, "If you don't say what it is you are afraid to ask, then I'm going to." She pushed lightly on his nose, withdrew her finger, and waited for his response.

Tom tried again, "I don't want to get into your personal life, but this little question keeps jabbing my logics." He drew a breath, then said, "Forget it, I have no business asking you questions that nothing to do with our new adventure. As you gainfully said, Albert, our job is to unravel the mystery of the universe."

Norma laughed.

Albert recrossed his legs and said, "Tom you sure know how to set the hook."

Norma said, "You guys, you acted this out, like you rehearsed this drama just to rile the female interest that all gals have. But, I'm not going to play this game." She stopped, looked at both their faces, inhaled, and really blew a cloud, then said, spelling it out. "I don't give a D-A-M-N." Inhaling too much smoke she coughed, and covered her mouth with her hand, indignant at Tom's, ring-around-the-rosy questioning of Albert.

Albert took command and stated, "No work today. This is a get acquainted, settling, relaxing, and go shopping day. Tomorrow we'll discuss our respective roles." He was attempting to take the pressure off Tom and said, "Why don't we have lunch at one of the eating establishments? I have to pick up a set of keys at Ardian Locksmith on the way. Then we will all have keys to the Scout." He glanced at his watch, then said, "in about three hours." They nodded in approval.

Tom got up and headed toward his bedroom, Norma one step behind, moved closer, and hooked her arm with his. Tom was startled then stopped. "What can I do for you Norma," he asked?

"Tell me," she said, "you guys staged that question for my reaction!"

He laughed sincerely, and said, "No. Sorry, Norma, I just backed out, although I would have liked an answer. You know when something bugs you?"

Norma said, with hands in air, "Oh, no, you don't. I'll see you later, Tom." She went to her room and Tom to his.

Albert had one advantage. Although it made his bedroom cramped, he had a half desk and chair.

Norma pulled her cigs from her deep-slit pockets, and tapped the bottom of her pack, only to find two left. She had a saucer to use as an ashtray, and set it on the bed beside her. She doubled her pillows and took a slow drag, blowing the smoke out very slowly. She stared at the ceiling which was desperately in need of paint. She let her mind wander, and her first reminiscence was of her lost husband. She thought of their courtship, their honeymoon. Everything was so natural, and then they told her that her husband was dead. Four years later, here she was in a scrubby little bedroom in Anchorage Alaska, on a dead-end street, named Downer. She smiled at the inference in the name--Downer.

"By God," she thought, "I still have one and one-half cigarettes left and two new friends." The past two hours with the two men had been sheer delight. She enjoyed clever and knowledgeable people. Her real adventure was just starting. She thought about what the cruiser would look like. She hoped she wouldn't get seasick. She neglected her smoke, and the ashes fell on her smock. No fire, she brushed the ashes off herself. She started to make a mental list of necessary personal items. She ground her cigarette out and told herself, "To hell with it." She was still tired from the long flight, and poor night, on this hard bed. She closed her eyes and cleared her mind.

The next thing she heard was muffled shots from a shotgun. She awoke, wide eyed. The rapping continued on her door. "Come in," she said, as she slipped to a sitting position on the bed.

Tom opened the door and said, "Is everything all right?"

"Sure, I'll be right there," she replied. She brushed her smock down, spreading out the wrinkles, and went to the bathroom. She splashed water on her face again and again. "Strange," she thought, "I haven't thought of Donald and those distant shotgun blasts in over two years." She combed her hair, checked her makeup and joined the men in the kitchen.

Albert said, "Ready, Norma."

"Uh-huh," she responded.

They scrambled in the Scout and motored to the first leg of their short journey, Ardian Locksmith. Albert excused himself.

Tom said to Norma, "Tell me to knock it off if you like, but are you having second thoughts about this venture?"

Norma answered, "Why did you say that? Earlier you hemmed and hawed with 20 questions about asking Albert, God knows what, yet you flat out asked a rather personal question."

Tom said, "I'm not sure, but I thought I felt a tension, a hesitancy, maybe, an awe."

She placed her hand, palms up in front of him, and said, "Do you read palms, too?"

He delicately put one of his hands under hers and very lightly rested his other hand over her four fingers to further expose her palm.

Norma was strangely startled. A warm hot surge flowed. She flinched, and her sensory muscles alerted her. She whipped her hand from Tom's tender grip.

Tom said in an apologetic voice, "I'm sorry, Norma, I didn't mean anything. I was just going to pretend to tell your fortune." He added, "It won't happen again."

She felt foolish, a grown woman. Her mind certainly hadn't reacted, those automatic contractual muscles had taken a split second command. She said, "Tom, that was indignant of me to react. Believe me, I didn't mean to imply, what I'm sure you did, and. . . ."

Tom interrupted and stated, "Albert's back."

Albert got in and dangled the set of keys and said, "Here, take these Norma. Now we all have a set." Albert drove to the Crescent Cafe and they elected to sit in a booth.

The waitress came and Norma asked before she greeted them, "Do you have cigarettes here?"

"Yes, we do, but you will have to pay for them, you know."

Albert took a twenty dollar bill out of his billfold and gave it to the waitress.

Norma smiled and said, "Thank you, Albert," then turned her attention to the waitress and said, "Please, as many packs as that twenty will buy, any kind, except Kools, and no menthols please." The waitress returned with Norma's precious cigarettes in a sack and inquired if they were ready to order.

They answered "yes" all around. The men waited for Norma, and she said, "I'll take the Spanish sandwich on rye bread, and black coffee."

Albert and Tom ordered sauerkraut and wieners, a small salad and coffee.

Norma said, "I would like a salad, too, thousand island."

Norma asked Albert, "Is this your favorite eating out restaurant, Albert?"

He smiled and stated, "I don't know, first time for me, too! I hope the food is tasty."

Norma asked, "Tom, do your charms, spells and incantations, give you any insight into the quality of our ordered lunches?"

Tom said, straight faced, "Absolutely. Albert's and mine will be good."

She asked, "And mine?"

He said, "A delightful treat?"

She laughed, "I'm not all that easy to please when it comes to food although, seriously, I did enjoy my breakfast."

Tom said, "I did take courses in the culinary arts."

Norma said, "My, my, did you know, Albert, we have a real chef?"

Albert replied, "It really doesn't surprise me. Both of you are seemingly blessed with more than one talent."

That really brought smiles all around. The waitress plated them with hovering majestic odors. Both men picked up their forks to attack, and were halted by Norma's gleeful, "See, see," she pointed at her Spanish sandwich. "They didn't respond. Hey, guys, look closely. It's sliced diagonally twice."

Tom said, "Oh, my God."

Albert joined in, "Just like a cat, always lands on its feet."

Norma smiled and picked up a quarter sandwich and sunk her teeth in a generous bite and remarked, "Mmm, this is fabulous." She put the bit of her sandwich under each of their noses, then withdrew it laughingly. The men's sensors were filled with the odors of kraut and failed to perceive its odor. They nodded anyhow and were busy filling their awaiting stomachs. Norma waved for the waitress, and caught her attention. She asked if she knew the ingredients of her Spanish concoction.

The waitress replied quickly that it contained onions, peppers, grated cheese, olives, and sour pickles all chopped and minced with mayonnaise.

"Thanks," said Norma, and put a buck in her hand.

The waitress refilled Norma's cup first and smiled a smile, that only other women could interpret.

Norma said, "Albert, you heard the goodies that went into this lovely tidbit. Do we have the stores to duplicate this sandwich?"

Albert napkined his lips, and answered, "No, I'm afraid not. We have mayonnaise and onions, but no sour pickles, pepper, olives, or rye bread."

Norma said, "I asked for a tub and was refused. May our kitchen be enhanced to house the makings of this wonderful sandwich?"

Tom asked Norma, "Isn't it a delightful treat?"

She laughed and said, "You just got lucky."

Albert said, "Tom was very precise in his wording. He gave no quarter."

She said, "Oh, you guys, you sure stick together."

The waitress made a bee-line for Norma and bent down and whispered something in her ear. Both men noted, in their fantasies, the feminine posture in that position. It spurred their hormones, and their eyes briefly met in mutual appreciation.

Norma said, "Well," and broke their trance.

"Certainly yes," said Albert, hoping she did not read what had just traversed as a common reaction in males.

Tom said, "Norma after this, don't ask my opinion on a subject if you take my magic frivolously."

Norma replied, "I didn't mean to hurt your feelings or your ego."

Albert said, "Norma, just a tiny word of caution. It's in your best interest not to criticize Tom too harshly. After all, he is the cook, and you want the substances to reproduce your sandwich."

Tom said, "That's not necessary, Albert. If she failed to take my prediction without a grain of salt, I'll keep my mystic powers under wraps."

The waitress was cleaning up the next booth and heard the entire conversation. She stopped her cleaning, and once again bent down and whispered in Norma's ear. The two of them laughed and the waitress was still snickering as she left.

Tom said, "What was that all about?"

Norma replied quickly, "Just girl talk."

Albert said, "I can imagine."

Norma really laughed and said, "No, Albert, there's a dark day in hell that you could imagine what she said."

Tom said, "I don't know. Do you want to me to take a guess?"

Norma said, "No, you just read me the riot act about my needless and wanton actions on subjects just like this. Besides," Norma said, "it just might embarrass you or Albert, or both of you."

Albert turned to Tom, and said, "Just what the hell is she talking about?"

Tom said, "Girl stuff."

Albert said, "Oh," then said, "Well we had a good lunch and a very interesting conversation. How about stopping at the store on the way home, and you two can negotiate over foodstuffs and expenditures for such."

Norma said, "Before we leave, speaking of expenditures, how are we to be paid--cash, check, week, or month?"

Tom joined in and said, "She's right, you know."

Albert said, placably, "I really apologize. No excuses. Just got caught up with day by day curriculum. Hope I haven't jeopardized a serious part of our working together, trust and such. How would once a month be?   I'll write you both a check when we get home. You can use the same bank I do, if you like."

Norma said, "Sounds good to me."

Tom said, "Great, let's go and do some serious grocery shopping."

Tom noticed that Albert had left a twenty for the bill, a tip of over five dollars. He smiled, and evaded all eyes.

Albert slipped some folded bills in Tom's hand and said, "Take this for your shopping." It was several hundred dollar bills.

As they left, both watched as Tom went over and whispered in the waitress' ear. She blushed and let out a startled gasp.

Tom smiled and assured her everything was all right. The waitress looked nervously at Norma, then turned away. Tom was still trying to comfort her.

Norma said, as she caught up to Tom, "What did you say to that woman?" For a moment there was terror in her eyes.

Albert witnessed it too and was so moved, at the change in the waitress, he thought, "Did you have the right to condescend, Tom?" Albert glanced at the waitress again and saw she was all right. He saw her mischievous smile.

Albert looked at Norma and somehow knew she, too, saw that secret smile.

Tom said, "I didn't mean to raise a ruckus, just testing." The three of them were standing near the door.

Norma said, "Okay, King's X."

Tom said, "Not fair."

Norma said, "Why not?"

Tom thought that one over and said, to the delight of Albert, "You don't have a King's X to give."

Albert said, "Tom, you never did tell me that part. Do you buy them, inherit them, or what?"

Norma said, "Thanks a lot, Albert. Well, your turn, Tom."

Tom had to laugh and said, "Oh, what the hell, I did use one King's X on Albert, although it was for your relief, Norma, not mine."

Norma said, "It would have been yours. You would have had to clean up the mess." Norma waited then said, "Or, Tom, I'll split the last King's X, but as right has it, as you said, Tom's King's X's ceases all actions, and revealment of truce follows."

Tom said, "Okay, what do you want to know? Just one thing for one King's X."

Albert started to speak, and Tom said, "No, no, it has to be from the one that calls King's X."

Norma said, "Looking directly into Tom's blue eyes, "What did you whisper to the waitress?"

Tom said, "By King's X law I must tell you but nothing is written or unwritten as to where. I promise to tell you in the car."

Norma looked at Albert for approval.

He smiled, and she said, "Okay, you got a deal." They left and scrambled into the Scout.

Norma said, "Albert, please don't start the car until we complete our deal."

Albert sighed and said, "I have to say you aroused my electrolytes. It's confession time, Tom."

Tom said, "It's no big deal."

Norma said, "What did you say to that waitress?"

Tom said, not looking at Albert, "I told her what she whispered in your ear the second time."

Norma responded too fast, "Bull shit."

Albert said, "Norma, is that what the waitress said?"

"No, she didn't say that."

Tom said, "No, she didn't." She said, "that magic wand was probably between his legs."

Albert said, "Good Lord," and grinned.

Norma said, "Well, I'll be damned. Yes that's what she said, but how did you know? Her hand was over her lips, and you couldn't have overheard her. Besides, it was just girl talk."

Norma thought, then said, "Tom, don't tell me you are telepathic."

Tom replied, "Yes, you might use that terminology."

Norma said to Albert, "Start the car."

Albert did as he was told.

Norma asked Tom, "Do you fly too?"

Tom smiled and said, "No, but I'm still not too old to learn."

Albert wheeled in the parking lot to a grocery called Shop Grocery on 6th Street, backed, and then said, "Well, here we are. Go get 'em gang. I'll wait here and do some pondering on our format for tomorrow."

Tom said, "Okay, slave driver. My partner and I will pick up the goods. Anything special I can get for you?"

Albert said "No, but thanks," then added, "The clouds are beginning to churn. We might be in for showers."

Norma appraised the sky as she was getting out of the car and said, "Might see some electrical discharges, too."

Tom said, "Come on, Norma. I'm a fast shopper."

An unpatterned storm front was moving in. The cool and warm fronts would have a convention. Low, long rumbling resounded in the distance. The storm was quickly shielding the remaining light.

Norma and Tom had a good time kidding each other on nearly every item. Tom rolled his laden cart with Norma on his left to where they had left Albert and the Scout. Both were gone. The thunder made the disappearance even more dramatic.

Chapter 3

Norma was visibly startled and looked around nervously and said, "Tom, I'm certain this is where we got out."

Tom did a full circle and excitedly said, "Jesus, Albert wouldn't just drive off. Something stinks. You stay here with the groceries, I'm going to phone for a cab and then call the police."

Norma said, "Hold it, Tom, no police. Think about what Albert would do."

He said, "Yes, of course, you're right. That was dumb of me, but I'll get us a cab."

Norma was still shaking as she dug out a cigarette and lit it up, then scanned the entire parking lot. She did notice two men walking side by side, dressed in overalls, who were walking in her direction. She bit her upper lip as they closed the distance between them. Both men appeared in step, with their heads down, wearing caps. She glanced to where Tom had headed and saw no one in sight. She focused on the two men. They were still coming on an intersect. She opened her purse, withdrew a small cylinder, and positioned it in her right hand. She gritted her teeth and muttered, "Come on, you bastards." Her left hand and arm was poised like a pent-up spring on the loaded shopping cart. She left-handed her cigarette. It glowed until it hit the asphalt. She was wearing a low flat shoe. It was semi-cleated, so that she could maneuver quickly on the parking lot. She pulled the cart tight to her body and glanced, at her tightly held canister. She noticed a can of jumbo pitted olives on top of one of the bags of groceries. Then she remembered how heavy it was. Her peripheral vision was exceptional. She captured just a quick partial image of Tom on her back left side. He had seen the two men closing on Norma and was hurrying. A can of something was flying, in a spiral spin, within his grasp. He caught it and rifled it at one of the men. The intruder screamed as it stuck and cut into his neck. Blood spurted, just missing Norma. She pressed the release trigger of mace, as the can struck the other man. Simultaneously, she hurled the laden wheeled cart, striking the other oncoming man in his kneecap. He cried sharply, expelling a high pitched, "You bitch." They both ricocheted, one, racing with a limp, with his hands squirreled in his burning eyes, the other racing with both hands on his neck in desperation. The two men displayed pain in their retreat with cursing and bellowing. A slight breeze swept the devastating properties of the remaining mace. Norma gave a thumbs up, and said excitedly, "Good shot, Tom."

He reciprocated with vigor, "Damn it, Norma, you're one hell of a woman. That was a dandy pitch. What the hell's going on?"

The rolling thunder continued to swell, then relapsed to a low hush, only to reverberate with moans of muted volume and rhythm. It was not cold, yet she trembled as she replied, "I wish to hell I knew. Now that it's over, I'm scared."

Tom waved as he saw the yellow cab approaching the lot and said, "Look at my hands." She saw them shaking.

The driver exited the cab and opened the trunk and helped unload their cart. Norma had walked a few paces and retrieved the can of olives that had a splash of blood on the can. They left the cart in the middle of the lot, bristling with flashes of light dancing when the black clouds swallowed its chrome frame. Although the back seat was wide they huddled, permitting body contact, each self-content, that the touch between them was an aura of safe haven. Tom stated, "I wish I was a drinker or a smoker. I'm still feeling giddy."

They arrived safely at their quarters, tipped the driver, and piled the heavy bags of groceries on the kitchen table. Norma quickly emptied the bags, putting the perishables away first. Some were left on the table in front of Tom. He grabbed a chair and just stared.

She saw the spooky look in his eyes. It hurt her to see a talented man become so unstrung. She hurried to a small bar Albert had evidently set up in the anticipation that one or both would welcome a drink upon arrival. She went back to the kitchen for sugar and lemon. She retraced her steps and mixed one teaspoon of sugar and two ounces of gin into a tall glass and added a squirt of pure lemon. She hesitated, then added a little more gin, and took it to the kitchen. She searched in the refrigerator and found a can of 7-Up, adding a little of the latter, and stirred generously.

She placed the tiny popping and bubbling drink in front of his glazed blue eyes. She said nothing. She moved herself behind him and grasped his right hand from his lap and guided it around the drink. She wrapped his fingers around the glass and lifted it until it gently touched his lips. She applied a pinch of pressure and hit his teeth lightly. He opened his mouth, and she watched him grimace during the reverberant libation. She forced his hand upward as he apparently finished the tasty concoction. She retraced her movements with his hand back to the table and had to pry his cramped fingers from the glass. She picked up and downed the little sip that was left. She licked her lips in appreciation. She had seen and personally experienced this delayed trauma and realized the power that it possessed.

As time passed, Tom blinked, coughed, and shook his head and shouted, "Christ, what was that?"

"That," she said, "was Tom."

"What," he said?

She said, "Tom."

He shook his head and stated, "Norma, we are in one hell of a mess. This is not time to play games."

She replied, "I'm not playing games. Do you feel better now?"

He nodded, and said, "Yes, yes, I do, but kind of giddy. I don't mean horsey getty-up. I feel rocky. What was in that?"

She replied, "Did it taste good?"

He nodded.

She continued, "It's called a Tom, officially a Tom Collins."

He said sincerely, "Thank you. What the hell are we going to do about Albert, and who were those two men in overalls--too much of a coincidence."

"I know," she said and added, "I wish we had pushed Albert for more information, but we had no idea this would be a dangerous assignment."

He responded, "We should have realized with all the bucks involved that there would be an adverse side. Someone would retaliate to gain information."

She said, "God they must have thought we already had some vital info."

Tom asked, "Norma, do you remember the license plate number?"

"No," she replied, then added, "I guess I was too busy trying to second guess those goons."

"Me, too," he said, and added, "We better go through Albert's desk and room."

She said, "This sounds so morbid. You go ahead. I'll finish putting the grub away, and then I'll join you."

He nodded, and was slow getting up. "By the way, that was a Jim Dandy drink."

"No," she replied, "It was a tall Tom Dandy."

He walked into Albert's room and went to the desk. He felt he probably wouldn't find anything in such an obvious place, but methodically proceeded, without a clue of what he was looking for. Norma entered the room, smoke rising in a trail that appeared to be like an elusive scarf on a windy day.

She asked, "Any luck, Tom?"

He turned and looked up, and answered, "No, I really didn't count on finding anything too accessible."

She said, "I just don't know. We have no leads, but, thinking about it, we do know Albert was a very intelligent man, We must put ourselves in his place."

Tom stood up, and crossed over to face her and tensely stated, "God, Norma, you just past tensed Albert."

She looked hawkish, "Christ, I didn't realize. This is spooky, but I feel he's still alive." She inhaled, blew a cloud, and looked about for an ashtray. She went to her bedroom to put her cigarette out.

Tom followed her and asked, "Are you a all right?"

"Yes, yes. I think so. I just wanted to put my cigarette out."

"Oh," he said, and looked about and commented, "Your room is as plain as mine."

She sat on the bed and placed her hands on both side of her face and stated, "Tom, we have to make some sort of plan. Maybe if we just talk it out, we can come to some sane conclusion and lay out a contingency."

"Okay," he said, and sat next to her on the bed. He asked, "Did Albert say anything when he negotiated with you on the phone?"

She thought, carefully recalling his conversation. "No, nothing, that would hint to his disappearance. How about you?"

"No," he said quickly, I ran it through my mind before I asked you. What about money?"

"What about it," she said, and added, "of which I have very little, remember. He was going to write out a month's wages for both of us."

"That's right," he said.

Norma asked, "Wasn't it in his desk?"

"No," he said, and trailed off. "It wasn't in his room either."

She clinched her hands together and brightly said, "Of course, there was to be a work place for all of us. His check book might be there and maybe some serious answers."

Tom got caught up in her excitement. He held up his index finger and spoke, "Yes, I believe you have hit on the next place to probe. Just for information, Albert said he was going to send his findings to Chicago on Van Buren Street."

She gasped his finger, then his hand, and a vibrant, nervous, warmth stimulated her body. She was conscious of terrific biological sensations.

He said questioningly, "Are you all right?"

She released his hand and slowly came back from somnambulism. Her black eyes had hidden granules of moisture that were perceptive to Tom and had him speaking again.

"Norma, you have beautiful eyes. They radiate and dance. Jesus, are you going to cry? Don't cry."

She pulled a crumpled tissue from her smock pocket and skewered her eyes, and replied, "No, I'm not going to cry, sorry. Did you say my eyes were beautiful?"

He answered, "Yes, I did. They were so, so. . .luminous I couldn't help myself. What was I going to say? Oh, yes," he sighed, "That question that I wanted to ask Albert, the one you and I sort of haggled over."

"Yes," she said, "What was it that bugged you?"

He said, "Before I answer, just tell me one thing. When did he call you? How many days ago, do you remember?" She responded quickly, "Of course, it was six days ago."

"I knew it," he said. "That was the same day he called me, that's my dilemma. How is it possible to have that 38 foot craft with a built-in electronic antenna, I presume it is new, rigged with all of the equipment, special ordered, when he only contacted us six days ago?"

She thought that over and joined in the mystery. She grabbed his arm, "Tom, that ship comes in tomorrow, remember?"

He replied, "It's not a ship, it's a yacht."

"Silly," she continued, "I know that, but the ship is named *Delta* and is arriving with the yacht as part of its cargo. He also told us the craft's name was *Magnetic*. He squeezed her arm and shifted a little on the bed and looked directly into her eyes. "By God, you're one hell of a partner. Let's call the harbor master and find out if she has weighed anchor."

She got up off the bed and said, "You men are always calling seagoing vessels she's.   I've never heard what I call a logical answer yet."

He replied, "I'll give you a quick one. How would it sound if they announced, all aboard, a him ship."

She grinned and said, "Thanks Tom, you've answered one of the mysteries of the sea. Really,

I like your answer. Anyone who said that he had curves and sleek lines, might be in the closet."

He stood up, turned, and looked down. Norma attentively followed his gaze. He was looking at the bed. Two very close impressions, were noticeably still present, the depressions still evident of how very close they were. He wondered what a hell of a time for sensuality. This was not the time for romance. They must find Albert. He broke his own tension. He matter-of-factly said,

"The information we were going to gather, from the universe was to be sent to a company where men dealing in futures trading are engaging in the most speculative type of gambling in the world. Speaking of money, how much do you have in your purse, or suitcase, or whatever?" He was counting the meager bills in his wallet.

She fetched her purse and took out her wallet. She counted out twenty-two dollars. She said, "Forty-two dollars."

He looked at the folding green in her hand and stated," I see only twenty-two."

She grinned and stated, "The smart gals always have some kind of poke for emergencies, especially when guys dump us because we won't put out."

"Oh," he said, sort of peevishly, "I suppose you have it tucked away in your bra."

She flashed her eyes and countered, "You suppose wrong, take my word for it, all I have is forty-two bucks." How about you?"

He said, "As a matter of fact, I've got thirty-six bucks, no hold outs."

"You men," she said smartly, and went on, "A poke is our life raft in this male dominated world. If I choose I would have kept my mouth shut. That my man, would have been a hold out."

He said sheepishly, "You're right, of course. I didn't mean to get your dander up. We do have food for a few days and, I assume Albert paid in advance for our quarters. But I don't know about the cruiser, but it will be on the bill of lading."

She walked, in short turns, and added slowly, "Even if he paid for the boat, the handling charges would exceed what we have."

"Damn it," he said, then snapped his fingers and said, "We have to find his checkbook. That would tell us what bank. Although I've only known Albert as long as you, I know he would want us to take every advantage, even to forge his name on checks."

Norma nodded and stated, "Yes, Tom, I agree. We know, at least we can assume he didn't carry it on him. He did give you a couple hundred dollars. He must have a safe place for his chronicles, his secret phone numbers, bank account numbers--well, you get the idea."

They both sat down on the bed again, almost in the same positions as before. Both looked into each other's eyes briefly, then quickly looked way, hoping each did not discover the physical bond that was evident in their eyes. All was quiet. They dared not shift their positions for fear the other would interpret it to be a negative reaction. They both enjoyed their close intimacy and had silly unsure looks on their faces but would not confront one another. Two mature grown-ups, caught up in one of life's wonders. each sorting the alternatives in their minds, of how to handle this pleasant, but frustrating situation. Pros and cons slashed through their memory banks.

Norma, with a natural grace, raised her left hand and laid it on the left side of Tom's face. He turned his head as he felt her touch, matched it, and looked into her lovely face. It was like it was in slow motion, so wonderful, so natural. He put his right arm around her. Both were pumping exciting messages as their lips touched, then pressed. Their bodies reacting, their arms hugged tighter.

The magnetic flux flowed between their hot, fiery responses. Each sensuous appetite absorbed the other's as their love was shared. Her body trembled, his quivered, each joining in rhythm. They sapped their bodies with love's majestic harmonies. Their lips separated and each feasted on each other's eyes, seeing the true resonance of genuine love. The purest affection for each other radiated with each move, as they undressed each other. No words were spoken. They were poised for what was to transpire. They timed undressing each other. Both looked at the other's loveliness, contentedly smiling, followed the coiling of warm arms about each other as they both bantered for breath, then they kissed tenderly.

She drew him to her, and he captured her in a sexual embrace. Their flowing sweetness rising in emotional spasms. The bouquet of mutual arousing, and the strong sexual awareness, blasted from their pent-up desires. A cataclysm of soft but violent moments, every inch of their bodies trying to absorb their ecstasy, if only to hold the rapture of pleasure for just another exquisite moment.

Tom rolled off her gently. It took all his strength, exhausting him physically with glorifying emotion. The fleecing of her pinnacle emotions, the excitement still lingered in her loins. They lay in perfect contentment. Both were thankful, that they had just experienced the most sensual saturation possible between man and woman.

Norma spoke first, "Tom, I'm not going to say I'm sorry about what happened. I want you to know no man has ever aroused or excited me like this before. I could have killed at that wonderful swelling of love. God, I'm so happy. That was atmospheric."

It was Tom's turn to relate as they lay side by side. "Norma, it was not my intention when I sat down next to you on the bed. By God, all premises were shut down."

He turned his head toward her and she did the same. She graciously rolled on top of him and said glowingly, "Tom, I want to express my gratitude, as a woman, who just had a voluptuous sensual awakening."

He said, excitedly, "Oh, Christ, your body and that word voluptuous." Their lips tasted each others, as their body parts enveloped. Their passion was resounding. Norma screamed in resonance. He pulled her closer and placed his hands on the sides of her face, comforting her through her spiral pleasure. She collapsed on top of him. He had to move his head to one side to breathe. Her scream was evidence of her complete fulfillment. He smiled and closed his eyes on this golden moment. They slept.

Tom thought he heard someone talking, a long ways off and was stunned to hear Albert say, "Tom, did you change rooms?"

"Jesus Christ," said Tom excitedly, and quickly added, "Good God, Albert, what happened to you?"

Norma had transposed from sleep, and saw Albert standing at the foot of her bed, his suit roughed up a bit. She screamed, and that startled Tom, who was embarrassingly sitting up in bed, facing his mentor. Tom had to help Norma cover her lovely breasts.

Norma asked in confusion, "God, Albert, where did you go?"

Albert grinned, and asked, "Did I walk in on my wake?"

Norma and Tom exchanged nervous glances. She said, "Oh, God, how this must look to you."

Tom said in a defensive voice, "Albert, I can explain everything."

Albert stated, "I'll bet you can, Tom. You both continue to amaze me at every turn."

Tom said, "For God's sake, Albert, I'm…. we're so glad to see you."

"Yes," said Norma and added, "you scared the pants off us when you disappeared."

Tom thought about what she had just said and gave her a queerish look.

Albert said, "I won't comment on that statement."

Norma flushed, and looked ridiculous.

Albert said, "Well, I'm glad about one thing."

Tom asked, "What is that?"

He continued, "When I walked in this room, I thought one of you was going to King's X me." He tiredly said, "I'm going to the kitchen and put a pot of coffee on, will you both please join me?"

They nodded and added, "yes."

Albert turned to leave, then stopped, turned around and asked, "What time is it?"

Tom checked his watch and said, "Five after two."

Norma asked again, "Albert, what the hell happened?"

Albert held up a hand and said, "I'll tell you both over coffee and maybe a sandwich. Robes will be all right. Don't get fully dressed."
As he left, they could hear him say, "It's comforting to know how protective you have become, Tom."

Norma laughed, as she saw the guilty look in Tom's eyes.

Tom muttered, "I wish he hadn't said that."

Norma still smiling, said, "You men, sometimes you're so silly."

Tom added, "It's not silly. You just don't understand.

She replied, "I don't want to, you better get to your room and get robed, I can't wait to find out what happened."

They both swung out of bed, graciously appraising each other's nakedness.

Albert had just about completed a baloney sandwich, slathered with mayonnaise and mustard. The coffee had started to perk. Norma reached the kitchen first and hummed, "Mmm that coffee smells good. You had us scared to death, mostly the latter."

Tom heard that as he came in wearing a thick maroon bathrobe and added, "Yes, that's true. She even past tensed you."

Norma put her hands on her hips and stated, "Sure, Tom, tell all, and I'll relate the sad tale of your behavior."

Tom said, sheepishly, "I'm sorry. I did it again. It's just that I'm so happy you're alive." He looked at each of them and was temporarily confused and added, and "apparently you're physically okay."

Albert, with his sandwich, slipped into the kitchen chair. He took a generous bite. Norma noticed and stated, "You must be hungry."

Tom said with conviction, "That's obvious."

Between bites Albert said, "Thank you both for your concern as to my safety." Tom's nose interpreted the aroma and moved to the perking pot. He asked Norma, "Want coffee?"

"Please," she replied and added, "What kind of sandwich is that?"

Albert listed the ingredients and she responded with a lackluster, "Oh."

Tom set out cups and saucers and did the pouring. Norma lit up, and smoked impatiently.

By quick nervous puffs, Albert alternately blew and sipped his coffee with pleasure. His eyes were intense. "Before we discuss my disappearance, there is a very serious situation that I walked into this morning."

Tom and Norma briefly locked eyes, and sheepishly waited for the chewing out that was going to be levied on them.

Albert said, still munching smaller bites, "I want you both to put yourself in my position. He turned and fired a question at Norma. "What would you do if you just walked in and you were in charge of this project?"

Tom quickly said, "Albert, if you're going to criticize, lay your questions on me. It was my fault. Norma had nothing to do with it." Tom's eyes shifted, then blinked. He just heard what he had said.

Norma stated, "Don't try to be so gallant. Both of us knew what we were doing. We're not teenagers."

Albert replied, fanning smoke, "Well, you certainly acted as such," and drank from his cup, still in a swirl of Norma's blue smoke. He added, "I don't want this to happen again, leaving our quarters unlocked. You both know how important secrecy is. All I had to do was turn the doorknob, and I was inside, home free."

Norma and Tom, felt instant relief. Tom asked carefully, "If I had locked up, you would have used your key."

Albert said calmly, "Not this time. They took everything that was metal and paper." He pointed to his right wrist, that no longer embraced his watch.

"Jesus Christ," said Norma, "What happened?"

Albert sipped from his coffee and related his story. "They came very fast through both doors of the Scout. One had a knife and gave warning that he would use it, if I did not cooperate quietly. They were white, in their late 20's or early 30's. They turned on the key and drove south on Tudor, then west."

Tom interrupted, "Didn't they blindfold you, or tie you up?"

Albert went on, "No need in hiding where I was going, and, since they were on each side of me, I couldn't move." He tapped his cup and said, "I'll proceed with my story."

"Sorry," Tom said, and looked across to Norma's vivid eyes. She missed his signal, and blew her last cloud, and ground the spent cigarette in her saucer.

Albert began speaking again, "They turned and circled Goose Lake Park that adjoins Alaska Pacific University."

Tom and Norma searched each other's eyes for any recognition of streets or buildings.

Albert noticed the temporary distraction and answered, to the amazement of Tom and Norma, "It's approximately eleven miles southwest of us. They found a lonely spot in the park, and took everything of value. They tied me up, then gagged me. Then they climbed in the Scout, grinned and waved goodbye. It took hours to untie myself. Then I had to hoof it back here. It was a long walk. No one stopped to pick me up."

Norma said quickly, "Why didn't you just hail a cab?"

Albert responded, "And if I did arrive here, and both of you were not here, no key to get inside, no way to pay the driver. . . ."

Tom said in speculation, "No poke, uh, no hold out?"

Albert replied, "Poke, holdout, what's that all about?"

Norma laughed and said, "A little joke about keeping a buck, hidden for emergency."

Albert said, "Thank God, I thought it might be phase two of King's X." They all laughed.

Tom said in earnest, "They stole your Scout, kidnapped, you and threatened bodily harm. Let's call the police!"

Albert said sternly, "Nonsense, we can't let the police in on our identifications, why we are here, and alert God knows who. I loved that four wheeler, too."

Norma quickly lit up, and nervously said, "God, you're talking about big bucks down the drain.

Tom joined in, "I don't know how much cash you had, or the price of your watch, but that Scout had to cost well over $20,000."

Albert asked for some more coffee, then took a long sip, and stated, "In hindsight, I should never have bought such a new and vulnerable vehicle."

Norma brashly stated, "You act as though it partly was your fault. What happened."

Albert answered, "It was, you know a new four-wheel drive, a high price up here. I should have realized that before I bought it."

Tom said, "Albert, I went through your desk and room. We thought after realizing our mutual finances, that we had very little."

Albert asked in a matter of fact voice, "If you both would have found it, would you have tried to forge my name?"

Tom looked at Norma for guidance, and both assumed they were on the carpet. She spoke up, "We both decided that was the best procedure."

Albert smiled and said, "I agree. I would have done the same. I carry my checkbook with me, they found it by accident and took it also."

Tom asked, "Do you think they will try to forge your name?"

Albert quickly responded, "I think not."

Norma asked, "Why not?"

Albert replied, "When they found it, they talked openly about it and decided to take it and destroy it later with the other things in my billfold that were useless to them. I remember one of them saying that checks are dumb, the percentages are poor. Yes, and with that news, I'm sorry, I had to renege on your month's pay. That will be taken care of this morning or afternoon. I want all of us to go to the National Bank of Alaska, on Fourth Street. I can get new checks and both of you can cash your checks and set up personal accounts," then added, "or joint accounts." That brought exciting, but nervous glances.

Norma blew smoke from a fresh cigarette and Tom looked about, avoiding all eyes. The smoke hung over them like a blanket. Norma broke the tension, "Thank God, you're safe, Albert. You gave us quite a start."

Tom added, "We thought that incident at the parking lot, when we came out to find you gone, was related to your disappearance."

Albert asked, "What happened?"

Tom detailed the complete scene and then added, "Norma was outstanding."

She countered," You weren't so bad yourself. That was a nifty throw with my olives."

Tom stated, "You just said my olives."

She said, "Well I requested them for my Spanish sandwich."

Tom retaliated, "I'm the cook, yet you don't hear me saying, 'pass my catsup, I'm stirring my beans, I'm toasting my bread.'"

Albert waved his hand in the air and said, "It's good to be home, it truly is." He smiled and asked, "What time is it?"

Norma responded, "2:25 a.m."

Albert said, tiredly, "Let's leave the dishes and hit the hay. It's been a bitch of a day."

Tom added his thoughts and said, "The audacious word of the day would be "bleak." He looked into Norma's dancing eyes.

Albert took note of the language and said in a humoristic manner, "Please lock up, Tom. I hope all of us have a good morning's rest. We have a full schedule today."

Norma said, "Sleep tight."

Albert went into his bedroom and closed the door. She purposely waited for Tom to lock up. She stacked the cups and saucers in the sink. Tom came up behind her, then hesitated. She felt his closeness, and quickly spun around to face him. Their eyes met in harmony. He tenderly put his arms around her and said, in a defeatist tone, "I can't keep my hands off you."

Unknown to Tom, she bit her lip and uttered, "Oh," then said, "Tom, what are we going to do. Albert is such a dear. We must have disappointed him as associates."

Tom added, "And as friends."

She said, "That too. I don't want to be the heavy in this affair, but by God, Tom, I'll use a King's X if I have to."

"Christ," said Tom, "Don't do that. You know how much I respect that remark." She hugged him, and he hugged her back. Norma said,

"Remember what Albert said at the table about walking in, and you and I were thinking of something else. He only chewed us out for not locking up!"

"Yeah," he said, and added, "One hell of a day."

She responded. "He is a sweetheart. That was his way of telling us to get a handle on the situation, but he sees possible flags or skulls or crossbones if we deter from our work."

He said, as he kissed her neck, "You would make one hell of a pirate, and yes, we know he is right." He looked into her eyes and said, "Can you now, in truth tell me it won't affect your contribution as a scientist?"

She pushed away, and strikingly stated, "You bastard, you loaded that statement with an impossible answer."

He said, questioningly, "You called me a bastard?"

She replied quickly, "A figure of speech only, but that was a loaded question."

"Good," he said and added, "That's the way I feel about you. What do you suggest?" He kissed her tenderly, on the other side of her neck.

She asked, "Tom, do you like old movies."

He answered, "You bet, comics, good strong lines and romances that are plausible. Why?"

She asked, "Do you remember an old comedy team in the 30's, with Wheeler and Woolsly?"

He kissed her on the forehead and said, "Yes, I saw some of their old films."

She said, "Well, that's my answer, Tom." She placed her hands on both sides of his face and asked with her eyes.

He thought for a brief moment and asked, "I don't know what you're driving at--give me a hint."

She said, "Okay, the scene was this. Remember this blonde was removing his clothes one at a time, and he was singing a song."

Tom's eyes brightened, and quickly and happily said, "The song was 'Just Keep on Doing What You're Doing, Although You Lead Me into Ruin,'" and they both laughed.

She said, "That's my baby, that's my answer, and don't stop."

Tom kissed her lightly on the lips, savored it, pressed lovingly and let his right hand gently pinch her left ear. He then let it drop, and slid it gently down her face. They held each other, neither wanted to let go.

She said, "As much as I love what you're doing to me, we had better get some rest, and begin earning our money."

He squeezed her and replied, "You're so right, yet sometimes, so wrong."

She answered, "I know. That's the story of my life."

He said, "One last good night's kiss."

"No," she spat back, then added, when he appeared rejected, "Yes, to one morning last kiss."

They did, and went to their separate bedrooms, each with the scent of the other's sweetness and love.

Chapter 4

Tom had set his alarm for 6:30 a.m. He wondered in confusion, and disbandonment what the annoying noise was. He attempted mentally to neutralize the piercing waves. The alarm won, and he fumbled, trying to shut it off. Then reality set in--his body functions were alerted. His first and dominant thought was Norma. God, what a lucky bastard he was. Swinging out of bed, he over reacted and tumbled to the floor. "God," he muttered, and thought how foolish he was acting. He then countered, "What the hell, I'm in love." He was an addict to the excitement. He got dressed and made a beeline for the bathroom. His next move was directly to the kitchen. He set up the coffee first, then got serious with potatoes, sausage, eggs and toast. He was busy, he took his preparations in earnest, and the exotic aromas were stimulating the others.

Norma, upon arising, failed to win the bathroom. She was about to savor, stronger odors, and turned to leave when Albert exited the same and greeted Norma.

"A good morning, to you, and thanks."

She turned and replied, "A hearty to you. What's the thanks for, thankful for being among the living?" She was sorry she said that and was about to tell him so.

He smiled and said, "This is a wonderful beginning, non-interrupted bathroom, No King's X, and those magical morning aromas."

She slid by without comment and closed the door. Tom saw Albert coming, poured his coffee, and they exchanged morning compliments.

Tom stated, "Eggs coming up, just the way you like 'em. I've added spuds."

Albert spooned his coffee and looked toward the direction of the bathroom and asked, "Everything savvy between our other colleague?"
Tom replied, with warmth, "You bet. I had one hell of a time concentrating on this preparation, but I did just that." He paused, then added, "Albert, it will not clash, or inhibit our abilities as a team; in fact, it will enhance it." Tom placed Albert's delicacy in front of him and was greeted with a nod of satisfaction. Tom had even pre-buttered the toast. There were generous portions of sausage and fried potatoes.

Albert said, enjoying his food, "Tom, you would make a hell of a good wife."

Norma cruised in with smoke trailing and in a perky voice said, "Good, good morning, guys."

Albert savored the sausage in his mouth and replied, "Good morning, Norma. My aren't we cheerful this morning?"

Norma replied, "Aren't I always?

Albert gestured, with his steaming coffee in a salute, "No you, yourself were quite adamant about not being a morning person. You said, I believe, you're a coffee, toast and cigarette gal, not necessarily in that order."

Tom laughed and filled Norma's cup with the morning succulents.

She graciously smiled, with both her face and eyes and replied, still emitting a smile, aimed Albert, "That was then, this is now."

Albert laughed and jokingly stated, "Is that a scientific fact?" He repeated what she said, "that was then, and this is now."

Tom said, "It sounded better when she said it."

Albert added, "A wonderful breakfast."

Tom grinned, and pointed to the plate, he was placing in front of her and said, "Equal portions."

She said demurely, "Thank you, Tom."

Tom pointed at her toast. She noted, and added, "That was sweet of you to quarter my toast, and butter it too."

Albert continued, "Congratulations to both of you on the selection of sausage. It was delicious."

Tom said, as he brought his plate in a condescending wave, "It's Italian sausage." Tom forked a generous portion of potatoes, chewing, and smiling, and said, "Norma, that wasn't a very literary answer, "'that was then, this is now,'" and quickly added, "Don't say, 'that's girl talk'."

Albert punctured his last egg with a piece of sausage, hesitated, then looked up and said, "Yes, you're right, Tom. It does sound better coming from her."

She proudly dunked her toast into her basted eggs, successfully guided it to her mouth without a drip, and confidently stated, "As a matter of fact, it has to with girls, only grown up, older girls."

Tom raised his coffee and was sipping and enjoying her defense.

Norma said clearly, "Menopause."

Tom's reaction was spilling his coffee down his face with traces in his lap.

Albert was amused with her answer and stated, "Menopause, in females, begins in the 40's."

Tom napkined his lips and said kiddingly, "That word, menopause, why the hell didn't they call it woman pause or female pause?"

Albert laughed and said, "I never thought about that."

Norma wasn't laughing. She countered a little louder, "Now listen up, males. Isolate those words, men-o-pause. Now if you interpolate the vowel correctly, it would be, and I'll spell it out, which she did, m-e-n-o-p-a-u-s-e. That justifies its meaning. Both men were about to speak, but she stopped them with a determined hand signal, then said, "That translates to 'slow down guys, we're getting older.'"

Tom and Albert looked at each other and laughed, and then shared their glee with Norma.

Tom raised his cup as he stood up in appreciation of her oracle and stated, "All well and good, but that palaver has no basis," and sat down.

Albert said, "Menopause is for gals in their forties to middle fifties."

Norma said diligently, "So?"

Tom questioningly said, "So, so, you're not 40 yet?"

Norma snapped at Tom, ""How do you know my age?"

Albert said, "Please pass the strawberry jam."

Tom did so, and said audaciously, "I don't know positively, but damn it, my powers, and they are never wrong, tell me that you are too young to be forty."

She took note of the compliment, and said, "Haven't you heard of pre-menstrual-menopause."

They both flushed and answered, "No." Albert and Tom were uncomfortable when feminine talk was the dominant topic. As with all males, not including physicians, they wanted to change the subject.

Tom asked, "Albert, what are we going to use for wheels, cabs?"

Albert quickly replied, "No, we'll cab to the bank and do our business, then I'll check out the used car lots."

Norma stated, "Don't forget to get extra keys made."

Albert replied, "Thanks for reminding me, Norma."

Norma had almost caught up to Albert in devouring their food. Tom chewed, pointed with his fork at Norma's plate, smiled and stated, "She likes my cooking, just a couple of bites left."

She said, with indignation, "Are you insinuating that I eat too fast and too much?"

"Oh, now," said Tom, "I approve of my concoctions being well received."

She replied, "Tom you are an excellent cook," to which Tom replied, "I know."

Albert laughed lightly and napkined his lips, and said, "There's a pay phone six blocks south of us. I'll need some change from one of you. I'm going to work off this dandy breakfast, and call us a cab."

Tom retrieved some coins and gave them to Albert.

"Thanks," Albert said, I'll have a cab here at 10:15."

They both nodded slightly. She lit up and blew a cloud that seemed to follow Albert as he pushed away from the table. As he rose, he was engulfed in a thin blue aura of smoke.

He said, "I'm off to call a Yellow."

Norma quickly added, "The Wonderful Yellow of Oz." They laughed in unison.

Alone, Tom asked, "Did you have a good sleep?"

She looked directly into his blue eyes and stated with enthusiasm, "You bet lover. You really sprung my mainspring."

He flushed, and shook his head, in shyness, then said, "I take that to mean you slept well?"

"Yes," she responded, and went on. "Don't be so stuffy, I'm sure you experienced a sexual release. It was dramatic."

He stared up and said thoughtfully, "Yes, it was, wasn't it?"

She stated, "Is that conveying by implication another re-enactment."

Tom said in a startled voice, "Jesus, you're something. Let's not waste a moment, race you!"

She yelled, "Your bed or mine?"

He yelled back, "Oh, God, yours."

Both ran and fell on her bed. Tom said, sitting on the edge, breathing a little heavy, "Christ, we're acting like kids, and I love every minute of it."

She giggled and responded, "Come on, kid, I'm going to depants you."

"Oh, Jesus," he said and added, "Hurry."

They laughed and made love, each belligerent in a war of sensations, yet so gentle and approving of their partner's sexuality. She suddenly sat up, startled and said, "Tom, Tom. Did you lock the door?"

He answered, "Oh, God, in the excitement, that wasn't on my agenda." He quickly swung out of bed and ran towards the front door and twisted the key and turned, he heard a noise of the vibrating door knob, then a knocking. Tom ran into his bedroom and grabbed his robe and put it on during the run back to answer the door. He ran his hand over his hair, blinked, and admitted Albert, who said, "All set with the oz."

He stopped and looked at Tom inquiringly. Tom sensed the stalemate and reacted quickly, stating, "I was just going to take a shower."

Albert accepted his reply and asked, "Where's Norma?

Tom said, with a sober face, "lying down, I think."

Albert moved toward the kitchen and said, "Well, get on with it. I'll have a cup of coffee, any left?"

"I don't recollect," said Tom.

Both men went to their appointed tasks and Norma walked into the kitchen in a regal manner, wearing a luscious, high-low emerald green robe. She had a strange smile on her face as she sat down across from Albert, and flippantly stated, "Wouldn't you know it, I wanted to bathe, but Tom beat me to it."

"Yes," he replied. "Sometimes he has no consideration. I'll talk to him about it."

"Yes," she answered slowly, "but don't be too harsh on him," and grinned, then added, "Any more coffee?

He replied, "Looks like I took advantage of you, too. This is the last cup."

She got up gracefully, hummed and stated, as she prepared a fresh pot, "Yeah, it's a man's world."

Albert shied away from that remark and stated, "Cab's all set. We'll go to the bank, do our business, I'll write each of you a check, then you two can set up accounts. Both of you can come with me, if you like, to buy a vehicle. If not, maybe you guys, excuse me, you or Tom, have some shopping to do."

She played dreamily with her spoon.

He said, "Norma, if you want to shop I'll get a car and pick you both up and we can see if our craft has arrived. I've already leased a slip."

She said, with interest improving, "You going to take it to the slip."

He looked at her disapprovingly and said, "It is called the *Magnetic*, or she." They both grinned.

He then said, "No, I'll hire a tow to the slip, have her fueled, watered and get her ready, comfortably stocked with stores, food, clothes and makings."

She brightened up, saying, "This is exciting."

Albert responded, "We'll have to put in some long hours at times. I want us to prepare for any unforeseen predicament. The sea takes no prisoners." Then in a sober candor he stated, "The poles will soon be reversing their polarities. I'll tell you, Norma," and he put his hand on hers. She held on to his every word. He continued, "Our fragile electromagnetic network is most vulnerable during this realignment of polarities. A violent storm on the sun, a dramatic electrical storm, simultaneously, would be disastrous for the world as we know it."

She could feel his concern through his touch. It frightened her. She said, wide-eyed, "Jesus, the way you are laying it out, the tone of your voice, you really believe this catastrophe will happen."

He withdrew his hand, sipped his coffee and stated in a spooky monologue, "This could be the final event." Goose bumps appeared on her arms, and she said in defiance, "God damn, Albert, I just got my life to its ultimate sensuality." She temporarily lost control and banged her fist on the table.

He stopped her pounding and grasped both her hands and said, "I'm sorry. Let Tom give you goosebumps."

She replied with honesty, "We have, and he just did." She got up and retrieved a fresh pot. "It smells good." She offered Albert some, and he held out his cup. She slipped into a chair next to him. He enjoyed her honesty, and the fulfillment of her professionalism. She said, "Thanks for being a dear about Tom, and we will work diligently to achieve your goals."

Albert said, setting his cup down, "I know you will, and both your and Tom's achievements are as high as your attitudes."

She smiled warmly and said, "All my feminine instincts indicate you're one fine boss."

He said, "Can you keep a secret?" He did not wait for an answer and continued, "Please do not call me boss. It's a quirk with me. Boss spelled backwards is SOB. It's sad, but sounds like slob."

She laughed and considered this bit of new information and responded, "You may have something in that reasoning. After all, Dog spelled backwards is God." She tapped a cigarette out of her pack, and gazed at Tom's eyes as he entered, freshly dressed, wearing a smile, and saying a pleasant, "Good morning, all. A shower is all powerful." She said, catching Albert's attention, "Tubs are nice too."

Albert winked at Norma, unknown to Tom and said, "You're in a good mood, with just a shower for your pleasure."

She giggled and covered her mouth with her cigarette, and emitted a slow cloud. Tom said stately, "I'm looking forward to a busy day."

Albert said, "Speaking of busy, all of the breakfast dishes are in disarray."

Tom blinked and quickly said, "Yeah, I got sidetracked. It won't happen again."

Albert said smiling, "I wouldn't ordinarily mention it, but have either of your lived with others in a small house trailer, apartment or boat?" They both shook their heads in a negative manner. Albert continued, as he lightly fanned the drifting smoke, "Living in close quarters can be tolerable or dreadful. We have to be neat, and respect one another. I'm sure we'll manage and have no serious problems. There is one thing we'll have to discuss. Norma, I have to be a stiff... smoking is a problem on board."

She blew a cloud and was about to speak.

Albert raised his hand and stated, "Please hear me out. I, too, was a smoker. I understand, but smoking will be discretely curtailed in your deployment of duties on, and in our home base, which will be on the waters."

Tom felt and saw grief in her eyes.

Albert continued, "First, when smoking in our working area, one hatch or vent will have to be open. On deck, no restrictions, except to have a caution for fire, of course. Second, purchase at least two smokeless ash trays, and stock up on batteries." He looked to Norma for her reaction. She inhaled, pulled it deep in her lungs, then let the smoke out, gently, and nervously tapped the ash off her smoking stick. She cocked her head slightly and relaxingly said, "The rules are a most congenial set of regulations. With neither of you being smokers, I thought you were going to King's X my smoking aboard."

Tom smiled, and rolled up his sleeves, and transported a few dirty dishes to the sink.

Albert said, "By the way, Norma, there's an electronic supply store located at 1207 36th Street. I'm sure they carry smokeless ashtrays and batteries. We are going to make a marvelous team. I'm going for another walk. What time is it?

Tom said, "9:42."

Albert continued, "I'm going to look through the phone book for used car dealers. I will be back before cab time at 10:15."

Norma quickly got up and went to her bedroom, returned, and caught Albert reaching for the door. "Here" she said, and transferred the coins in her hands to his.

"Thank you, Norma," he said.

She replied, "My thanks to you for understanding and being so liberal about my smokes."

Albert raised his voice so Tom could hear, "Tom, the cuisine was grand."

Tom edged out of the kitchen toward the front door. He nodded in appreciation.

Norma was in a delightful mood, and asked Tom, "What kind of magic did you unfurl on those hash browns?"

Tom relished the compliments and replied, "The key is using just the correct juice of the bacon and sausage fat, with a smidgen of milk, a dash of salt and pepper, and fresh spuds. Sometime maybe I'll prepare lyonnaise or fracoma potatoes."

Albert said, with a hint of a smile, "If you were mix-sexed, what a trophy you would make, for a man who loves intrinsic and exciting foods." Tom actually blushed.

"Whoa, Whoa," said Norma, and added, "Tom has the right sextant for measuring celestial bodies."

Albert applauded, and said, "Norma that was an excellent repartee."

Tom turned his head away and said, "Gaud."

Albert and Norma laughed. Tom refrained from all eyes and did not answer. Albert waited, then asked, "Tom that word gaud, did you say God, or" and spelled it "g-a-u-d." Albert waited again, then said, "No mind, but one is a Supreme Being, while the other gaud, is an ornament or a trinket."

Norma laughed and said confidently, "I can assert to which he is."

Albert could not resist and said in jest, "That may be true, Norma, but only Tom can quantify his celestial parts." Albert and Norma enjoyed the moment with merriment. Tom had to grin, then he joined them.

Albert said, "Bye, all. Got a date with a phone book."

Tom grimaced, faced Norma, and said indignantly, "Christ, do you like to watch me squirm. Men don't talk about man parts, it's a code, it's uncouth."

She replied, "You men are so silly about ordinary things."

Tom sternly said, "I don't think of my thing as ordinary. No man believes that of his thing. Do you think it's ordinary?"

She lightly laughed, smiled and said, "I think its organoleptic and orgulous."

Tom and his masculinity, accepted these gracious oratoricals, and capriciously asked, as he rose with their cups and saucers, "Want to give this thing a hand?" Tom thought of what the hell he had said, and quickly stated, "Belay that question."

She also thought of what he said, and asked, "Would you like me to translate what you so condescendingly said?"

Tom said, smilingly, "No please let it come out wrong."

She asked as she handed him her cup and saucer, "That word belay, is it a seafaring lingo. A feminist might relate that to being laid in bed." She felt flippant and continued, "and that hand thing, are you asking for a hand job or classification of its performance."

Tom was blushing red, embarrassed, and nervous, and said, "I'm just a tinkle away from a King's X, on this bedroom subject. I'm not as comfortable, or liberal, as you are."

She lowered her voice and said, as they washed and dried, "You, Tom, are a typical male, within the context of sexual conversation. Men negate talking about it, to men, except to boast. On the other hand," she paused, "I'm not going to mention hands. Women feel more comfortable in discussing all aspects of sex, without feeling distressed. You guys, you think sex 24 hours a day."

Tom retaliated and spoke, "Was that a typical female pretentious performance <u>of pretending</u> to enjoy the act of sexuality."

She smiled in remembrance, and took the kitchen towel, slipped it around his neck, and gently wiggled it, until she really got his attention. Then she gave it a little tug.

"Oh," he said.

She said determinedly, "When it comes to sexual engagements, it's 101 percent genuine. I take extreme pleasure in having sex." Then, she added, to his enjoyment, "with you." She pulled and snapped the towel from his neck. Their lips softy met and tasted one another. Tom's wet hands coiled around her. She shivered, but not from the moisture. They held each other tightly. They both started to talk at the same instant. Tom said, "Go ahead."

She said weakly. "I want to, but we don't have time."

Tom gave her a quick squeeze, then responded, "That was my thought too, let me just hold you awhile longer. I know we haven't time for pleasures. You know, I used the plural, but this is a fine second feature."

She said in no uncertain terms, "You know how to treat a lady."

Tom replied, "There's really only one way, in my book, and that is honestly."

She touched his chin, directed his awaiting lips to hers, and pressed firmly in frustration. They uncoiled slowly, and she looked into his blue eyes and stated, "It will be more powerful the next time, when we have time. Tom, just think, you and I have at least one full year's contract."

He smiled and watched her dancing black eyes and said, "God, just like being married, in fact, better. We can't be separated, for at least a year, as you said." He paused, then added, "Some marriages only last weeks or months."

She gave him a quick and unexpected full kiss that delighted the hell out of him. She said coyly, "So there, too."

He said, "Thanks for that unexpected kiss, mate." Tom grasped both her shoulders gently and stated, "We'll be on the waters before long, and Albert, indicated much of our work will be on the

Magnetic. We had better refresh our memories and learn sea terminology."

She smiled and said, "That's a grand idea. Excuse me, Tom, I have to go to the head." Tom smiled as her eyes said goodbye.

He said, as she trailed away, "I'll be at the galley, finishing up."

Albert returned and found Tom finishing drying. Tom asked, "All go well ashore?" Albert replied, "Ashore?" Tom retaliated, "Norma and I thought we would brush up on seagoing lingo."

"Good idea, where's Norma?" Tom answered, "In the head." They both laughed, and Albert stated, "You both have no comprehension of how easily you have molded into an enterprise, getting on the right foot with soft waves and a safe harbor."

Norma entered the kitchen and overheard what Albert had just said. She stated, "I want to compliment both of you. Aye, Captain."

Tom joined in and said, "Aye, aye Sir, I'm ready to roast barnacles, fore and aft."

Albert brightened up and stated, "If that's the Captain's plate, I'll have you keel-hauled."

"No," she said quickly, "When you keel-haul him he can grab some fresh barnacles from under the hull for future torments."

Tom laughed and said, "Thanks a lot, shipmates."

Tom was startled as Norma grabbed him excitedly and spoke daringly, "Let's you and I get tattooed." She couldn't stand still, with anticipation.

Tom looked at Albert, for a life raft or any help, and was denied. Tom was ardent, changed the subject and said, "Cap, when you keel-haul me, I'll collect fresh barnacles on the starboard portion of the hull, then I could serve them on the half-shell." He laughed.

Norma stated, "That's not funny, Tom," and Tom replied lightly, "I know."

She said, "Now, damn it, Tom, answer my question, better yet, my request."

Tom asked, "A real tattoo?"

She joyfully said, "Yes, yes, I'm sure there's a shop in Anchorage!"

Tom asked, pleadingly, "Albert, do you want a tattoo!"

Albert laughed and replied, "Don't drag me into your adventure, besides, I have one."

Norma said smartly, "See, see, our Captain's been ordained already." She released Tom, and strolled up in front of Albert and asked, "What design are you sporting?"

Tom said, "Good God, ordained, sported, what's with your judgment to the scarring of the skin?"

Albert coughed politely and spoke, "Tom, everyone has his or her opinion. It is quite delicate, like a wig, colored hair, fake fingernails, all choices."

Norma said, "Thank you for laying it out to Tom, but you evaded my question."

Albert stood a little straighter and said with dignity, "Double cobras."

She looked at Tom, then back to Albert and said, "Interesting. Why did you choose that design?"

Albert responded, "I don't believe that has anything to do with this discussion. I prefer not to answer your question."

She looked down, took a small step back and stated, "I didn't mean it personally. I thought your input would assist my choice of design." Silence, then Tom said, "Norma, tattoo's are personal monograms, like a special girl, animal, our flag, and Mom. Well, it's a damn private selection."

She responded, "Why hide it. You don't buy a picture and put a cover over it. Another thing, are you ashamed of it, is that why you hide it. Give me some answers, Captain." No response. She continued, "Captain, where the hell is it?"

Tom grinned and was happy to be a spectator, instead of the victim.

Norma cultured a special smile, Albert witnessed it, and she said, "Look, we're going to be shipmates, and as <u>you</u> so diligently stated about living in close quarters on a boat, smoking and tidiness."

Albert said, "Give me one reason. What has that to do with my tattoo?"

"Yeah," sided Tom.

She replied, "You heard him. Our Captain just said, one reason and he will relinquish his negativity as to the location of his cobras."

"Oh, oh," said Tom.

She smartly faced the two men, and with complete confidence said, "Okay. Listen up, I'll set the scene."

Albert glibly said, "Oh, God!"

She flashed a warning not to be interrupted, and said, "We are aboard the *Magnetic*, the seas are rolling whites, and we're being tossed about. You, Albert, have just finished taking a shower, and had a towel around you. The craft lurches, which sends you sprawling, separating the towel from its coverage. I was coming from mid-stateroom, and was hanging on looking down at you and screaming." Both men were intrigued by her exciting story, Tom, more so. "I didn't scream because of your nakedness, but unknown to me I see double cobras, and scream again at the unknown. Then I'm thrown to the deck, because of the shock, and fall on top of you and the cobras."

Albert said, "Good Lord, what a tale of adventure. Did my cobras bite you?"

Tom started laughing and pictured the fictitious scenes. Then he said, "She's got you by the short hairs. She gave you one reason," and continued to enjoy Norma's success, then stated, "Captain, you should have said, one valid reason. One little word, cost you your vanity."

Albert said to Tom, "I don't particularly approve of your phrasing, I'm no stick in the mud. The very thought of that scene, and vivid picture, you have validated, the right by concession will reveal the hidden nest of the cobras."

Tom said, "Well, where the hell are they?"

She cried out, "Oh, no, you don't," and continued with a perky voice. "Better yet, Captain, show time."

Tom snickered, and Albert laughed at this wholesome conversation and subject. Albert finally said, "No, Norma, you'll have to conjure up a storm, and have me toweled, and then magically tumbling without my towel."

She stood up to him, and said, "Well, Captain, the deckhand and the cook are waiting."

Albert said, "Okay, no big deal, it's on my thigh."

She shot back, "Which one, Captain?

Tom horned in, "The left side."

Norma gave him a hellish look, then faced Albert again.

Albert said, "He's right. I don't know how Tom knew it, must have been a guess."

Tom said, "I knew. Did you know that when a cobra is excited it expands its neck in the shape of a hood by its anterior ribs."

She said, "Captain we have a cook who's a walking encyclopedia."

That hurt Tom's feelings, and he said, "I don't spurt off to try to be smart, I thought the information, if not known, would be helpful."

"Sorry," she said, "I do enjoy the interesting knowledge."

Albert said, "You both are right. Knowledge is power."

She said, "The cobra is in the genus of Naja."

Tom responded, "Yes, that's right."

She said to both of them, "Wasn't there a discussion about two heads are better than one?"

Tom asked, " Are you talking about heads on board or two heads of the double cobras."

Albert said, "This all is double nonsense. The facts are there are two heads on the *Magnetic*, but just one complete bathing facility."

"Oh," she said.

Tom asked, "What builder?"

Albert responded with delight, "Bayliner. There are two staterooms, a large salon, galley, fully equipped, even a microwave oven. She is powered by two 210 H.P. diesels inboard, dual controls at flying bridge, and lower helm." He smiled broadly, and stepped close to Norma and said, "I have a surprise, just for you."

Norma became excited, she loved spontaneous, unexpected gestures.   Albert continued,

"Stereo, C.D. player, dual cassette desk with remote control."

She said, "Great," but rather lacklusterly.

Albert said, "Tom, I had a larger water capacity and holding tanks installed, from 88 gallons to 100 gallons and from 36 to 46 gallons. You will be delighted with the galley, freezer, six cubic foot refrigerator, twin stainless steel sinks, full size stove and oven, with a rotisserie."

Tom said, "Hot damn, you have my full attention."

Albert related, "The craft is trimmed in teakwood, even in the galley flooring."

Tom stimulated, said, "This will be like living and dining at the Ritz." He trailed off, locking his eyes into Albert's and stated, "Aye, Captain, don't question my knowledge, but I know you have a treat you have yet to reveal. If for some reason, you are saving Norma's surprise, I'll say no more."

Albert looked at Tom and said, "You are spooky. I believe you do have the divine gift. No, I'm damn sure of it."

Tom grinned and said, "I do know what it is, too, I just know."

She said, "Come on, Cap, tell me, a mirror in my stateroom?" Both men laughed.

Albert said, "What the hell? What is it, you wished for?"

She stared at Albert, her eyes grew in size, and she screamed, "Oh, my God, a bathtub."

They both clapped and said, "Yes, yes. She said with just sprinkles of tears in her eyes, "You guys. How was that arranged?

Albert said, "In truth, the Bayliner *Magnetic* I ordered, the 38 footer, the bath was part of the complement. It was only in telling it to you that I held back."

She happily said, "Where is this jewel located in the Bayliner?

Albert said, "From where, fore or aft?"

She said, "From the back aft."

Albert stated, "From aft to salon to mid-stateroom to your jewel molded tub."

She screamed again with delight, and moved and wrapped her arms about Albert, and kissed him on the cheek. She squeezed him generously and said, in elation, "Dream's do come true, not only in fairy tales."

Albert hugged her back and stated, "Captain's orders, no soaking during working hours, after that, you are on your own."

Tom laughed and accepted Norma's embrace. They hugged, then she released him slowly and asked, "When did you know, Tom, about the bathtub?"

Tom blinked and said, "Just a few minutes ago, it just happens. I can't will it."

Albert said, "Maybe Tom picked up when I told you I had a surprise for you."

Tom quickly stated, "It was about then; it just popped in, spooky like."

Albert winked at Norma, and said jestfully, "Imagine if you were married and your mate had incantation powers?"

Tom looked into Norma's eyes, and then Albert's in search of an answer.

She broke the tension saying, "Just another male script to impress me."

Albert said in a determined voice, "No, Norma, I swear on my double headed cobra, and that is as strong as Tom's King's X's, I didn't tell Tom. He has never seen me undressed. You had better come to a sober realization that your Tom has exaltation powers."

Tom reported, "Yes, he's right, and don't you forget it." He paused and analyzed what he had just said, then added, "please."

Norma said with confidence, "We'll see."

Albert said, "Our craft also has a freezer in the galley, navigational equipment, top of the line, and all hydraulic steering. The main salon has a wet bar with an ice maker. The tinted windows will protect and gamma waves. It is a magnificently built craft, with all bronze fittings. Well, you'll see soon enough. It's quite luxurious, yet soundly built. We'll be spending most of our time, in the salon, lower helm or the flybridge, with the exception of Norma and her tub." They enjoyed that summary. He continued, "At Seattle, I had some fundamental instruments installed in the salon. I must purchase some notebooks and scratch pads. For now, all our work will be done by hand."

Norma flashed a message to Tom and Tom crunched.

Albert continued, "We will decide on a computer in a few weeks. Meanwhile, we will use pens and pencils in our input, so all of us will have hands on each other's information." Norma and Tom laughed.

Albert wanted to know, "What's so funny?"

Tom said, laughing, "Just don't and I mean, do not ask me."

"Well," Albert said, as he turned.

Norma threw up her hands and said, "There you go again. Things get tough, or spicey and you males bail out and ask questions after."

Tom said, "I don't think that's quite correct. You're mixing two different slogans."

She snapped back, "Back off Tom, or answer the Cap's question." Tom put his index finger to his lips.

Albert said, "I recklessly told both of you about my tattoo, and its position."

She said, "Okay," then laughed. She pointed her index finger at Tom, and with her other index finger, crossed them several times, indicating kids' way of shaming. She continued, "Tom in his haste blurted out what could be interpreted as an up front sexual question."

Albert said, "My god, not again. Don't get into details. What is the phrase or word that has you both deliberating, and I'll try to handle it." They both snickered again, and she said, "Tom got his thing, being a co-pilot, to his hand."

Albert thought about what she had said, and gave Tom two finger shames.

Tom announced, "It was simply a dumb play on words, in a lousy arrangement."

Albert asked, "For the record, did you intend to convey that statement in a sexual context?"

Tom said, "Hell, no, I just said 'want to give this thing a hand, the thing being me, not the thing.'"

Albert was amused, and said, "There was an old, but good science fiction film called, 'The Thing,'"

"Yeah," she said, "I remember it on T.V."

Tom said, "I didn't see it."

Albert said, "No matter, I will attempt to, and remember, I said, try to substitute or delete, the extremities of our arms." Tom and Norma clapped politely, and grinned.

Albert said, "Just for all to know, I've decided to lease a car rather than buy. It just makes more sense."

Norma lit up, and blew a satisfying billow of smoke.

Albert continued, "I am allocating chores for both of you. Tom, the mystic and the cook, will be responsible for purchasing, stowing and preparing meals. Norma, all toiletries, soap, tissues, laundry and batteries, flashlights and all safety equipment."

"Aye, aye, Sir," she said and continued. "Did I hear correctly about extra water in the holding tanks for my tub?"

Tom and Albert laughed and Albert said, "No, I can't take credit for the larger tanks. We may be on the water for long periods of time and our drinking water is precious. Our water heater is only eleven gallons and we really have just one hundred gallons for all of us."

Tom spoke up, "The sea is nice to bathe in," and received an elbow form Norma.

Tom said, "That longing question that I so wanted to ask, before your kidnapping, but didn't, I'm asking now."

Norma blew a blast of smoke, interrupted Tom, and said, "Tom, why ask? If you have telepathic powers, use them."

Tom replied, "Damn it, I told you I don't have these turn on powers. They come spontaneously without a will on my part. Now Captain, how the hell did you order, receive, and transport, a 38 foot yacht with customer equipment, in less than nine days? Whew! I finally got that off my chest." He then added before Albert could speak, "It seems impossible, it's spooky."

Albert laughed and she looked at him with wonder. He said, "I could see that would make an interesting point. It's really simple. I have been retired for about a year. It had been one of my goals to purchase a new yacht that was suited to my retirement needs. I researched and talked to many men of the sea. A veteran Coast Guard sailor summed it up best. The Coast Guard crafts are 38 footers, they cut and trim in the calm and rough seas better than any other. I selected a builder with a reputation for detail. Bayliner filled that slot. I've been waiting half my life for such a craft, and nearly a year after I ordered it. Then I received an intriguing offer of continuing my work, with adventure and an eminent salary. But unknowingingly, at the time, you both have been and are the crowning jewels. Friendship and respect is the eminence of life."

Norma was so impressed and humbled that she said, "Captain, I'm sorry about mouthing off about your tattoo." Tom said, "And forget that nude scene, without the towel."

They laughed and enjoyed each other's flexibilities, and candor. Norma blew her last ball of smoke, killed her cigarette, and asked, "One of those extraordinary heads is a bath, what color is it?"

Albert answered, "Like the color of the drapes. We'll save small surprises. By the way, Norma, you will be pleased to know in the forward stateroom is a clothes hamper. I don't know the color."

Tom asked, "What is the fuel capacity?"

Albert replied, "Three hundred and four gallons, I wish it was four hundred. I've maximized on many pieces, but unfortunately, the larger tank was prohibitive. She weighs over eight and one-half tons without fuel, water, and goods."

Norma said, with reluctance, "You guys, shipmates to be, have straightened out my thinking on using the word 'she' for boats and cars. Feel free to call the *Magnetic* she, she sounds like a regal sort of gal."

The men clapped in appreciation, and especially at the thought of changing a woman's mind.

Albert asked, "What time is it?"

Tom told him and asked, "You going to buy a watch?"

Albert said, "That's on my mental list too. I'm saving the best surprise of all when we are physically aboard the ship." He started to walk to the kitchen sink.

Norma stated, "You can't hang a statement like that on a line, without a clue."

Albert said, as he filled a glass of water, the two had trailed him, "Oh yes, I can, I'm the Captain, and you have your orders. The Yellow Cab is due in a few minutes.

Tom looked at Norma and said, "He said the best surprise of all, that must be a dandy."

She said, "Tom, tune in your telepathic mind and find out the Cap's secret."

Tom said, "I haven't the slightest idea. You can bet your pants, oh, oh, bet your last cent, he is enjoying the mystery and that only he knows the solution."

She said, "Damn it, Tom, I'm more excited than ever. If I only knew his secret."

Albert overheard parts and looked out the window to see a Yellow Cab pull to the curb. He said, "The cab's here, let's go."

They locked up and were walking to the yellow conveyance. Norma skipped up close to Albert and asked, "Please, just one clue, you have me rattled." He stopped and whispered in her ear, then said, "We have a full day ahead of us." Albert told the driver, "National Bank of Alaska on Fourth Street."

Tom whispered in Norma's ear, "What did he say?"

Albert was delighted with their secrecy. She whispered in Tom's ear, "*celestial bodies*." They both looked at each other in awe.

Chapter 5

The trio cabbed to the bank, and Norma paid the fare without incident. Albert identified himself to the Assistant President, and loosely detailed the loss of his checkbook. Albert introduced his colleagues, and they both set up accounts. They drew cash on their first check, and tucked the money in their purses. The three huddled, and Albert said, "I have to cash a check. Both of you have mutual assignments." They both nodded positively, and he added, "Why don't both of you cab to the electronics supply at 1207 36th Avenue. Norma can pick up her smokeless ash trays, batteries and flashlights, and any other safety equipment, such as first aid kits, etc. Then both of you cab to Shop for Groceries at 700 West Sixth, and Tom, you pick up your stores. Oh, yes, and don't forget to save the receipts so that I can reimburse you both. I'll pick you up in the parking lot."

Norma said, excitedly, "Albert, that's the place where tragedy nearly happened twice."

Albert replied, "I know, the chance of lightning striking again is in our favor. However, we are on the alert now."

Tom asked, "How will we know your car?"

Albert replied, "You won't, I'll find you. If you should hear two quick horn blasts--twice, that will be a signal, just in case." He looked down at his naked wrist and said, "I'm buying a watch first, then leasing a car. What time is it?" Norma said, "11:15," and Albert replied, "I'll be in the parking lot at 1:15, maybe then we can get some hamburgers to go, and check on the arrival of the *Magnetic*.

Tom said, "Sounds good, watch yourself."

Albert laughed and replied, "Yes, I'll be sure to buy a watch."

The trio broke up. The sun was playing peek-a-boo with billowing cumulus clouds. There was a soft roll of thunder. Tom and Norma had new resilience in their bounce, with bucks in their pockets. They kidded one another, and each put positive input on the other's assignment. Tom reminded Norma to save the receipts at the electronics supply store. They cabbed, it was a short distance to their destination. Inside she bought safety gear, her ashtrays, extra batteries, and first-aid kits. They discussed, and agreed that the *Magnetic* probably did not include an emergency flare gun and loads. She bought two sets. Tom independently agreed without comment. She only smiled. She conferred with Tom and picked out two hand held mathematical computers. Tom picked up one more and gave it to her, making three. While paying attention to her comments on articles to buy, he was mentally listing the goods to buy at the grocery store.

Tom asked the young clerk if he would call for a cab. The clerk replied a little brazenly, "Sorry, can't tie up the phones."

Norma stepped up and said, "He's with me," and gestured to all the purchases they had made. No response from the clerk. She threw the receipt on the counter and said, "I've changed my mind." The youth's smile faded. Norma said, "Make it fast, junior. Is your boss in?" The youth replied, "Yes, I'm the boss."

Tom said, "Forget this puny purchase, seeing that we've just incorporated. There are other stores in town that cater to their customers."

The young boss apologized, and said, "What cab company?" Norma replied, in a vindictive manner, "Yellow, and tell them to high tail it." He did as he was told.

Tom was mad, not because he was slighted, but because the clerk's arrogance towards Norma was intolerable.

Norma saw a Yellow Cab pull up through the glass doors, and hand pointed to Tom in that direction. Tom flashed a cruel smile at the proprietor, and said grimly, "Alex would roll over in that cheap gray box if he had just witnessed his son's insolence."

The kid's mouth dropped and meaningless prattle emitted from his quivering lips. His face became twisted, and he spat out, "How did you know the color of the casket, you weren't at the funeral?"

Norma smiled, and gathered in two large bags. Tom had captured the other two. The young boss was dumbfounded. She said, "Tom, you're spooky, was that tale true?"

Tom replied, "You can bet your buns on it."

She said soberly, "I've got to take your gift, or your cross to bear, more seriously."

Tom said, as they entered the cab, "It's scary some times. I'm a witness in these incantations, and they are always vivid, and in color."

Light drops of rain fell on their cab and the cabby engaged the swipes. She grabbed Tom's arm and said, "Damn it, Tom, I can't get those two words out of my mind."

He asked, "What two words?" She looked at him in disgust and repeated, "What two words?"

"Yeah," replied Tom, weakly. She countered, "Just the biggest surprise, our Cap said, don't you remember, Swami, Good Lord!"

He looked serious and stated, "Yes, of course, the clue you literally carved out of him, *celestial bodies*." She replied, excitedly, and beat on Tom's knees. "Yes, yes, we're dedicated, and fairly smart. Let's examine those two words every which-way, to solve this agonizing mystery."

He said, "Yeah, it's agonizing to you. Why don't you just let it happen when it happens. You women are occupying a whole different galaxy."

She thought that over and was going to retaliate, but he remarked, "I'm mentally and chronologically preparing a list of goods to buy."

She touched his chin, and flashed her warning eyes and stated, "Good God, how in hell can you think of food when this mystery has our welfare at stake?"

He quickly remarked, "What makes you think it will be to our welfare? More than likely it is a unique apparatus, involving our new experiments."

She said, dejectedly, "You could be right. Maybe, just maybe, that smile meant something special."

Tom replied, "Albert is as religious in his work, as if it were his wife. Remember, he was retiring, yet the yacht he ordered was named by him, *Magnetic*, and he had a radio magnetic antenna built in."

He opened his hands, palms up, and continued, "That's as capricious as you can get." She thought about what he said, not wanting to believe the validity of its soundness. Tom asked, "What is your favorite dish? I will lay in extra stores. Damn it, I should have asked that of our Captain."

The rain increased and was cascading about them. Occasional rolls and welts of thunder and bolt lightning was on display. The driver flicked to a higher speed on his swipes. He pulled to a stop, evidently at their destination. Tom paid the driver, and had him wait, for Norma to gain possession of an empty cart. She rushed back, rain soaking her hair and face. They hurriedly transferred the bags in the trunk to the cart and both half ran, wet, but happy, into the grocery store. Tom wiped the water from his face with his hands and secured another empty cart. Norma had to rush to stay close behind him. The rolls of thunder could be heard inside the store as the storm intensified. Tom would read the label, and whisk a can of this and that, into his cart. He stopped, smiled, and turned back to look at her and stated, "Norma you look good enough to eat, wet or dry."

She replied, hair hanging straight, "I'll take that as a compliment, seeing as how you're a gourmet cook. Hey, here's some smoked oysters."

He scooped up seven of them to her delight. They left the canned meats, mustard, and hurried to the spice and coffee aisles. Norma said, "This is my favorite area in a store, except the bakery, Mmm."

Tom said, as he carefully selected his perfume for his savory concoctions, "This is the catalyst of all great feasts." This was where Tom's spent the longest time in making his selections. Norma wished she had a cigarette and that Tom would hurry. Her wish came true on one score. She looked up at the directory and shouted, "Tom, take a right, this is my cigarette alley, and I'm stocking up for those long moonlight cruises."

Tom said, just so she could hear, "Yeah, some romance, a cigarette billowing smoke in one hand, and a smokeless ashtray in the other."

She looked about and said seductively, "You set your scene, here's mine with yours. My hands may be busy, but you're on top of me. How's that sequence?"

Tom thought that over and replied, "Lay in the cargo," and laughed.

She inquired, "Is that a play on words?" and Tom quickly answered, "Doesn't matter, it sounds delightful."

They moved to another aisle. Norma felt something faintly touching her neck, and then it hit her lips and nose. She screamed and released her grip on her laden cart, and beat her hands against her lips, still screaming shrilly.

Tom had just picked up a jar of sweet pickles and the intense scream startled him. He dropped the glass container, shattering it. Pickles and sweet juicy syrup drenched his pant legs. Carts were coming from both directions, being propelled by those who had heard the screaming above the intermittent volleys of thunder. An Odonata, lay at her feet, its immense eyes staring, it's long slender body quivering, and its short antennae oscillating. The glossy texture of the twin-wings were gyrating. A small stream of sweet pickle juice made its way to the stunned dragon-fly. She looked down at the large flapping winged insect.

They heard a female voice in back of her, apparently excitedly talking to another, "Why did that man throw the pickles at her?" She heard a man shout, "Are you all right, was that guy, trying to hurt you?"

Tom also heard that nasty conversation. He shook one leg at a time and stepped closer to her. A large, red-headed, bruiser of a man, a good 200 plus, moved toward Tom and shouted, "Hold it, Buster." People with their carts attempted to get close to the scene to see what was happening. The short, balding manager, was desperately trying to get to the area of the accident. The carts and customers kept him at bay. She bent over to inspect her downed intruder. Another woman, blocked in further back, only saw Norma bend down. She screamed, "She's going to faint." The fat-jowled red-head grabbed Tom around his waist, pinning Tom's arms. The hulk shouted, "I've got him, call the police."

Tom uttered, "Jesus."

Norma was transfixed on the flailing fly and picked it up carefully. She dug in her slash pocket and retrieved a crumpled up tissue. She placed the bewildered dragon-fly on a large can of relish, and ever so lightly patted the churning wings. She blotted the sweet pickle juice off its flaps. She then righted his four-wing body and watched the recovery of the dragon fly.

The manager finally squeezed and shoved his way through saying, "Sorry, I'm the manager. Sorry." He was unaware that in flaying through the crowd and congested chrome carts, he had suffered a thin, but long cut, just below his elbow. The blood was oozing, then he inadvertently hit it again on a sharp projection from another cart. Out of breath, the pudgy manager swung around to see what the hell was happening and he splattered drops of fresh blood on the red-headed man's face. Feeling and seeing blood on him, he shouted, his arms still clamped about Tom. "He has a knife, he's cut my face!" Tom struggled and the manager helped to subdue him, flinging more blood on the grim, red-headed man, and the angered Tom. The carts and spectators tried to get closer, jamming each other tightly. The wail of a siren stopped.

Norma's was unable to look at Tom who was being held by two men with splashes of blood on them. She was transfixed with the antics of the double-winged, dragon fly with the bulging eyes.

There was shouting from the front of the store. "Police, step aside, please!"

Norma looked up and shouted, "Let him go, he didn't do anything."

Redhead yelled, "By God, I'm not going to let him get away with cutting me." The manager raised his arm and saw the blood from the cut. He cried out, "God, he slashed me too!"

Norma barely caught a glimpse of the bug-eyed dragon fly as it became airborne and then disappeared. The two policemen finally made their way to the struggling trio. One of them told the Redhead and manager to release Tom. Red- head resisted and said, "He cut us both, lock him up. He attacked her, too."

The manager said, "Look at this mess."

The officer in charge said, "Let me see some identification." He also asked Norma if she was all right.

She replied, "Yes, thank you, this man is with me. My screaming caused him to drop the jar of sweet pickles."

Tom withdrew his wallet and slipped his Ohio Driver's License out and placed it in the officer's outstretched hand. He read the brief description, compared the photo, then handed it back.

The curious spectators kept pushing tighter. The aisle in both directions was jammed. A heavy roll of thunder was heard and had the cart holders looking up. A man yelled from an adjacent aisle, "Was it a rape?" A woman's hysterical voice arose. "Women aren't safe even in a grocery store anymore." Another voice chimed in, "Quit shoving." Still another, "Go to hell."

The officer in charge asked Tom, "What happened?"

Norma broke in, "That question should have been directed to me. I'm the one who screamed."

Tom hunched his shoulders, and the policemen said, "Yes, Ma'am, what happened?"

Norma replied, "This four winged insect touched my neck, then skirted across my lips. I didn't know what it was, at first. But it was big, and terrified me. I slapped my face while screaming, and Tom dropped the pickles."

Tom said to the cop, "It was an Odonata, that attacked Miss Moon." The cop pulled his head back, frowned and asked, "What in the hell is an Odonata."

They both replied, "A dragon fly.

The policeman asked, "Where is it?"

Norma looked at Tom, and Tom just stared back. She said, with hesitancy, "It flew away."

Tom added, "She's right, you know."

The officer, not liking what he heard, stated, "No, I don't know, I thought you said you slapped it."

Norma's eyes were flashing and she replied, a little sharply, "I did. It fell to the floor, and was quivering, and rivulets of sweet pickle juice flowed to one side of its wings. It lay there thrashing in desperation.

The other cop said, "Why didn't you step on it?"

"God," said Tom.

"What?" said Norma.

The cop who should have remained silent said, "I mean move it with your foot, out of the juice."

"Well," said the cop in charge.

"So," she said, then went on, "Oh yes, I dried his wings and turned him upright, and it flew away."

The officer in charge said to Tom, "Lean again the aisle and spread 'em." The second cop frisked Tom and had him empty his pockets. Tom said softly, "Christ," but did what he was told. No knife was found.

The redhead spit in his handkerchief and was wiping off his face. The second cop looked up into his face, took his hanky and cleaned off a couple spots of blood on the redhead's sweaty face. He turned to his superior, pointed to the redheaded man, and stated, "Ed, not a scratch on him. Must have come from the store manager's cut."

"Yeah," was his partner's reply.

Tom whispered excitedly to Norma, "I have never, in my lifetime been involved in strange situations until I met you, two days ago. I've had three life threatening episodes with you." He stared into her black eyes and added, "Are you the wicked witch of the North?"

She countered, a little louder than necessary, "You're the one who fumbled the sweet pickles. You did a hell of a lot better with that can of olives. You, yourself, admit to spells of incantations, you're a sorcerer and a dowager!"

A woman stacked five or six deep screamed, and shouted "Blood." The man ahead of her shoved his cart forward in the excitement. He struck the buttocks of a well endowed female who couldn't turn around. Being jammed, she shouted, "God damn, you." The man behind the gal he had rear-ended said, "I'm sorry."

Ed, the cop in charge, pointed and indicated to his partner to take care of the situation. His partner looked back at Ed and said, "I wish I had four wings.

The shoving and the bumping of carts and curious shoppers suddenly caused items to fall from the shelves. The storm and thunder were becoming more promiscuous. Shouts and angry voices were gaining prominence with "Back off, screw your sister, you bastard." It was getting out of hand. The manager shouted at the two cops, above the quarreling din. Food stuffs could be heard striking the floor. Another woman screamed as a can hit her open-toed shoes.

Ed turned to his partner, and stated nervously, "Get down to the end of the aisle and start moving them out." Ed realized no one could hear him without a bull horn. The heat from the crowd was evident on their desperate faces. Their clothes were beginning to stick.

His partner snapped back at Ed in a loud voice, "How the hell am I going to get through that, Sam?"

Ed lashed back, "I don't give a damn, just do it!"

Norma realized their dilemma, waved, and crooked her finger.

Ed saw her gesture and squeezed over and gently shoved to get next to her.

She stated to the cop, "Be explicit, be firm. Tell the first person, 'police business, and tell him/her to hold the cart steady. Then climb up and into the cart. Tell the next person the same thing. Step from one cart to the one next to get you to the end of the jam."

Tom, standing next to Norma, overheard her assisting the boys in blue and added, "Watch where you step, some carts have pastries and little tots in them."

The frustrated cop didn't even thank Norma, but he ordered the first person to hold the cart and he climbed aboard it and stood up. He stepped to the next cart with people yelling. One shouted, "I'm going to sue this damn store," a baby crying could be heard faintly above the noise. The crowd was becoming hostile. More canned goods came tumbling from the shelves. One man added to the debauchery and screamed, "It's an earthquake."

Ed and the manager tried effortlessly, faces sweating, to calm them down. Ed looked down and saw his uniform dotted with blood. Although it was hot, Tom put his arm around Norma to protect her. They were all squeezed in together.

Tom said over the clamoring, "You and that dragon-fly sure caused a rumpus. It was probably a boy dragon-fly, and was just kissing your lovely neck, then your lips."

She replied loudly, "You're a melancholy mongrel!"

Tom answered, "Is that the male terminology for a bitch?"

"Yes," she shouted," to which Tom replied, "Oh."

Ed's partner had opened up one stretch of the aisle.

A familiar figure had followed the officer, and was searching for his associates.

Norma's back was to Albert, in Tom's arms. Tom saw his Captain, and waved him over.
Albert squeezed his way to their side, and stated loudly, "I might have known you two were the instruments of all this chaos."

Norma, upon hearing her Captain's voice, turned slowly out of Tom's arms to face Albert, and said in a chipper voice, "Hi, Cap."

Albert said, "Good Lord, how did this all come about?"

Tom chimed in and stated, "A male Odonata kissed her neck and lips."

Albert said in disbelief, "A dragon-fly did this. My God, it looks like the aftermath of a riot." Then he added, "Speckles of blood on carts and on the floor."

Norma said, "I know."

The manager and Ed were talking and Ed had his pad out taking notes.

Albert said, "How about our goods, are you about finished shopping?"

Tom said, "No, not really, another 20 or 30 minutes.

Albert said, "I'll take Norma to the car and we'll wait for you there."

Norma replied, "That's a great idea. I have a full cart from our other shopping spree, and damn it, I want a smoke."

Albert said, for Norma's sake, "Can we leave you alone to shop?"

Tom quickly replied, "Certainly, I didn't cause this uproar."

Albert said, "Damn, my shoes are sticky." He sniffed, then said, "but sweet smelling," and Tom replied that he had spilled a jar of sweet pickles.

"Oh, said Albert, then stated, "Well, I won't say, hurry, but please be careful."

Norma laughed, and said loudly, "Tom, don't forget my cigarettes!"

Tom replied, "Yes," waved an okay with his fingers. He captured his cart and moved down to the intersect aisle that contained rice, macaroni, and spaghetti.

The manager was puffing and racing after Tom. "Please stop." Tom did. he said, "Be sure to tell the clerk at the checkouts to charge you for a 22 oz. jar of Heinz sweet pickles."

Tom nodded, and he scurried away shaking his head.

Norma stopped Albert as soon as they exited the store. She raised her hand, resembling a stop, and pulled out a pack and tapped the white stick out and set fire to it. It was in her mouth in less than a second, she inhaled deeply and sighed. She stated, "God, I needed that," as the smoke streamed from her mouth. She then blew another cloud and gave the patient Albert a tug to get him started. He then lead her to his leased car. The rain had stopped except for an occasional drop.

Stock boys and other employees were picking up the spilled canned goods on the floor. Ed's partner related his finding of blood on two carts, indicating that the manager had cut himself. His uniform was stained, and his hip pocket ripped, and his hat had been stepped on. They wrote up the incident, still bruised, and no one to lock up.

Tom completed his shopping and told the clerk to add the price of pickles to his bill. He then asked, "Were the Heinz pickles on sale?" The clerk replied, "I'll have to check." Tom said, "Never mind."

The young clerk stated that was the most excitement here since the big earthquake in the '70's. Tom asked, "Were you working here at the time."

"No," was the clerk's reply.

Tom thought that it was possible that he hadn't been conceived when that disaster occurred.

There were low rumblings of distant thunder as Tom wheeled his bagged cart to the parking lot. Three out of his four wheels were working, the fourth jitterbugged crazily. Tom was wearing a smile, and searching for his friends. An unfamiliar car wheeled and braked. He saw a familiar female hand waving out of the passenger window. Albert exited his leased 1992 Jeep Cherokee Briarwood, 4 x 4, and opened the tail gate, and the two stored their goods.

Tom said, "I had to pay for the jar of sweet pickles that broke during the ruckus."

Albert said, smiling, "That will be fine. Just give me the receipt and I'll make out a check to you when we get home."

Tom asked Norma as he slipped into the back seat, "Anymore double winged gargoyles."

"No," she said and countered, "You smell sticky, but sweet."

Tom said to all, "This is a nice leather interior."

Albert said, "Yes, we'll stash our stores and then check on the harbor."

Norma said coyly, "Tom that was quite an experience, you're exciting to be with."

Tom replied, "I had nothing to do with the kiss of the fly."

Albert said, "I recall a show called <u>The Fly</u>."

She said, "Yeah, it was good too, the remake was the best."

Tom chimed in, "I didn't see it."

She said, "Tom, that big heavy set red headed man sure had you pinned."

Tom answered, "What he surely had was B.O."

Albert laughed and swung left, stating, "I'm going to have to re-think about turning you both loose in the midst of the good folks of Anchorage." He paused, "On the other hand, you'll be easy to find. Just locate a riot, and move into the axis of it, and I'll find my two good friends."

Tom said, "I had absolutely nothing to do with that, I'm innocent on all charges."

Norma stated, "Tom, if I were alone, nothing would have happened, but place a male and a female together, and screams are heard, the worst is believed."

Tom said, "The dragon fly was a male."

Albert said, "She has a valid point, and she is right. The same is true, if you are in the company of someone who commits a crime. You then are an accessory after the fact."

"God," said Tom, then added, "I'm just a damn accessory, a part of something."

She said, "That's right, Tom," and laughed with Albert. She lit up a fresh cigarette, and smiled. She loved it when the conversations were based on her assumptions. Tom inhaled a portion of Norma's cloud and coughed. She said, "Sorry about that."

Albert said, "Here we are on Downer Street," wheeled and braked.

She turned in her front seat and looked into Tom's eyes and stated, "Thank's Tom for your arms."

She could sure make him feel good at times after she took him down. He felt he was the puppet, but she controlled the strings. He smiled back at her. Norma crushed her cigarette out and slipped out of the Briarwood. They all took on pack horse duties and settled at the kitchen table.

Tom asked, "Anyone for coffee? My treat." Albert declined, and Tom got up and started his chore.

Albert sighed and said, "No need for you to tag along. I'll check up on the *Magnetic's* arrival, and if she is present, I'll have her towed to birth 28 at the "White Tern." Neither commented, their eyes not wanting to meet. Albert broke the silence, "Well, I take that to be no? Can I leave you two alone without your getting into trouble?"

She lit up, blew a long stream of smoke, and asked, "What is your definition of trouble?" and awaited an answer.

Tom butted in and said, "Captain, you should belay that last remark."

Albert thought it over and was about to speak.

Norma said, "Aye, aye, Captain. We'll be delicately fastidious."

Albert said, under his breath, though audible, "Good Lord."

Tom said, "She's speaking for herself, but I'll keep an eye on her."

She retaliated, "Thanks guys, I see it's still a man's world."

Albert said, "Bye now, be sure to lock up."
More good-byes followed him.

Tom turned to his coffee, and asked, "Please lock the door."

She replied, as smoke filtered about her like a net, "I can't, that would be a dereliction of duties. The Cap gave you a direct order."

Tom said, "Since when do you walk the straight and narrow?"

She skidded her chair hack hastily and quickly grabbed a startled Tom.

She said, sensuously, "Since you and I locked horns, Lover." He baited her and replied, "Is that what we did?"

Norma said, "Shut down the coffee, mate."

He queried, "Are you serious?" then quickly added, "Don't answer that, I know, Christ yes, your bed, yes, yes." They both laughed and hustled and threw themselves on the bed. Tom smiled and asked, "Let me shuck your clothes. Just like peeling a lovely peach."

She replied, as she bounced off the bed, "Be my guest."

He said in anguish, "Christ, you got me so excited."

She said coyly, "I know, and it shows." They both laughed. She had a divine satisfying smile, and her eyes flashed of pleasures to be. He had her standing in just two remaining pieces of cover. He looked at her in anticipation and said, "Oh, Jesus, " and wound his arms around her, and tasted her sweet lips lightly, then heavily. He was shaking as he unhooked her bra clasp. He stepped back slightly, and withdrew her breast cover, like a matador. He stared in awe at her hemispheric loveliness.

She adored watching him, his excited eyes, as he attempted to cover it forever in his memory. She stood still, drinking in his enthusiasm, and relishing the grandeur of the power she possessed, as a woman. She wanted to touch him and relieve his tension.

He repeated the same words, as he delicately used both of his trembling hands and slowly pulled down her lacy panties, "Oh, God, Oh, God."

She waited. She touched his lips with her fingers, and very softly said, "Tom, it's my turn, let me shuck you naked, just like peeling a lovely banana."

He muttered, "Oh, Sweet Jesus, hurry, hurry."

She did. The only garment on him was his bulging briefs. Standing naked and beautiful before him, and teasingly pointed to his hidden pulsating manhood that was straining for freedom, she said teasingly, "What have we here?"

He said, pleadingly, "Hurry, Norma and find out, I can't take it any more"

She gracefully pulled them down in one motion and marveled at his bulb of life. He stepped out of them, while grabbing her, and both fell to the bed. They pleasured one another in ecstasy. After, they lay in sublimity, they fondled, and caressed each other. They wondered at the magnificence of each other's body, and enjoyed the elixir of life. She spoke softly, "Tom, the solace and what we say to one another, before and during our romantic bash, must be our secret. Let neither of us demean what we do. It is too precious. It's a whole new wonderful world, making love."

With a smile, Tom said, "You're red, white and blue. If we didn't deviate, I couldn't function as a literate male. The two worlds, as you say, must be hailed as two fragile increments, and added coyly, "Enough of that palaver, let's fantasize again. As they began exploring, "What's your fantasy?"

Tom smiled and said flippantly, "Two naked, two hundred pounds of female flesh. Ouch, damn," he shouted in earnest.

She playfully finger kicked his manhood. He said, suffering, "Now look what you've done!"

"My God," she said, holding her hand to her mouth in disbelief.

Tom said dejectedly, "Don't ever do that when a man is fully aroused, didn't you know that?"

She said sadly, "No, I didn't. I'm sorry, Tom, I was just fooling around."

He sharply stated, "Norma, let me tell you a true tale. The hospitals during World War II were full of injured young virile men, thousands of them, all having various injuries. But sexual wants and desires were always in working order. In some cases five nurses tended to the wounds of fifty men. Nearly every time a nurse walked by any of the men's bed, they saluted her. "You follow me?"

"Yes, I know what you mean, Tom," she continued.

"Well, it got so bad, and the nurses were getting so self-conscious of what they were doing unintentionally that they had a strategy session.

The next day one of the nurses went to the P.X. and purchased a box of No. 2 long pencils, with erasers, and passed one out to every nurse on duty, although most used pens.   Some of them carried them in their hair or clipped to their pocket or necklines.

Norma laughed, and he continued, "It's not funny if you were the guy under that tent."   She continued laughing, and finally said, "I know, no disrespect, but picture fifty beds appearing as tents."

Tom laughed and said, "Yeah, if you were the female," and they both laughed. He continued, "Well, if she saw an obvious conspicuous tent, she would go to his side, quickly pull up the sheet with one hand, make sure his manhood wasn't hiding. Using the other hand, she would whip out that pencil and strike his manhood on the head.   It would immediately wilt, as you just witnessed. She would cover him and he no longer had that yearning sex drive."

She asked, "Why do you call your thing a manhood, instead of a penis?"

"That's easy, that reminds me of a pencil."

"Oh," she said in delight, and added, "Are you a refugee from a striking pencil?"

"Yes, once, it wasn't World War II. Tradition had prevailed, and the nurses passed it down as an endowment. It hurt like hell, but it never failed, and it was the nurse's quick fix for a man in that condition."

"God," she said, "you are a wonder, one more question all females would have to ask, how long was it detained after being struck by a pencil?"

Tom smiled, and she said, "Come on, Tom, we have no secrets!"

He pointed to his and said, "See how pathetic it is?"

"Yes," she said, and added, "I am sorry, but you haven't answered my question."

He answered, "Well, let's put it this way. You put him out of commission. I can't perform right now, and I am truly sorry."

She said, "I know and I am truly sorry."

He replied, "Yeah, you're not the only one."

She said, demurely, "How long will it be out of order?"

He gave a disgusted sigh and stated, "Jesus, now I'm some kind of broken machine."

She thought that over, grinned, and said, "Just a broken erector set," that brought a grin, then a laugh from Tom. He said in jest, "You might as well put a bad order tag on it."

She laughed as she thought how it would look, then said, "Tie it with a blue ribbon."

"Yeah, said Tom, "you couldn't put a yellow ribbon around it cause that goes around the old oak tree." They both snickered.

She said, "A blue ribbon is nice around your weeping willow tree," and pointed to it in jest. They hugged one another and had a great time nuzzling, and enjoying their intimacies.

Chapter 6

This was a happy drive for Albert. If all went well, he would capture one of his dreams. The almost invisible scar on the left side of his neck was itching, most likely due to his anticipation of seeing and claiming his yacht. He was humming as he pulled into the ocean dock road from Harvard Loop. The presence of birds, and the smell of the salt water became stronger. He rolled down his window to fill his lungs with the perfumes of the sea and the beauty of the gulls, bobolinks, albatrosses and ospreys. The winds, though light, were coming seaward, and made the presence even nicer without the manmade pollution of Anchorage. He rationalized about his two new friends, and his new awaiting craft, but there was one link missing. He had no mate with whom to share the realization of heaven on earth. He thought how Radiant Norma and Tom had become. He absent mindedly pressured the accelerator in his excitement, glanced at the speedometer, grinned and eased off. He braked and saw the many sized crafts of the sea, riding soft shelled, like babies in a mutual cradle. Filled with enthusiasm, he made his way to Anchorage Harbor information, of Knik, and Turagaun Arm. In the hallway was a large, self-standing "V", advertisement inverted utilizing Old Time West block letters. It was ornate and an eye catcher. He stopped and read it, "Chilkoot Charlie's Rustic Alaskan Saloon, we cheat the others, and pass on the savings to you. Two dance floors, with stages, and live entertainment. Horseshoes, pool, video's and pull tables. Special-Chilkoot Charlie's Sourdough Ale."

He thought, "Good grief, I would like to visit such an establishment and treat my two new friends. He was feeling a tiny bit sorry for himself and wished he could share all this magnetism with someone. He straightened up and fielded the directory. Main floor, Harbor Directory, Room 106, Anchorage Federal and U.S. Courthouse and government office complex covered over three square blocks. He entered and was submissively startled by the woman at the desk. She observed his sudden sequester, and asked, "Are you all right?"

Albert felt foolish and was taken aback when his body language was so obviously interpreted. Rarely did he ever stammer. "Miss, Miss," then looked down at the name tag and continued, "Miss McAgneti."

She laughed softly and said, "Yes, my last name is difficult to pronounce, Irish and Italian heritage. It's pronounced envisioning the "M" and "C" as "Mick", then add Agneti. Just like the word gent, then add an "i".

He stated, "I'm terribly sorry, I'm usually quite good with names. It's just that your face and hair remind me of someone I once knew."

"Oh," she said. She wore a yellow and black velour sweater, embossed, and trimmed with scallops. Her hair was a light, but fervent brown, and was well groomed. She had striking features and welcoming eyes. He admired her hands and the lovely cultured nails. He said, "Your nails are lovely, they look like each of them has been calibrated. Do you do your own?"

"Yes," she answered and added, "I believe they are as much an extension of our body as hair. What can I do for you?"

He stated, "My name is Albert Dryfuss, I have a bill of lading signed by the agent for cargo on the freighter *Delta*." He handed the slip to her and their fingers touched, and just lingered for a second. It was like a hot poker stinging him, without the heat. He said, looking at his nails, "I really should pay more attention to the extensions of our bodies."

She read and said without looking up, "It's quite acceptable for men to have manicures."

He said, "I wish I were more receptive to my nails and contours."

She raised her brown eyes and looked him over demurely, then scanned his hands and nails. She said pleasantly, "I can tell usually by one's nails if they are healthy and take care of themselves."

He asked sincerely, "Really, how so?"

"Well," she said, setting his paper down, and gave him her full attention. She then added, "Well, first, I look to see if you bite them. The shape of your fingernails is important. They might be rounded or convex, like a face down spoon, with a characteristic thickening of the nail bed. This condition is called clubbing. If ever so slightly, it is of no consequence, but when obviously convex it may reflect chronic disease. Clubbing is not really understood, but might have something to do with an increased blood flow to the fingers."

Albert interrupted her with, "My God!"

She continued, "In contrast to the spoon-up nails, most people who have that shape of nails have a chronic iron deficiency. When the nails are very thick, a fungus might be present. The color of the nails is revealing, if they are pale, you may have anemia. A blue means not enough oxygen in the blood." She stopped, smiled, and then said, "I'm sorry, I didn't mean to go on, but our nails are the reflection of our physical being."

He stepped closer, and offered his hand, and laughingly said, "Please, how do I stack up on a range of one to ten?"

She had glanced at his nails, but now took the opportunity that was presented to her. She took his hand gently, looked, and then turned them over to look at the nails. She said, "Mr. Dryfuss."

He quickly said, "Please, may I call you Laura, and please call me Albert." She replied, "If you like," and he nodded.

She continued, "Albert, you eat well. Your nails are well nourished, not too thick or thin. Just about all you need is a good manicure to shape them." They still held hands and she said, "Our fingers have 14 bones in each hand."

He thought that was an interesting and intimate phrase. She suddenly released his hand, and resumed a businesslike attitude, and stated, "Yes, Albert, everything looks to be in order. She thumbed through a large ledger book, flicked a page back, then added, "The *Delta* notified us of its anchorage late yesterday. The cargo is probably, or part of it, is being unloaded at this time." He smiled upon hearing that bit of information.

He asked, "Are you folks doing business longhand?"

She laughed and said, "Well today, anyway, our computers are down." She handed him back his invoice and he took it reluctantly. He said, "Laura, I noticed your fingers. Is that beautiful ring your birthstone?"

She replied, "Yes, it is."

He said, "I was captivated by your knowledged of our nails, I will hope to be more attentive to mine. Believe me, you might do me a great favor. I'm new in Anchorage and my craft will have to be towed. Do you know of any one that caters to this task?"

She smiled, looked at her watch and said, "Yes, you might be in luck, if he's free. Dan Satters, with two t's, is honest and capable. He could be at Larry's having coffee. Do you want me to call?"

"Please do," he replied. She tapped in Larry's number and asked for Dan Satters. She took time to look at Albert while she was waiting. She said, "Dan, there is a man here named Albert Dryfuss who has a 38 footer and needs a tow." She asked Albert where to, and he responded, "White Tern, Slip No. 28." She repeated this information to Dan, and she told him it was coming off the freighter *Delta* out of Knik Arm. She listened and then said, "Yes, I'll tell him," and she disconnected. She said, "Please sit down, he'll meet you here."

He said genuinely, "You must permit me to take you to dinner for what you have done. It is most kind." There was a pause, and he stated, "That is if you are free to do so."

She quickly countered, "No, it's not that."

He said, "I have two colleagues, we could make it a foursome."

She said, "Well, why not!"

"Grand," he said, and continued, "You mentioned that each hand, that is fingers, has 14 bones."

"Yes," she replied.

He said, "It may be a little of fate or something close to it, my slip number is 28."

She radiated a wonderful smile and said, "Yes, that is what you told me to tell Dan."

He said, "That's not the only coincidence, it absolutely boggles my imagination."

She looked apprehensive, and stated, "You've really got me interested."

"Fine," he said, then added, "When I tell you, I will be waiting with anticipation for your input."

She said, in empathy, "You have my undivided attention."

A man in his mid-40's, square faced, deep black eyes, with a slight shine of a beard, walked in. Dan was wearing blue jeans and a thick colorful fuschia shirt. He nodded to Laura, and the two men shook hands, and called each other by their first names. Dan asked, "Are you going to tiller your craft?"

Albert asked, "Are you free to do it now?"

"Yes," was Dan's answer.

Albert hesitated, then said, "I did want to take the wheel the first time, but my car is here."

Dan said, "I see, is this your first boat?"

Albert replied, "Yes, I've been waiting just about all my life, but rationally, it's been a little over a year."

She said, "I would drive, but I still have a little over two hours to go."

Albert said, "Thank you, Laura, that was kind of you. I'll just let you handle it, and meet you at White Tern off Knik arm."

Dan looked at Laura, then back to Albert and stated, "I have to charge for my full crew in any event. If you wish, I could have one of my crew drive your car, and I will tow with a partial crew, with you at the wheel of your boat."

Albert's face lit up and he smiled and said, "Wonderful."

Dan stated, "I still have to charge for the full crew."

Albert stated, "Of course, I'm obliged," and then asked Laura, "May I have your number and call you around six and make arrangements?"

She had already written it down, handed it to him with a smile, and said, "Don't forget about the mystery, and my input."

Albert said, "Not on your life, or mine, for that matter. Thank you again."

Dan asked, "Are your tanks dry?"

"Yes," Albert said, and added, "She's being, or all ready, to be unloaded."

Dan remarked, "I think not, they probably wouldn't be able to tender it."

Albert replied, "Yes, you're probably right. Thank you Laura, I'm looking forward to our foursome."

She smiled as the two men left her office.

Dan stated, "If it's sports fishing, you've come to the right waters. "

Albert replied, "It is indeed fishing."

"I wanted you to know up front, for a full crew, and turning over my engines, the minimum is $500.

Albert replied, "Thank you for your prudence. The price sounds equitable."

Dan said, "I'll need your car keys, Albert, and description."

Albert answered, "Yes, of course." He told him the make, color and the location of the car.

Dan said, "Wait here just a minute." He saw one of his crew and hollered, "Tony, over here." He gave instructions and passed him the keys and told him to wait on the dock, and they'd all come back together. Tony raised his index finger to his right forehead and took off toward the parking lot.

They boarded Dan's 46 footer. The gulls, bobolinks, and sea fowls were everywhere. Many were perched and others were land lubbers rather than in flight. The exciting noises of birds, whistles, and surf, where the sea slaps the land was a glorious blend. The air was caressing, a slight breeze was charged with odors of man, fish, birds and rotting kelp. Albert followed Dan to the wheel house, Dan shouted, "Rick, Rick."

A young, good looking, man joined them. Dan stated, "Release our ties. We have a tow out of Knik arm to White Tern. Rick, this is Mr. Albert Dryfuss, the owner of the tow."

Albert stuck out his hand and stated, "Call me Albert. Happy to meet you, Rick." They locked hands and Rick nodded to Dan, and then scampered out. Dan pointed northwest through the pilot house windows, and stated, "That's the sea gate, it's a wide beam, you don't have to use it, but its safer, lots of sea room, if you're coming or going in rough waters." Dan watched until Rick gave him a clear wave. He had engaged his twins, in neutral, then reverse, turning them forward. The powerful craft instantly responded with over 500 hp at his command. He kept his speed low, in order not to congest the waters so close to other crafts and the seashore. Albert appreciated the sound and competent seafaring man.

Dan stated, "The ships lay out a few miles, anchor, and unload to smaller crafts. It's actually cheaper. They need no pilot boats or docking charges. And they do dispense their bilge pumps. Some of the older and foreign registered aren't seaworthy. They are so run down and patched, they leave oil and their hulls are literally their skins. Peeling them would devastate the environment, if those haunts were to go dockside." Dan depressed his speaker on his radio, and dispatched, "Wimpee to Delta." The third call, they connected, they talked, and Dan ended with an over and out. Dan said, "Well, you heard. They have a savvy skipper. Your boat is laying starboard of her. The winds are light, but steady. He anchored the Delta adjacent to the winds and tides. He's had two tie lines, fore and aft, letting mother sea swells, maintain the distance between his hull and your boat. He was careful in protecting your craft. Your boat is no threat against his ship. It must be a big one, well over 500 tons is my guess."

Albert asked, "How much longer?"

Dan replied, "You will be able to see her after we take a full port."

Albert was about to see his fairy tale come true. The Bayliner brochure was tastefully done with specifications and colorful pictures, but to see the real motor craft and feel the McCoy, was a pleasure he awaited with anticipation and sensual gratification. They felt the sways as the Wimpee throttles accelerated and was brought about to cut the incoming waves.

Dan smiled, handed Albert his binoculars, and pointed. Albert pressed them to his eyes. He saw his dream, white and proud, bouncing with the sea shine, and using the ocean for its cradle. The sun refracted the bright stainless steel rails. The fly bridge looked to be suspended, due to the wrap around tinted glass that separated 360° from the main hull of the vessel. Dan was correct, two tie lines, had her cocooned, forward and aft. Albert licked his lips and his eyes were dancing. He removed his glasses to wipe the moisture from inside the lens.

Dan said, "Let me have those binoculars, I want to scan your new boat, take over the wheel." Dan said, "She rides like a feather, she has lines, and the strength of a cutter."

Albert swelled up a little and stated, "It's trimmed with teak, except the floor in the galley, which is teak-parquet."

"Jesus," said Dan, then added, "If she's looks half as good inside as outside, you've got quite a gal!"

"After the tow, it would be my pleasure to show you about."

Dan smiled, "You're on. Will you let my crew take a squint? I would especially like my son, Rick, to see your craft. It will heighten his passion that someday he will be the captain of such a vessel."

Albert said, spreading out his hands, "You didn't do too badly, Dan, this is a fine craft, nothing to sneeze at."

Dan replied, "I suppose. The Wimpee is a working boat, and I will be happy when I can really call her my own."

Albert felt empathy for his new friend. He dared not ask how much of the boat he did indeed own. The wave in Dan's voice, and the far away look, indicated to Albert that Dan was trying to walk on water. The Wimpee was but 100 yards from the *Magnetic*.

Dan stated, "I'll come starboard to stay clear so you can jump on your transom platform."

"Yes," replied Albert.

Dan said to go aft, and he would back the Wimpee toward your aft.

Albert hurried out. The swells were light and Dan eased the Wimpee against the waves, cutting his engines slowly, the closer he came to the *Magnetic* aft. Albert had a short jump, and waved in Dan's direction. Albert went below and opened the doors in the rope locker. He returned to deck side, made his way forward and tied the end of the tow rope to his bow pulpit. Rick was aft on the Wimpee, and waved to Albert to throw the rope end to him. Rick grabbed it and secured it to the Wimpee's aft. Albert moved swiftly, and released the two lines back to the old ship. Albert looked up at the high stained steel bow, its massive weathered hull, soaring above him. The choppy waves tore the *Magnetic* and began carrying it. Dan reversed the forward throttle slowly, until the tow rope was taut, and eased off slightly. Albert was at the fly bridge, sitting in the helmsman seat, and grasping the stainless steel steering wheel. He adjusted his seat and permitted his eyes to rove the chrome gauges, compasses, depth-finder and complete instrumentation. He looked down at the remote control spotlight, spot, and flood beam. He smiled--at long last he had found his home. He could hardly wait until the time when he could engage his 420 h.p. diesels. He admired the radio, wind screens and the dual trumped horns. He wondered again how his craft would measure, and absorb, with the rolls of the water. It all really came down to this, so far, excluding his near disaster at Goose Lake Park, his expectations far exceeded, his negatives. A smile crept, then completed itself across his face, as a sea fowl had just christened his boat. He visualized two sea horses, twisting on a bright swivel. They had small raised brocades of green stones. The little horses were dangling from Laura's ears. "Damn it," he thought, he could see them so clearly, why didn't he compliment her on her choice of earrings. They certainly weren't a cheap trinket. He thought for a moment, and then justified his mind for playing games. He didn't want to accept the fact that he was more interested in thinking about Laura. Just to hold the wheel of the *Magnetic* was almost too good a fortune for any one man to capture. He then thought of his two companions and the speed with which both Norma and Tom had tendered to each other.

This was unlike Albert to daydream. He should have turned on the radio, and checked the compass as to their heading. This was not in accordance with being a scientist, but he felt wonderful. He noted the heading and switched on the radio. He could see the coast clearly and was not far from his harborage. The swells were changing, and the screaming sea birds were prominent. The smells became stronger. Dan skillfully maneuvered his boat, and Rick leapt on the dock with the tethered rope, and tied a black wall hitch to the slip's tie down.

Rick ran down the dock and received a line from the Wimpee and tied it to an empty slip that barely accommodated her size. Dan and his son returned, and physically pulled the *Magnetic* tightly in her mooring. Both were admiring her sleek lines.

Albert said, "A masterful job, you two work well together. Come aboard." Albert gave them the grand tour. Dan and Rick enjoyed what they saw.

Dan said to his son, "A craft such as this, could someday be captained by you if you work hard and want it bad enough. Mr. Dryfuss had to work hard nearly his whole life, but now you're standing on his dream come true."

Albert joined in and stated, "That's the truth. It's all up to the individual and if you want it badly enough, you will succeed."

Rick replied, "You both sound like it's a lark."

Dan quickly stated, "Only if you want it bad enough. Patience is one attribute that makes the wanting and getting easier."

Dan asked, "What's that strange antenna?"

Albert, not liking to deceive his two new friends said, "Oh, that, just an untried unit for finding fish. I have no idea if it will perform."

Albert wrote out a check for $600 as Rick was waving. Albert turned around and saw his other crewman, Tony, making his way toward them. Albert handed the check to Dan and said, "I left the payee blank, is that all right?"

Dan nodded, and captured his check. He started to address his crew and noticed the extra $100. He turned to Albert and stated, "Why, our deal was for $500."

Albert smiled and replied, "You're correct, the tow, and your crew was exemplary, I want to thank all of you, and if for any reason, I am in a position where I should need you, you can well expect a call."

Dan handed Albert his card, and both men engaged in a tight handshake. Dan stated, "Welcome to Anchorage."

Albert acknowledged his kindness, Tony handed Albert his keys, and said, "Nice leather." Albert and the gliding birds watched the Wimpee and the crew reverse, then head out. They gave him a final wave.

Dan turned, called his son, and said, "Rick take the wheel, and wake me after you dock her."

Rick replied excitedly, "You bet." This was the first time Dan had handed over the control of the Wimpee, for an extended run, and most of all, the docking. Rick gripped the wheel more firmly and with renewed authority.

Albert finished tying her down and wanted to go through her slowly, but time was not as accommodating as his love of life at this period. He turned back to admire her one more time.

The dock began trembling, a sudden and unexpected compression clamped his right shoulder. He quickly swung around in bewilderment.

"Where's that pirate?" said Ruth Dander, owner of the White Tern. She pointed to the waters and added, "Damn him, he polluted my waters."

Albert said, "Good day, Miss Dander."

She countered, "It was that scum of the ocean that utilized my slip, without payment, and defiled my waters." Tiny bits of food had escaped from her busy mouth, due to her aggravation.

Albert then took note, and discovered she had a partially eaten sandwich at her side.

He was undaunted as to what the makings were. Albert raised his hands waist-high, in a push position, and said, "Now Miss Dander, everything is all right."

"Right," she spat, and added. "He had no right without my permission, can't that scurvy pirate read my signs?"

Albert noticed an awkward strutting bird with a long strong beak, ending in a hook, its nostrils opening from round horizontal tubes on each side of the base of its bill, moving behind, and toward his excited slip lord. Having webbed feet, but lacking a hind toe, it waddled steadfastly. Albert apprehensively was pointing behind her and shouted, "An Al, an Al."

She cut him off, and snapped back loudly, "His name is Al, I'll get him and haul his." She screamed as the albatross clamped his strong hooked beak into her unfinished sandwich, which included one of her fingers. She screamed again and the albatross flapped his wings in distress while tearing and downing his captured meal.

During the debauchery, she thrust forward toward Albert, who reached out in haste to steady her. His right hand caught her left shoulder, unfortunately his left hand missed and fell to the cleavage of her dress. Being strong and stout, she attempted to right herself and pulled back.

Albert's left hand was not as fast. The end of her second scream ended in a sharp high shrill rip. The noise and confusion grew louder as the bird rose and screeched, and circled to escape the mayhem.

Wide-eyed, Ruth looked down and saw her familiar pair of sphericals completely exposed, splashing light from the water, giving the illusion of enchanting movements.

Albert stared at the twin marvels and said, "Good Lord." He looked down again at them and was mesmerized by how they looked with the sea shine dancing on them.

She had never seen them exalted with the wondrous flashing designs.

Albert again uttered, "Lord," and sheepishly handed her the ripped garment which he clutched.

She finally acknowledged his gesture and took the cloth and tried to cover herself. It didn't even come close. It would have been a shabby bandaid covering nothing or very little of her endowment. She straightened her shoulders, thus further extending her two virtues, and let loose of the ripped end that haplessly fell in front of her.

Albert thought of being gallant, but he only had one handkerchief. She stated with dignity, "Please do not leave the premises without seeing me, Mr. Dryfuss."

He questionably spoke out loud, in frustration, "I've seen you already."

She did a complete turn, the dock bounced in rhythm with her mighty spheres, as she strutted, her brown hard nipples in cadence with her body orchestration. She heard, but did not see, a horn blasting in short quick intervals, as if applauding her feminine rendition.

The gyration of the dock subsided the further she distanced herself from Albert. He had to steady himself a moment after that precious display. Another bird glided down and landed and picked up a tiny morsel, digested it, strutted around looking for some more of the same and left quickly.

Albert wondered at the albatross and the extrinsic exposing of Ruth's lovely bosoms and then grinned at the audacious splendors of life. He wanted to fuel and water the *Magnetic*. He would have to face Ruth and get the key for the diesel pump. "Lord," he thought, the incident still fresh in his mind. "She was certainly well endowed."

He followed the path that Ruth had taken, but stopped and turned to look. Yes, it was still there. No dream. He smiled, and continued to assimilate the grandeur where the sea shared its permanence with the land.

Ruth was in her stout oak swivel chair and wearing a blouse that clashed with the rest of her attire.

Albert loudly cleared his throat and she recklessly whirled in her creaking chair. He said, "Miss," she interrupted and barked back, "Call me Ruth, it isn't as though we haven't met, and I'm sure you've seen more of me than most men."

He quickly, uneasily stated, "I have, indeed, and please call me Albert."

She asked, slouching down in her chair, "Were those pirates in your ward before they escaped?"

He cringed, thought that over, and replied, "As a matter of fact, I did pay for a tow to your harbor."

She snipped, "They defiled my waters, upset the birds, and utilized my slip without so much as a dime, or thank you. He is a son of a bitch of a scavenger."

He held up his hand and stated sternly, "Ruth, he and his crew are not what you perceive them to be. The captain, in my opinion, is a fine seaman, fighting to keep his boat in his name for his son's future."

She whipped back, "Then you condone what that crossbones did?"

He wrinkled up his face and stated in defense, "He had to enter your waters to deliver my boat to its slip."

She pushed the chair back, stood up, her bosoms shaking. She saw his eyes and chirped, "What are you looking at?"

He blinked, and replied, "A remarkable endowment of femininity."

She smiled, "I knew you were a gent, the minute I set my eyes on you." Her smile faded, and she spoke, "That would have teen tolerable, had he not oiled my water and helped himself to my largest slip."

He said with prudence, "What is your charge?"

She countered, "We have yet to discuss other damages." She then added, "Are you assuming services rendered by the Wimpee, I did capture the boat's name."

He said, "Yes, of course, what is the luminosity of your charges?"

She grinned, "You gents sure use high falutin' words, but I, too, engage in the English language, the brightness of my compensation, let's see." She sat down in her swivel, it groaned, and she laughed, then laughed all over.

He was taken up in her antics, and feeling like a child about to be chastised, laughed nervously.

She stately held up one finger, crooking it with a finger of the other hand, and said, "There is also a matter about my blouse that you destroyed."

"Lord," he said, then added, "Now, now, in all honesty, that was between you and the albatross."

She retorted, "Now, now, it wasn't the albatross who ripped my blouse, and everything else, with your hand, and exposed my femininity to your captivating eye."

He replied, "Oh, good Lord."

She continued, "If that damn bird could testify, I'm sure I would win a case on rape charges in 50 states."

He thought about that and unwillingly had to agree with her synopsis. He countered, "You know, Ruth, I would never invoke my advances woefully on any woman."

She said, "I don't know you, but in most instances, you have been a gent, so far." She crossed her legs and asked, "The question is this, what am I willing to accept for your exposing me in front of all to see, the replacement of my debauchery, and your hired pirates?"

"Good Lord," he said, "you make it sound so....so."

She barked in "sordid."

"Yes," he said. "Thank you, I want to take on fuel and water. I'll need a key, then I'll pay for my goods."

She recrossed her legs a little higher than necessary. He couldn't help but notice, and observe. He was very uncomfortable, to say the least.

She stated, "I know we both have our priorities. Mine is payment in full." Having said that, she pulled out a little drawer and fumbled inside and captured a key with a tag on it, and held it teasingly. This got his attention, and she stated, "You look and act like a gent, but you and I know what you did--you bared me from the waist up. I'll let you decide what compensation you feel is adequate."

She held the key out for him to take. He did so and she whirled in her chair, braking with her feet and bringing her to face her desk.

Albert left her office unpretentiously.

She spoke as he left, and he heard her say, "Beauty is in the eye of the beholder."

He thought he was certainly on the spot. He fueled his craft, with diesel and water. He couldn't stall any longer. He returned and she did not hear him or want to turn around in her chair. He dropped the key with the tag on the desk. She turned to face him and he noticed that her skirt had edged up on her thighs. He desperately attempted not to focus on such sights.

She asked, "How many gallons of diesel?"

He said, "Three hundred even."

She said, "I generally check, but I'll take your word this time."

He thought as soon as he left she would check the meter.

She pulled a ledger and thumbed through it. She doodled on her bad and swerved again to face him and stated, "The diesel is $382.25. Seeing you're a paid up at my harbor, the water is free. You know who to make the check out to, let's see if you make it out for the correct amount."

"Damn," he thought, "How did he get involved in a situation like this that required the most delicate taste of payment. He timidly laid it on her desk and said, "Thank you, Ruth, it's been a capacious day."

She replied, "See you, Albert. May your sea chest be free of barnacles."

"Whew," he said as he was outside her door. He thought, "My god, what if he had repeated what she had just generously said to him?"

She picked up the check and noted the amount. A sweet smile swept across her face. She knew he was a gent, and she continued to smile at her adventure. She remembered the look in his eyes. She had never seen her breasts as beautiful before. She was happy now that the incident happened. She sucked her finger and thought, "That gent knows how to treat a lady."

Chapter 7

It was beginning to darken. He turned on his headlights and was pondering his decision not to have a telephone installed. What if Tom had already started preparing dinner? Neither of them knew of his plans for them to all dine out. The heaviest possible objection, was that he was neglecting his primary course. He did know, he was damn contented. He had to repress himself from looking forward to tonight's dinner engagement with Laura. He grew excited and knew that Norma and Tom would probably tease him upon hearing about his new date.

The revelation of Ruth's protuberance, and his new yacht, and the awaiting to solve the mysteries of the universe, and Laura, with Norma and Tom, was a culmination of joys in his life, all squeezed into one sequence. He braked at Downer Street and glanced at his watch--five after five. He walked lightly on the porch and grasped the door knob quietly, and gave it a turn. It was locked. He smiled, took out his key, and engaged the metal mechanism nosily. He entered and heard light laughter coming from the kitchen. He walked in on them wearing a smile.

Norma said, "Hi, Cap. How'd it go?"

"Yeah," said Tom, and added, "Did you capture your cruiser?"

Albert hastily slipped into one of the chairs at their breakfast table, and stated, "Indeed, the *Magnetic* is a masterfully built craft. I fueled her with diesel and water. I also met seamen who were eloquent in the performance of their duties. I've so much to tell you, and yet so little time."

Norma said, in a concerned voice, "What happened?"

Tom said, "It won't take me long to fix dinner, I thought of two courses."

Albert butted in, raising his voice and using body language. "Hold it, Tom, I have exciting news for both of you."

That caught their attention. Tom and Norma's eyes flashed signals of anticipation.

Albert continued, "I'm taking both of you out to dinner on me, to a very exciting establishment." He hesitated and she said, "I didn't bring any evening dresses with me."

Tom said, sternly, "No monkey suits for me. Where are we going?

Albert asserted himself and stated, "The famous Chilkoot Charlie's Rustic Alaskan Saloon."

She lost some of her enthusiasm and said, "God, a beer hall?"

Tom sided with Norma and said, with a grin, "What's on the menu, hard tack ala carte?"

She snickered and Albert had to hold back a laugh an stated, "Hey, this place is a classic, you can even pitch horse shoes."

She uttered, "Good God," and turned to Tom, and stated, "Tom, if you ever take me out to dinner, and we wind up pitching horseshoes, you'll regret it on those long lonely nights at sea."

Tom said in defense, "God, I didn't do anything, now Captain, look what you've done."

Albert had to laugh, then said, "Listen up, you two, it's a unique place, lovely dinners, two dance floors, live entertainment, and two stages. It sounds exciting as hell to me."

Both of them perked up, and messages of fulfillment were in their eyes.

She said, "Why didn't you say that in the first place, Cap? Sure, I'd love to go. Just think a dinner, show, but most of all, to dance with my two best men."

Albert said, "You could, and should hold an ambassador's job, what a tribute you present."

Tom said, "What palaver," and was sorry he said it.

She hooked an elbow in his side.

Tom shouted "ouch," and added, "Cap, you certainly have exuberance, but I have a feeling that you are beating around the bush about tonight."

She caught the keenness in his voice, and asked, "Okay, Cap, if Tom's magical mind is working, and I've no indication it is, what is the mystery? Is it that you don't dance?

That made Albert laugh, and he replied, "No, Norma, I'm average, but there is one favor you can do for me."

She thought that over and responded wisely, "Within reason, Captain."

Albert digested her response and slightly flushed. He said, "A woman notices many details that men never would realize. I'm asking for one of those special qualities that you possess."

Tom got up out of his chair and wildly pointed at Albert and stated, "Oh, God, our Captain has a mate."

Norma wanting to fathom Albert's question, looked up, and exchanged glances with Tom, turned to Albert and said gleefully, "You old Sea Dog, or should I say Sea Horse, you got yourself a filly."

Albert said demurely, "I thought a foursome would be more compatible, this is our last night on shore for awhile."

Tom said teasingly, "You're a fast worker, who's the lucky gal?"

Albert brightened up and said, "Her name is Laura. She is lovely and I'm calling her at six to set up our adventure for tonight."

Tom said smugly, "I told you so."

Albert said, "Norma, it's peculiar you said sea horse. That was the trinket I was going to ask you to perceive. When I met her she was wearing a unique set of earrings. They were sea horses, encompassing small green gems. Norma, I wish you would look at her earrings, and see if they are screw-ons or pierced."

She replied, "Sure thing, Cap, but odds are on the pierced ears, they're in fashion." She looked at Tom and asked, "What do you think, Swami?"

Tom said, "You certainly came out smelling like a rose."

"How's that," she replied.

He explained, "When you called our captain, a sea dog, then followed up with sea horse."

She looked indifferent, and said, "so."

Tom countered, "So what is the mate to a dog? It isn't a poodle."

She laughed and said, "I like that word when I hate a gal."

Albert said, "Thank you for that revelation. I have some other news I would rather not disguise, but I must be up front. Oh, Lord, I mis- spoke. I must be candid!"

Both were smiling at his misgivings. Albert continued, "The owner of the White Tern is a wholesome woman, by the name of Ruth Dander. An incident on the dock, at my slip, will undoubtedly be rumored. It would be proficient of me to tell you the true version."

She said, "The way you are backtracking, it must be a goody."

Tom stated, "My vibes aren't in tune, but I agree with Norma. Let us in on your so-called incident."

He went on about the albatross and the baring of her breasts. Albert said, "It was funny, but it was an accident."

Tom was partially in a hypnotic state and said, "My God, two women within hours." He laughed and added, "You said, I believe, she is a wholesome woman!"

Norma piped up, "You men and boobs. "

Tom repeated the old joke about keeping abreast of things and they laughed.

Albert stated, "The sea shine was on them, and it was a fascination to behold" I would just about bet with assurance that she too was taken by the dancing lace."

The men smiled and looked at Norma for further comment. She looked at both of them, several times, then stated, "No comment, but I will say this, I've yet to have the opportunity to sea shine my breasts."

Albert, still smiling, just shook his head and stated, "It was a wondrous sight."

Tom said, wide-eyed, "God I wish I was there." Tom's fascination with the story had beguiled him. He meekly turned his head and saw Norma's burning eyes. He realized just then, that this was a major mistake, for they had made love at that time. That dumb inference, indicating to her their love making was second fiddle. He reached and put both his hands on her warm face. Her first thought was to slap them down, but she let him wiggle. He looked pleadingly into her eyes and said, "What I said was indecent, I swear. I just got caught up in the fantasy."

Albert obviously felt the tension and guessed what had prevailed. Albert nodded his head at every word Tom was saying and sighed when Tom finished with the word "fantasy."

She keenly saw the real terror in Tom's eyes and his wanton look. Then she realized that it was just a fantasy. In her book, fantasies are just that, have nothing to do with your partner, or your love, they just are seasoning, or spices, that add to the pleasure itself. She had to let him off the hook, but predication must prevail. She gently caught his face with both of her hands, and made certain his eyes didn't wander. She said, "You did hurt me."

Albert said, Oh, no." He got up and stated, "I'm going to take a shower," and left the room.

This hesitation made it even worse on Tom. That word "hurt" was like a red hot poker. He started to defend himself, and she hushed him. She stated, "I'm not going to say, 'Don't let that happen again,' we're not married. But I love you so."

"God," said Tom. She responded, "Shhhh," then added, "Please try to be more complacent and assure me that the next King's X is mine, deal?"

He shook his head, and replied compassionately, "Deal, yes, a deal, my hormones and my mouth were out of kilter.

She said with a laugh, "I'll split it down the middle, I'll take care of your hormones."

Tom smiled and stated, "I may talk, and respond, a bit slowly, for awhile, in order to not let my mouth precede my brain."

She said, softly lowering her hands and moving in for a kiss, "Be yourself. Be the man that you are, and reflect what you really mean." They kissed and she said, "I like you just the way you are."

He replied with a smile and said, "You're right, you know."

She nodded in agreement. He said, "I love you," and she responded, "you're a classic." They kissed again, reaffirming their affection. She stated, "Hey, we'd better get ready, I'm starved."

Tom remarked, "You're always starved."

She responded, "That's my boy." They headed for their bedrooms.

Albert was singing in the shower, his voice a poor modulator--the tune, "It Had to Be You," was under frequency and the rendition was atrocious.

Tom bridged himself between Norma's door jams and said, "The love bug has captured our mentor."

She replied, "She must be an eye-catcher. we'll soon know."

Tom asked, "Do you women kind of dissect one another and give and take points away for various reasons. Just how the hell do you evaluate other girls?"

Norma replied, "It all depends. If she is better looking, the man at her side, her hair do, and naturally what's she wearing." He asked what about her cleavage? Norma replied, "What about it?"

Tom said, "Never mind."

Albert came out of the bathroom sporting his blue, short cut bathrobe." As he passed, Tom said, "You can use singing lessons."

Albert smiled, and said, "Thank you, Tom, I'm glad you're among the living."

Tom moved in the hallway and Norma shut her door.

Tom said sullenly, "Boy, I sure screwed up, part of it was your fault, you know."

Albert replied, "Nonsense, you acted on your own volition."

Tom argued that wasn't altogether true, the tone of your voice, and the essence of your portrayal of the eye-opening experience, had laid a hypnotic scene on me."

She shouted, "I heard that, Tom."

Tom said, "I'm going to shower."

Albert stated, "A sound idea, Tom."

Albert had to hurry to dress and hike to the phone booth to call Laura. He called through the door and asked her to lock up, as he was going out to telephone, and set the time at 7:30.

She yelled, "Yeah, hurry up, I'm starved. what part of town does she live in?"

He replied, "I have no idea, but I will possess that bit of information shortly, bye."

She said, "See you, Cap." She quickly undressed, robed, and silently slipped in the bathroom. The steam was prevalent and she could just make out the back of Tom's wet head above the glass shower door. She slipped off her robe, and quietly stepped inside the steaming, hot shower, much to Tom's delight. She said, "Would you like your back washed?"

He turned and handed her his washrag and soap and stated, "You're a shine-on harvest moon," and kissed her sweetly.

She gripped his shoulders and turned him back around and said, "No, Tom, I'm Norma Moon."

He said, laughing, "You're a super nova."

She began briskly on his back. She smiled to herself and asked, "You're not getting excited are you?"

Tom said, "Let's say it this way, I'm occupying more space since I have a wonderful guest to share my shower."

She laughed and said, "Turn around, Tom."

Tom asked, "Do you want to wash my chest?"

She replied as he did, "No, I want to validate your last statement."

The laughed and she said, "My true blue-eyed, Tom."

He said, "I wish we had time."

They slipped their arms about each other's body. He felt and saw her satisfaction, and said, "This isn't bad, Miss Moon."

She said, with intrigue in her voice, "If you like, you can make it a full moon!"

"Oh, God," Tom said, and obliged them both. They hugged and kissed and moved ever so slowly, unabated by the fresh water streaking over them and the moderation in temperature. Their bodies were completely conformed to the extremities. The pulsations slowed and she dropped the washrag that was snuggled around the soap.

She said, "More hot water."

He said, "You can reach it easier than I can."

She laughed and answered, "Of course," then added, "There isn't any more."

Tom said, "It's just like me, it's absconded." They laughed. He said, "Shut it off, Miss Half Moon."

She said, "Mr. Spoones, that was lovely."

He placed his right wet hand on her left nipple, and made a small circle and said, "I want our song to be 'Shine on Harvest Moon.' Both our last names chime--Spoone, Moon, and I love you and. . . ."

She kissed him smartly and replied, "Yes, Tom, that's nice and it will be our song," then added, "We had better get out of this amorous chamber, I have to dry my hair for Charlie's Saloon, and my love, Albert will be back any second."

He said, "You're right, you know."

She replied, "I know," and stepped out of the shower and began toweling. Tom stepped out and found a towel. They both watched each other drying, pretending not to notice.

Tom said, happily, "Miss Shine On, shower cages have an advantage over tubs."

She replied, "Mr. Shine a Poo, just wait until I get my tub, then I'll let you be the judge."

Tom said, "Oh, God, I'm looking forward to your tub as much as you are."

She laughed and said, "Tom, you're beginning to fantasize."

He looked down and laughed, "It's a good thing men wear clothes, his barometer in effect is his fantasy, he hasn't a chance when naked." He then added, "I was in fantasy land when you made those protrusive statements about you and me in the tub."

She laughed, and said, "Keep that good thought, Tom."

"Yeah, thanks," he said. "It's easy for you to say."

She trailed off, "Not always."

They smiled and robed, and he followed her to the hall where they each went to their own rooms. They heard the front door open and close.

The Briarwood holding the trio was traveling East on Tubor, then he turned left on Cranberry, then north toward's Laura's home on Diamond Street. She asked, "Albert, Skipper, how about revealing the mystery of those two eloquent words *celestial bodies.* I have to admit it's a garter snapper."

Both men laughed, and appreciated her selection of words. Albert replied, "I am just a little surprised that you haven't solved the nonplus."

Tom said, "The contrivance is compounding my mind."

She said, "Yeah, I know Swami, me too."

Tom said slowly, "Something in my cranium has been swirling, and in effect has to do with our new adventure. Captain, and I don't mean your dream boat." He thought about what he had said and added, "Let me restate the latter, I don't mean the *Magnetic*."

Norma looked back into Tom's blue eyes and approved of his rescinding of dream boat.

Albert didn't answer.

She said, "Albert, is it true that you have been hiding some scandalizing tidbit?"

Albert responded, "Yes, but not in that context."

Tom said, "I know."

She said, "Damn you guys, you're in on the poop, and I'm left out to dry."

Tom said in defiance, "No, that's not true, even I don't know the whole mystery, but it has to do with a name or names. Only the Captain knows."

Albert responded, "Tom, you are a wonder. Yes, the crux is the unbelievable, your pragmatics are astounding. The first presents were her wonderful hands and nails."

Norma interrupted him, "She really got to you."

Tom tapped her on the shoulder, and she captured Tom's leer-look. He didn't want his Cap to wind down just yet.

Albert continued, "Laura, explained to me how our nails are our extensions of our bodies.

She related to me that our hands, that is our fingers, contain 14 bones."

Norma and Tom both said, "So?"

Albert hastily said, "So one of her hands and mine, of course, two together is 28 bones. The same number as my slip, that is protecting the *Magnetic*."

Norma and Tom looked at each other in puzzlement.

"That isn't what I picked up, it has to do with two names."

Albert responded, "Yes, but the amazing facts are yet to be told. Even Laura is excited about what I will reveal tonight. I told her I, too, was apprehensive of her input."

"God," said Tom, "Are you going to propose?

She shouted, "Oh, my God, is that true?"

Albert laughed and it sounded hollow, then stated, "Nonsense. No, no, as I have stated, and Tom, the magician, has related to, is a show stopper."

"Damn you," said Norma, and added, "That's two closely guarded secrets that you have denied us." She added, "I thought we were all friends?"

Tom quickly inserted, "Norma, that's close to a play on words, and just a wee bit of blackmail."

She said, stubbornly, "Like hell it is, I would like to know. This is driving me crazy!"

Albert clamped the wheel tighter and said, "All in good time."

She said, "And that's another male lined statement. How do you know it's a good time? Not for me it isn't, I'm frustrated."

Tom said, "She's right, you know."

Albert said, "Never in my whole life have I had to rephrase so many consensual sentences. Norma, for your insight, half of the mystery will be revealed at Chilkoot Charlie's."

Tom smiled and said, "Not bad, Cap."

She said indignantly, "That still leaves me just two little words."

Albert was enjoying his mystique and replied, "Those aren't just little words--they're astronomical."

Tom said happily, "Good work Norma, you squeezed another clue out of the Cap."

She said, "That's better than you've done. Why don't you sashay up a charm, or magical rite, or whip up a spell, and solve those two massive words?"

He sheepishly said, "I can't as you just said, whip up an answer."

She countered, "It just seems, you know, everything else. But when I want one little thing."

Tom interrupted her and stated, "You want two big things."

She thought that over and asked, "Are you in fantasy land again?"

Albert said, "Good Lord," then added, "Norma, please don't forget about the ear observation."

Tom said, "That was a rather cold statement."

She said, "Yes, it was, at that."

Albert said reluctantly, "I'm not going to change my statement, just act upon it, please."

"Okay," she said and continued, "I told you I would, and I didn't forget, but ear observation is so, so."

Tom butted in quickly and stated, "Sterile."

"Thank you, Tom," she replied.

"Gracious," Albert added, "You two really keep me on my toes."

She laughed and said, "or ears." They all enjoyed this pun and laughed.

Albert braked, and looked to the back seat. Tom gave him a thumbs up. Albert nodded, and smiled.

Norma and Tom discussed the third clue, the word "astronomical."

Tom sided and said, "Why didn't he just say enormous, stupendous, or. . . ?"

She stopped him by clapping her hands excitedly, and said, "Tom, you've cracked a portion of the mystery."

Tom asked, "What did I do?"

She obliged him. Albert picked that special word, with the relationships of space and what it embraces. His three little words or big words, all relate to our galaxy. "You're finally using your bean."

He said, "You could have said brain! I do know we are in for a wonderful spectacular experience, one that cannot, and I repeat, cannot be explained."

She sternly stated, "God, Tom, don't lead me down the dark tunnel."

Tom said, "That comes close, yet spooky, it must have something to do with it."

She said smartly, "Stop it Tom, you're giving me the willies. I'm going to get in the back seat with you. Besides, they should be together."

Tom said, "Good idea."

She just slipped in and closed the door. The overhead light came on, and revealed a good looking, well groomed woman.

She said, "Thank you," as Albert closed her door.

Tom said, "I'm Tom and this is Norma. We're pleased to meet you, Laura."

Norma said, "Me, too. What do you think about going to a saloon?"

As he slipped behind the wheel, Albert asked, "Introductions?"

Norma said, "We beat you to it."

He pulled away from the curb and Laura answered, "I really don't know?"

Albert asked, "What was that about?"

Tom volunteered, "About us varmints and two beautiful saloon gals."

Albert laughed and stated, "If it's not our cup of tea, we'll find another harborage."

Norma said, "Aye, Aye, Captain."

Laura said, playfully, "You have a contented and loyal crew."

They laughed and made small talk until they arrived at Chilkoot Charlie's Rustic Alaskan Saloon. It was lit up brilliantly, and grand in size. They entered, and felt they were transcending time to the late 1800's. The props looked to be authentic. The sporting women were beautifully and brightly dressed. There was an attractive bar, with the attendants wearing beards and colored garters on their rolled up and billowing white sleeves. Most wore wheelbarrow mustaches and wide bright red suspenders. It was busy, the flow of men and women appeared to be having a good time.

The music streamed from the stage, as a sporting, or dance hall girl, swirled up to them, in her billowing dress of that era, and asked "country western, or old time?"

They looked at each other's eyes for an answer. Albert said, "Let's try old time."

The young dance hall gal said, "This way, is four a stay."

Tom added, "Yes."

The girl inquired, "Dinner," and Albert replied, "yes." She added, "We have a rootin' tootin' menu," and lead them to a large round oak table, with old fireman's chairs, and highbacks, with ornate wood scrolls on the back. Bright red and white checkered tablecloths set the scene. The billowy dress said, "Your waiter will be right with you, enjoy!"

They thanked her and the men did their chair act for the women. The dance floor was peppered with scented sawdust, a five piece combo was playing "My Blue Heaven" on stage. It was a beautiful rendition. The foursome was smiling and enjoying the mystic and the atmosphere.

A decked out, Western waiter glided up to their table with a smile, and a plush towel embossed with the saloon's name on both ends. He cheerfully said, "My name is Charley. What might be your pleasure?."

Albert said, "Let's begin with a pitcher of your sourdough ale, menus and one sheet of blank paper, 9" x 12" for me, please."

Charley said, "I'll be right back and please don't shoot anyone till I get back."

Laura said that she had been in Anchorage for about three years but that this was her first time at Charlie's.

Albert said, "Those tortoise earrings are stunning."

Tom said, "They are, you know."

Laura smiled a thank you, and said, "I don't mean to be bold, but I haven't been myself since you left me dangling. I believe, your words were to boggle my imagination."

Norma ironically noted, "Yes, they set the stage, and as you said, Laura, they leave you dangling."

Tom said, "It's a wonder."

Albert asked Tom, "You know?"

Tom replied, "Just about."

Norma talked to Laura and said, "Albert pumped us up too, but sometimes Tom can communicate, you know, spooks, and things."

Tom said, nominally, "Maybe it's a gift, but it's spooky, like telepathy, it's time travel, but it just happens. I have no control, or calling."

Albert looked at Laura, "Yes, he certainly has mystified me more than once. I'm just beginning to realize the powers that seem to drift, or will themselves, through his mind."

Charley returned with a full, large pitcher and four glistening mugs on a large round platter. Charley did an "S" with his heavy platter and tactfully made a safe landing. Albert thought that was the act of an artist. The large pitcher was topped with an inch of white froth, and deep below it the effervescent was busy bubbling. The four menus were under his arm. He whipped them out one at a time, and using his thumb, opened the first page, all in one masterful motion, to each patron. They clapped for his sanguine performance. Charley reciprocated with a warm smile as he twisted and twirled his long waxed, elegant mustache. He had started with one of the ladies first and ended with Albert, only his menu his was covered by a large piece of paper. Albert especially enjoyed this type of entertainment.

Norma said, "Where do these talented actors come from?"

Tom joined in and looked at Albert, "That commands a healthy gratuity."

Albert pulled out his old Parker pen and stated, "He will deservingly have such a tip," and added, "Tom, will you pour?"

Tom smiled and said, "Delighted." He made a big deal of just filling the mugs. Tom failed three of the four. He pretentiously poured and the white foam boiled over and ran down the side. The girls smiled, and flashed that message between them, that only women command.

Norma finger pointed at Tom's mug, and kiddingly said, "See, he poured his to perfection, and slopped all of ours."

Tom looked at Albert, then without a word, changed his mug with Norma's.

Norma said, "Well, Tom, 50 percent of nothing, is better than 100 percent of nothing."

Norma slid her mug in front of Laura and retrieved Laura's mug for herself, then stated, "Now, Tom, I hope you learned a lesson by this act?"

Tom smiled and said, "Yes, I'll let you pour the next time."

Norma said, "That's my boy." They all laughed.

Albert said, "One last ritual, then an explanation or exploring of that admiration that strikes reality, but now a toast."

Norma said before Albert could continue, "Cut it in four pieces."

Albert and Tom burst out laughing. Albert finally stopped, and related a quick anecdote of their first breakfast with Norma as cook. Laura enjoyed the humor as a woman would perceive it, and flashed another one of those female messages; that probably meant good for you, one for our side.

Albert said, still smiling, "A toast." They all lifted and gently touched the heavy mugs. Albert said, "Salute," and Norma said that it would be kind of difficult now, and they all grinned.

Albert said, "No, you know just clink the mugs, one more time." They did, and then Albert added, "May each of us remain blessed as we are this night."

They smiled and all took long sips of the frothy, bubbling ale. The girls reflected their remarks in unison with Mmm, and then looked in each other's eyes. The men started to wipe their mouth with their sleeves, then both napkined them. They flashed a warning signal between them that could have said, "Watch it."

Albert said, "Laura, I can hardly wait for your reaction."

Laura looked at Norma. Norma hunched her shoulders and Laura said, "You two are wonderful, but I can certainly see your tactics, and making it most difficult to just get on with it."

Norma clapped and said, "Really, Laura, I've known you less than an hour, but I follow my instincts. If I could have been blessed to have a sister, you would be at the top of the list."

Laura reached out, and warmly patted and then squeezed Norma's hands.

Norma said, "Listen up, Cap. Do it, now"

Laura said, "It's getting close to nail biting time, Albert."

Albert poised over the paper and asked Laura, "Please pronounce your last name."

She did, emphasizing the Mc, and ending like agent with an 'i.' She pronounced it "Mc-Agneti," with a smile.

Norma said, "So."

Albert asked Norma, "Please spell it and I'll write it down."

Tom said, "I hope this isn't a seance with you being the medium."

Albert smiled and replied, "That's your sophistry."

Laura spelled it out with a small "c", McAgneti. Then she looked about to see if there was any response.

Albert wrote it in large block letters, but made the "c" the same height as the rest. He reversed the paper so they could read and positioned it in front of each of them for a few moments.

Four rustic lanterns hung from a ship's wheel above their table, each lantern had been wired for an electric bulb that made up the chandelier over their heads. Albert took a long slow sip of his sourdough ale, then patted his lips, and stated, "There it is. The challenge is yours." They questioningly turned their heads, their minds reeling. Albert said, to break the silence, "Can't any of you see the extraordinary coincidence?"

Norma startlingly said with a gasp, "God, well, I'll be damned."

Albert said, "One down and two to go."

Norma suddenly started to laugh, a little trace of hysteria, made it sound eerie.

Laura said, "Spell it out, Albert," then she looked at the paper again, and covered her mouth with both hands, then said, "No wonder you looked so peculiar when we met."

Tom said, "My mind is on hold, let me in on the secret."

Laura said to Tom, "It's not a secret, I can't explain it. It could be called unknown revelations. My last name spells Magnetic, if you take the "c" and put it at the end, the name of Albert's motorcraft."

Tom said, "God, that is dandy. The odds of two people in the world with names as unique, it is an omen, a good one. It must be some extraordinary power. Do you believe I'm talking serious stuff? Believe me, the meeting of Laura was not by accident."

Tom had their complete attention. His voice and sincerity was a bit scary. "Some day, the secrets of the very power of magnetism will be unfolded. Magnetic forces are the only proven link in our galaxy with humans and the spectrum."

Tom stopped when Norma nervously shouted with eyes blazing, "Stop it, Tom, stop it!"

Albert asked Tom, "Are you all right?"

Norma had grabbed Tom's hand and was squeezing it with determination.

Tom said, "I'm sorry, excuse me, I felt I was in a tremendous field of energy. God, it was scary, I was in position and had the power for an instant to solve any problem or secret of my choosing, I had to command, and the knowledge was there for the plucking."

Norma said, "Tom, maybe we had better go home."

"Oh, no, "it's gone, but one crazy phrase was phenomenal, clear," he said.

Albert said, "What was it Tom?"

Tom looked at Norma and said, "Shine on Harvest Moon."

Laura said, "That is queer unless there is a link."

Norma said, "But there is, we made it our song today. My last name is Moon and Tom's is Spoone."

Tom said, "I need a drink, I did something with that song."

Albert said, "I am concerned about you, but what possibly could you do to a song, written so long ago. Nonsense."

Tom said, "I honestly don't know."

Norma captured the eyes of Laura, and stated, "They say, and I do believe, that love is magnetism."

Laura looked a little shy, but alertly looked into Albert's dancing eyes.

Tom adjusted his menu slightly, and said, "Let's see, what is the most expensive dinner in this honkey-tonk saloon."

Albert said, "Be my guest."

Norma said, "We are your guests."

Albert scanned the menu, then asked Laura, "See anything exciting?"

She replied, "I've had enough excitement to last a lifetime today."

Tom looked at Norma, then they both looked at Albert.

Laura said, "I would like to see your cruiser sometime, that is, if it is all right."

Tom and Norma grinned and awaited that answer.

Albert looked at his two colleagues and responded, "It would be my pleasure."

That flashed messages, and stirred Norma and Tom in relation to the secrecy of their venture. Albert asked Laura, "Are you fond of the sea?"

She replied, "Very much so, years ago I mastered at the University of Washington in Marine Biology." She sipped her ale, and patted her lips slightly, then added, "But that seems a long time ago."

Tom said, "The assimilation of knowledge is still in place in your mind, all you need is to stir it up."

Norma said, "As Tom would say, he's right, you know."

Albert chuckled and said, "Yes, he knows."

Tom nodded in agreement.

Laura asked, "Are you here for business or pleasure?"

Tom and Norma again let Albert field that hot question.

Albert said, "You might say a little of both."

The smiles on Tom and Norma applauded his delicate answer.

Tom said, "I've decided on my order."

Norma said, "Yes, Oh Swami, what is your palate telling you?"

Tom said, "Duck Veronique," to which Norma replied, "Sounds good."

Tom replied, "If they prepare it with the skin, and the correct wine and grapes to match, it's a delightful treat."

Norma said, "I'm going to order the roast duck with orange rice stuffing."

Tom grinned in appreciation of her choice.

Laura said, "I'm on a seafood kick." She laughed, then added, "For the last 20 years or so, I've chosen barbecued crab, and chicken baked with lemon butter."

"Mmmm," said Norma, and added, "that sounds grand."

Tom asked, "Well, Cap, what's your delight?" He looked at Laura. Norma and Tom looked at each other. Norma said kiddingly, "The dessert follows the gourmet meal."

Albert reddened and said, "Good Lord."

Laura asked, "What is your?"....she paused, and Tom and Norma waited in anticipation for that last word. She added, "selection."

Norma grinned admiringly at Laura, indicating a thumbs up.

Albert missed the eye messages, and stated, "Savory sea food on scallop shells." There were smiles all around. Suddenly they all looked with startled eyes, and a tenseness, gripped all of them. The musicians had started the beautiful rendition of "Shine on Harvest Moon." Tom and Norma cupped each other's hand, and each transmitted a warm loving squeeze. Tom and Norma rose. Tom said, "Excuse us." The first time they held each other while dancing, and it was to their song. She put her hand on Tom's chin and guided his head to search his eyes and said, "You're magic." He said, "you're magic-gal." They kissed, then glided across the floor that could have been billowing clouds.

Charley, the waiter, came to their table wearing the same smile, with a pad. Albert relayed the dinner orders.

Laura's hand covered his wrist, and she said, "Let's dance, it's a lovely piece."

Albert replied, "Yes, by all means."

They rose, and they held hands to the dance floor. The fragrance of sawdust mixing with the women's perfumes, and the men's colognes, made it wistful. The five piece band was exceptional. Their orchestration sounded like an eight or ten piece instrumentation. They eagerly wrapped their arms about each other. Then their cheeks touched.

The room was filled with an aroma of incense. They danced as one, in the flume of ecstasy.

Norma whispered, "You're a lovely dancer."

He gave her a hug and replied, "You're the star of this pair."

She said, "Is there anything you aren't skillful at?"

"Yes," he replied quickly. "Sometimes I talk before I reason."

She laughed and stated, "You're right, you know." They both enjoyed the sweet blending of words.

Albert and Laura complimented each other's rhythmic glides. Albert asked, "Have you thought about our name connection?"

"Yes," she said.

Albert prodded, "And,"

She replied, "I'm a true believer in fate."

Albert wondered about her reply. He asked, "Have you been married before?"

"Yes, I'm happy I came tonight. Your friends are gracious."

He said, "We met just a couple days ago, and they're just like family to me," then added, "They keep me on my toes, too."

The band swung right into "Let Me Call You Sweetheart," not missing a beat. She said, "I'm in a phase of my life that has many paths, and I know I will have to choose--just one."

Albert could not respond to what he had just heard; the statement left many questions unanswered. He also thought that this might be a crossroad in his life. She ever so gently squeezed him and he realized the sweetheart dance had ended. The band had just played a short melody. Their fingers laced each other's as they recaptured their table. Tom and Norma had just returned. The music welled up with ponderous drums, followed by an announcement, that there would be a fifteen minute break.

Norma turned to Laura and stated, "You dance as one."

Laura smiled and was slightly embarrassed.

Tom said, "Let's touch our drinks one more time and that will seal us as one." They did so with wholesome smiles.

Laura said, "I see why your new friends are like family, all of you have made my presence very comfortable."

Albert said to Laura, "And you have fulfilled my imagination."

Their waiter, Charley, emerged and said, "Succulent time." The flow of vibrant odors had mixed with each fragrance attempting to emulate its rivals. Charley whisked each lavish dish close to the owner's nose. Both girls said in unison, "Mmmm," that brought surprise and laughter.

Tom cut a slim slice of his Duck Veronique, savored it. He then said, "Yes, yes," and plucked a seedless grape and popped it into his awaiting mouth.

Albert, foregoing sampling, said, "Tom, our dinner has passed its test with the most prudent cuisine teacher."

Laura selected, forked a small bite, dipped it into a cup of lemon butter, and guided it to her mouth with an "Mmmm, delicious."

Tom said, out of the blue, "Heaven."

"Yes," Laura remarked.

Tom positioned his empty fork toward Laura, and said, "May I?" She nodded and he selected a small portion of herb barbecued crab and he dipped a corner of it in her lemon butter, and his mouth quickly stole it from his fork. He smiled, and chewed.

Norma said, "Well."

Tom replied, "No doubt it's good, but just a dash too much of Rosemary."

Albert said, "That's my boy! You want to plunder my dinner, too?"

Tom said, "Sorry."

Norma cut a piece of roast duck, then selected a spoon, and joined it with her orange rice, tasted the morsel and added, "Good, Mmm."

Tom finished chewing and washed his down with a swig of ale.

Norma prepared a small bite, off her plate for Tom and he captured it with a "Thank you, Hon."

He chewed, and said, "Good, yes, could use just a bit more honey."

They ate and made small talk. The band returned and their first rendition was an old smoothy, "Red River Valley."

Tom napkined his lips, leaned back, and used both hands, and patted his tummy. "Nice," he said. Then he looked and caught the pleasurable eyes of Norma. He dared not search Albert's face. He turned slightly to face Laura, and said for all to hear, "Please, Laura, your circumstances appear devastating to you, but with time, it will heal, as all of us know."

There was silence, not even the tinkling of utensils. Norma had lit up, and blew out a net of blue clouds that matched the sentiment at that moment.

Albert said, "Good Lord, what are you babbling about." Then he looked into Laura's eyes, then back to Tom, and asked a little harshly, "Why did you say that?"

Tom looked at Norma and said slowly, "I made a judgment at the spur of the moment."

Laura said with her hand up gesturing, "Albert, it's all right."

Norma whispered loudly, "Your one weakness has risen like a ghost on such a festive night."

Tom brought both of his hands from his lap, and pleaded his case, and stated, "Damn it, I'm not going to say I'm sorry, Laura. I've never laid my eyes on you until tonight. But on the dance floor, you sent out waves of desperation."

Albert said, "That's nonsense."

Laura put her hand up to calm him, and said, "I was having a wonderful time, then for a moment the realization struck, like a trauma."

Norma saw the brief terror in her eyes and said, "My God, what happened?"

Laura said, "Let's not spoil our wonderful party."

Albert tried to look through Tom.

Tom stated to Laura, "I'll keep your secret, if that is what you wish."

Albert looked disturbed, stared at Laura, and asked, "You want me to take you home?"

She tried to smile but failed to pull it off, and said, "No, no, I don't want to go home."

Norma said to Tom, "This is our dance."

He answered, "Yes," and avoided Albert's eyes.

On the dance floor, Norma asked, "Just what the hell was that all about?" then added, "I could see the terror in her eyes."

He stated, "It came to me while we were dancing, the horror was in her mind, and I intercepted the pain.

"God," she said. She didn't deny anything, but what right have you to openly exploit her fears?"

Tom said as they danced, "That's easy to answer."

She said, "Well."

He said, "You heard her say not to spoil her evening."

She said, "But Tom, you've already ruined the evening."

He thought that one over and then said, "Listen, Hon, I'm torn between two poles, I admire and respect Albert's date, Laura, but I also respect Albert as a man and a friend." He spun her around as the band musically switched to another old but wonderful tone, one of Eddie Cantor's Favorites, "Making Whoopee." The male singer was good.

She said, "Then you didn't do it to show off?"

"God," he said, "You really don't know me. I did it for Albert."

She kissed his neck.

He smiled and looked into her eyes, and said, "Think about it, do you really want me to deceive Laura?"

She replied, "I wish you hadn't used that word 'deceive.' No, but you're right, it has to be her decision."

Tom said, "I'll tell you one thing that is important, as to why I spoke up, I took a calculated risk."

She asked, "How's that?"

He said, "I'll say it once, don't laugh, but I cherish our relation, and you are precious to me. I'm still scared that something unforeseen will take our love away. I'm just so damn happy, I would like Albert to have the same chance of happiness. The want of it is that Albert was the interjectory of our meeting."

She hugged and kissed him firmly, and said, "God, how did I get so lucky. You're such a romantic and thoughtful."

He stated, "I also took the risk that she might reveal her grief and Albert would help her in her difficulty, therefore bringing them closer together."

She said, with a wholesome laugh, "Tom, your quivers and bows are showing, and so is your uniform."

He asked, "What kind of uniform?" and she retorted, "A prince charming."

The Making Whoopee tune had just ended. They returned to their friends, just in time to hear the start of another old, but grand song, "Springtime in the Rockies," flowing from the raised podium on the stage. Norma said, "Hey guys, do you realize I've just lit up one time all evening, that's a record for me."

Tom said, "I'm proud of you."

Albert said, "I applaud you for relaxing your habit."

Norma reached across, placed her hand over Laura's and stated, "Tom has kept your secret from me, even though I pestered him ardently." She gave her a squeeze and retrieved her hand, then added, "Your nails are lovely."

Norma replied, "Thank you, Norma."

Albert asked if they wanted another pitcher of ale, but all declined. Albert excused himself and left for the restroom. Norma quickly asked, "Laura, I'm not really a snoop, just a woman's curiosity. Maybe Albert can help you."

Laura replied, "You've all been so nice, but this is my problem. It could be worse, I guess. I haven't even called Linnel, my only brother in Oregon."

Norma stated, "The genesis of the tempo, and the rhythm of time, relegates that many times, can free moments of time, to years of memories, so time is really no consequence."

Tom said, "She's right you know, you and Albert could very well be perceived as a married couple. You both have that comfortable halo about you, like magic. It's really magnetism."

Laura said, "It's not my place to burden or involve someone I just met."

Tom said, "Says who?"

"Yeah," said Norma, then added, "Reverse the incident."

Laura asked, "What do you mean?"

Norma explained, "I don't pretend to know if you have feelings for Albert, but what if he inherited your conflict, wouldn't you want to know about it?"

Laura said, "You guys sure know how to squeeze a gal."

Tom replied, "I do a lot of squeezing in preparation of our meals."

A little grin appeared on Laura's lips.

Tom continued. "And besides, I know what it is."

Laura asked, "Do you mind telling me discretely?"

"Certainly," Tom replied, and took a pen from his lapel pocket." He pulled the paper from Albert's area, and folded it in half, thus screening his upcoming message. He began to doodle, finished, folded and handed it to Laura.

She opened it, and gasped with an "Oh, how is it possible?"

Tom said, "Let's just say, like Doris Day, it's magic."

Norma said nonchalantly, "He did it again, I take it."

Laura nodded and handed the paper to Norma, as Albert recaptured his chair.

Norma took one glance and said, "Jesus," Her voice hung, and she looked nervously at Laura.

Albert reached over and asked, "What's this?" No one said a word. He looked at the paper and saw a picture of a hand gun with lines coming out of the barrel and big letters spelling out 'Bang'!!

Chapter 8

Devlen, 5'10", was a little bit fat, and at 52 years, was approaching being a slob. His jowls hung, precluding a tarnished silver spoon in his mouth, since the first day he blinked. His hair looked to be on fire, a bright red, with his enlarged ears, looking like an earflap on a winter hat. His smile was that of an apostate, angel or the ruler of hell, whichever was prevalent, or the topic, at the moment. His last name was Dawson, some called him D.D. Most called him Mr. D.D. He was sprawled in a masterful upholstered chair which had maximized its extensions. His fat round belly adorned a white, white, silk shirt, along with his size 46 white stretched slacks. His taut belt was the result of the killing of an alligator for its skin. A large silver buckle portrayed two sea nymphs embracing each other. There were four small, but prestigious rubies on the bold relief. His small black pupils blinked obsessively, as he demonstratively stated, into the cocked neck phone. "Yes, I took care of it, I fired her. Yeah, she has no recourse. I told her, it was a hand down, budget cutback." He laughed, then said, "She bought the cutbacks. They were authorized in Form 906." He paused and laughed again, "I know there is no such form." Laughter emitted from both ends. He stated, "Too bad the computers broke down, and she had our material, that is sensitive to our business."

He listened, then said, "No, she didn't check anything out. I'm sure she won't connect on anything today." Yeah, she'll be concerned about finding another job. Everything here is as smooth as a baby's. Yes, okay, see you soon."

He smiled, and cradled the phone. He lifted a thick short stout glass that held a swallow of straight whiskey, and rotated the heavy tumbler. He tossed, grimaced, and filled his gullet. He shook his head, and emitted a satisfactory sigh. He loved the good life, and he knew it was getting even better. He was smart enough to know that the more money he amassed, the deeper he was becoming involved in the operation. He would prepare a back-up, and watch his backside. Honor among thieves was a Hollywood glib.

Albert didn't wait for the check. He took a Hamilton and a Jackson and tucked most of it under his nearly empty plate. He picked up the paper with the anagram on it of the shooting gun, then tore it three times, dropping it on his plate. Albert looked at his two friends, one at a time, then asked, "Tom will you and Norma please take a cab. I would like to take Laura home now, if she wants to go."

Laura, trying to smile, said, "Yes, that would be fine."

Tom perked up for Laura's benefit and vivaciously said, "You bet, Cap, I wanted one more dance with Miss Moon."

Norma stated, "What's this Miss Moon affair? Call me Norma, Mr. Spoones, after the knives and forks."

That brought a little cheer which quickly faded. Laura, Norma, and Tom exchanged gracious remarks. Tom said, "I wish this one time I could predict your destiny, but smooth waters, is our mandate for you, Laura. Norma and I have a wonderful new friend."

Albert smiled benevolently.

Norma said, "My Tom has said it all."

Laura graciously moved to Tom's side and had to stretch to kiss him on the cheek. Albert's and Norma's hands touched; both squeezed leaving messages unanswered. They left. Norma, was still nervously attempting to tap a cigarette out of her crumbled back, and several fell on her plate.

Tom asked, "Do you have another pack?"

She said, a little upset, "Yes," while trying to get the last fresh cigarette out successfully.

Tom covered her busy hand, withdrew the pack, retrieved the lonely smoke, and said, "Open up."

She did and he picked up a souvenir book of matches and lit her bobbling, smoking stick. She inhaled deeply, and slowly released the blue carbon particles into the air which shifted like a waving net. Norma said, "Just two comments. I needed this cigarette, and secondly, this was the first time you lit me up."

Tom grinned.

She read it, smiled and stated, "Thanks for not reminding me that my last statement was out of context."

Tom said, "I know you are concerned about Laura and Albert, now that he has the choice to get involved further, or let it go."

She stated, "I know, but the impossible coincidence, and she seems approving, and not deserving."

Tom said, "I agree, but it's like this."

She watched, as he tried to capture a handful of her smoke.

He then looked at her and stated, "You see, the significance is, if I could capture your smoke, it would be short lived."

She responded, "That's true, but not to try would be unconscionable."

He replied, "Yes, you're right."

She paused, then remarked, "I expected your usual answer."

He then replied, "You still know little about me, I prefer, not to jest when a serious situation is the topic, or on the agenda."

She stated, "You're top flight."

Tom replied, "I'm having a slight wave of second thoughts, by revealing Laura's situation."

Norma asked, "How so, Tom?"

He turned to squarely face her and replied, "What of my first thought, was to reveal Laura's perplexity, so that Albert would play on his consciousness and mutually bring them together, like iron filings, to a magnet?"

She attempted to look deeper into his blue eyes, and responded, "It seems to work. They are alone, and the Cap seems very concerned."

He leaned over and kissed her, then said, "Yes, and what if that big little word 'if' because of my meddling, he does get involved, and Albert is seriously harmed?"

She replied, cocking her head, "Tom, you did what you thought was right for both of them. What happens after the fact is a whole new ball game." She paused, pointed a finger at him, and continued, "You are now a spectator, not a participant."

"Thanks," Tom said, and continued.   "I mean it, and what the hell?   If the Cap gets in trouble you and I will bail him out."

She responded, "Sure thing, so now I'll give you a choice, dance with me, or call a cab, and we'll work off that lovely dinner together."

He said quaintly, "Shine on Yellow Harvest Moon."

She smiled broadly, she had him back, and said, "I'll walk with you to the lobby. I don't want to be the last one at this table."

Albert braked at Laura's residence on Diamond Street. There was little substantial information passed by either of them. He said, "I'm sorry Tom upset you, and abbreviated our night. What was that drawing of a gun?"

Laura responded, "First, Albert, I had a wonderful time, so good in fact, I did forget my immediate problem. As for Tom, he did what he thought was right, and I'm flattered with your concern."

He stated, "I am concerned, I was taken the first time I saw you."

She reacted, and replied, "You really don't know anything about me."

He countered, "I do know how I feel about you; the knowing will be the frosting on the cake."

She laughed lightly and said, "You have really perked up my spirits. I was fired at work. There was a letter on my desk."

He asked concerned, "For what reason?"

She replied, "Some reference to budget cutbacks, I'll have four weeks severance pay but no job."

"My God," he said, keenly sorry, and put his arms around her. She relaxed and leaned closer to his protective hold. She tilted her head upward, and he responded to her gesture. Their lips touched, then pressed, exciting messages surged through their bodies. They tried to snuggle closer and he kissed her eyes lightly.

She responded with a warming hug, and asserted a pleasurable kiss. She asked, "Do you want to come in?"

"Yes, I would like that very much," he replied. He hurriedly got out and opened the car door.

She thanked him with an unexpected kiss. Their hands found one another and they headed toward the door. She retrieved her key from her purse.

The ride home in the yellow was exceptionally quiet, both were mindful of Laura, Albert's consequence, and the anticipation of making love. Tom broke the silence, "If my intrusion is a possible, we'll have considerably more time."

She smiled, patted his hand, and stated, "I'll leave the timing to you, Swami. I hope you're right about our Cap. If we do get caught, I'll gladly resend you to center stage."

He grinned and responded, "It will be worth it."

She said smartly, "Is that another male's response, that you envision greater pleasures than I, or is it a female's?"

He laughed lightly and said, grinning, "I'll tell you another secret." He put his finger to her surprised lips, then continued, "Anticipation for guys, is as close as that sensual, lavish moment."

"Yeah," she replied. "That's all you guys ever think about."

He responded, "Not true, we love, good food, too."

She smiled, then smartly stated, "Have you ever thought, how selfish your gender is, always taking."

His blue eyes danced, and he replied, "No, Miss Moon, I give you every millimeter. It's a give and take situation."

She laughed generously, and said, "That was a marvelous rebuttal."

Tom said, "Speaking of butts, we're here, after you." He paid the driver, keyed, and went inside, and made certain that Tom locked the door. He reached out to keep from crashing into her.

She placed both hands on his face, surprising him and presented him with a smart smacker, then looked into his blue eyes and hastily stated, "Hit it, Tom, you're in center ring, in my bed." They playfully ran and shucked their clothes carelessly. Norma won, and said, "You lose, please get me a glass of water."

Naked, he turned and heard her say, as he left the room, "Think about me. I'll be able to visually verify your lust." She doubled up the pillows, and lay waiting for his return. She clapped graciously upon his entrance and said joyfully, "That's my boy." They laughed, and she captured the water and sipped it.

He was excitedly waiting, close to lip biting, till she playfully finished her drink. They nestled in each other's arms. All the parts fit with acquiescence. She said seductively, "You're still my Lancelot."

He said, spiritedly, "I may well be, Guinevere, but don't tie a flag on my lance."

She laughed and elbowed him in fun, grinning, and stated, "It would look cute!"

He grinned and said, "We've already been through that attaching to!"

She said, proudly, "That's true, it wouldn't fly now anyhow."

He looked down and said, "You're right, you know."

They kissed and she said, "Tom, is there any chance we could get a tape or recording of our song? Wouldn't that be lovely, during our comings and goings?"

He quickly responded, "That's a peachy, dandy idea. Our song, Are You a Harvest Moon, Miss Moon?"

She pretentiously said, "I would have to say yes, in respect to you, you seem to reap my harvest."

They giggled, and he playfully patted her soft thigh. He said, "The most exciting response for me was in the shower when you indicated you like to make a full moon."

"Yes," she said, "that was generous of me. If you take that out of context, we would only have pleasure once a month, a full moon."

Tom quickly retorted, "There is one month that has two full moons, according to the celestial calendar."

She suddenly screamed with joy, Tom was temporarily terrified. She accidentally struck him soundly in the chest, and excitedly finger pointed at him like scolding, shouting, "Jesus, Tom, Tom . . ."

He nervously interrupted her with, "You sound like an Indian drum!"

She cried out, "Damn it, I know Albert's secret. Oh, God, he is a sweetheart."

He sat up on the bed and exclaimed, "What is it?"

She said, "Sorry I hit you, but you said it yourself."

He searched her beauty, and the mystery, and asked, "How?"

Still bubbling, she countered, "The moon phases, the full moon, the celestial calendar, with the celestial bodies."

He said, "I still don't know what you mean."

She asked, "Do you want me to spell it out, it's a male hidden covenant." She paused, then went on, eagerly and smartly. "It took a woman to break it."

He said, "You've made your case. It must be a wonder, you called our Cap a sweetheart, I'm just a little jealous."

She said, wearing a smile, "Don't pout, he was thinking of you, too. Okay, Tom, what does the word celestial mean, in definition?"

Tom pondered and pulled up the sheet a little higher, then spoke: "Worldly, supreme, heavenly, so....?" She happily stated, "Tom, the second word is exactly what it means, bodies, like male and female bodies, don't you get it?"

Tom swelled up, and a wondrous smile captured his face. He said, "I'll be damned, what a romantic!"

"Yes," she shrieked, "Our own stateroom together."

Tom said in the festive mood, "No more pretending." They hugged and kissed, and Tom said in jest, "You're a meteorite bed partner."

She responded, "You have a heavenly body." They hugged again, and lay in each other's arms."

Albert looked about the clean, homey, small apartment. The scarves and curtains were bright orange with brown and black flecks. He selected an overstuffed chair.

She settled in a matched one, directly across from him. The distance was approximately ten feet. He had declined a drink, and stated, "This is lovely, so bright."

"Thank you," she replied.

He continued, "I don't mean to be nosey, but the government job you had at the harbor."

"Yes, what about it?"

He stated, "I held a few short-term jobs for the government. It's a little unusual to be pink slipped, unless you are in charge and the heat is on."

She stated, "I've never been fired in my life, until now, and it hurts. Excuse me, Albert."

He nervously shifted in his chair. The coffee table was within reach. A folded paper lay next to a bowl of colorful artificial flowers. He stretched, and plucked the paper, opened it, and began to read its contents. "Dear Miss McAgneti: Due to necessary reduction of staff, your position has been terminated. The lack of trade importing, has compelled us, with regret, to reduce our budget. This notice is in effect immediately. The usual four weeks severance pay will be in addition to time served." Signed, Devlen Dawson. The top left was also typed, Form number 906. "Jesus," he thought, "what a cold impersonal dispatch." He folded it back.

She returned, and stated as she sat down, "I'm sorry. I'm better now."

He said, in a sullen tone, "I took the liberty of reading," pointed, and continued, "What kind of man was Dawson to work for?"

She perked up a little, at his interest, and replied, "I only saw him once or twice a week. He seemed to be all right."

He said, "I didn't know ship trafficking was on the decline in Anchorage."

She replied, "Actually, it has increased in tonnage."

He said in question, "The letter stated the contrary."

She remarked, "I know, maybe they balance exporting with importing."

He thought and said, "That's possible, what are you going to do? What about your major in marine biology?"

She said, "The master's that I hold would qualify me as a teacher and maybe a professorship, if there were any. I would have to be evaluated, a 90-day course, then tested. I have already looked into it, the government has dropped its subsidy at the University of Alaska. The closest, and still maintained, is at the University of Washington in Seattle. I haven't contacted the University because I wanted to stay here in Anchorage."

He said, "You certainly are well informed."

She half-smiled and said, "When I first arrived here, nearly three years ago, I put in an application. That's why I am up to date as to what prospects are open in my field."

He asked, "Do you have any idea what your recourse will be?"

She looked grim, and said, "No, I'm still a little dazed by the finality of my job."

He asked, "Do you qualify for seniority?

She replied, "I doubt it, at least none in this state, and three years in government employment is nominal." She gave her best smile, and stated, "Albert, we have talked about me all night, what about your fishing?"

He said soulfully, "I must confess, I'm in a discreet situation, Laura. The word fishing was used discriminately. I am in a legal business, but very confidential. But I can tell you this, my conquest in life has been, and currently is, the search of knowledge."

She asked, "Will I see you again?"

He hesitated, his reply was most crucial and she was waiting for a determination. He saw a disturbing look in her eyes and finally stated, "The timing is like quicksand." He rose, she followed his lead, and he moved to embrace her. Their arms emulated, both holding each other vigorously. They relaxed their tense hold, permitting their waiting lips to taste the flowing warm sensations, that sensuously invaded their bodies. He said, "Thanks for the life raft."

She smiled back and replied, "It's the same one you gave me earlier this evening." They kissed again for the want of excitement. Their receptors were drowning in polished sensitivity.   H   e spoke in a spellbound, yet coherent voice, "I can't let you go." He was squeezing her a little harder than he realized. He said, "Let's sit down. I want to ask you some serious questions." His eyes were bright and focused as they recaptured their chairs. He soberly asked, "Laura, what ties have you right now!"

She accommodated what he had stated, and sullenly said, "None."

He stated, "I'm hoping you don't think me intrusive. I want, I truly want you to come with me, until you resolve, and rectify your life." He held up his hand and sensitively stated, "I must say this, I had no revelation I would meet, let alone desire you when I made that tactful promise to Norma and Tom. I told them that I have a generous surprise for them on the *Magnetic*. I presented them with a bewildering mystery, to unravel two words, celestial bodies. Norma and Tom have become romantically and hopelessly in love, in a matter of two days." He laughed with freshness, and continued nervously, "My surprise is to give them a stateroom together." He waited, then added, "I hope I haven't embarrassed you?"

She replied, "They look comfortable, I'm happy for them, and no, it does not embarrass me, but I do have a couple of questions."

He said, "Yes, I expect you would," then paused and added, "I cannot renege on my promise, there are only two staterooms on the vessel." He looked at her in wonderment and quickly added, "I'm sincere, I wear pajamas." We would be together, and you'll have time to brush up on your maritime and marine biology. And the most compelling reason of all, I think I love you." Revealing that fact, they both got up and met in each other's warm embrace.

She said, "We've just met, we're strangers."

"Nonsense," he stated briskly, and went on, "I perceived you care for me, as I do you. I wasn't intruding in your life, or taking advantage of you. I respect your privacy."

She replied, "I do like you Albert. I really don't know if I love you."

He countered, "I thought it would give you time for solace with friends, and not be alone. Then you can go forward on a firm path."

She said, "You are caring."

He said truthfully, "I'm also shy. I've never said that before, but it's true. I must tell you, I'm on the adventure of my life. It would be the adventure of a lifetime with you.

She said, "I couldn't, and won't be a bit of cargo, without being purposeful."

He shut her off, "Laura, you're not imposing, your background would be of infinite value in our goal. Yes, Laura, I need you even more than my fascination."

Still holding one another, she stated, "Yes, I'll sign on Albert, only if I can enhance your cause, whatever it is." They matched lips again and firmly protected one another.

He said, with a relaxed sigh, "We'll leave in the morning, hopefully before noon."

She stated, "Are you sure?"

He said with certainty, "I'm certain, as I am about to unravel one of the mysteries of life on the universe." They kissed again and searched each other's eyes for guidance.

She said, "I can't believe what has happened to me in a matter of hours. I've been fired, and now am committing myself to the sea, with a man, I haven't known 24 hours, and obviously, sharing his cabin."

He stated, "This episode that sprung me out of retirement, is so compelling that I would have wagered heavily, if I was told another person would be involved other than Tom and Norma. I wouldn't foolishly change my integrity, and a vast amount of money. Laura, we are not risking or committing, we are fulfilling our lives, The pluses far outweigh the minuses." He paused and kissed her neck.

She said lovingly, "My God, Albert, I'd walk with you through the internal cavity of hell, if you wanted me to."

"Oh, Jesus," he said excitedly.

She laughed shyly and said, "Oh, Lucifer."

They both enjoyed their closeness and hugged even tighter and emitting intense, violent emotions. The electric pulsations of positive to negative was the clincher. She guided him toward the big bedroom.

He stopped her and said, "You don't have to!"

She nodded, and replied, "I know, I want to!"

He was at a loss for words, only expectations, they made love sensually and finally spilled into epicurean. Contented, they lay side by side. His left arm was under her head. A sheet covered their waists. Albert had directed his index finger, and was circling and massaging the bright, springy rim of her enhanced nipple.

She watched him with fascination. His touch was that of an artist's brush creating a master. Lulled into rapturous, emotionally sensitized by his antics, she asked, "I have an interesting question, Albert."

He replied, "Sound away" as he brushed with light pressure strokes.

She asked, "Please clarify celestial bodies."

He laughed lightly, and replied, "That's easy. Celestial is just what we experienced, heavenly bodies are their bodies."

She said, "That was clever."

He stated, "We engaged in kidding, enjoying humor. It absorbs some of the stark, rash realizations of life. The universe, as we know, it is becoming very fragile and dangerous."

She said, still enjoying his busy finger, "Life has always been unforeseen." She laughed, captured his eyes and said, "Like this extravaganza today." She smiled and sent warm signals, then added, "and tonight." He gave his attraction a slight fondle then pressed it. She blinked.

He rolled half on his side, and pulled the sheet up to cover her. He said, "What I'm about to say is as serious as it gets, it's dumb for me to tell you now, I never use that word dumb. But I want you to know what could possibly happen. Have I your full attention?"

She looked bewildered and said, "yes."

"Laura, there have been eight stages since our planet has mushroomed naked single celled embryos, emerging in saltwater. Thirty to forty million years are time zones in which our earth has been lambasted. Each strike has obliterated thousands of species, and nearly destroyed all life. My studies relate that these zones of devastation, are in relative time zones that are directly related to the change of our magnetic poles."

She asked, "Do you mean the North Pole is reversing its polarity?"

"Yes, that's exactly what I mean. The poles, or axis of earth, are the exponents of a massive electrical magnetic fields, therefore the most vulnerable to any infusion of matter from space or earth, electrical or otherwise, is inducive. All of the energies on our surface are bombarded into our atmosphere. Transmission lines, radio, TV, and electromagnetic fields and magnetic lines of force, are more violent than ever previously measured. Life would be tenured without the magnetosphere shielding us. The magnetosphere does not rotate as our earth does, but remains fixed. One side always faces the sun. Any given spot on the surface of the earth is in a constantly changing magnetic field.   The solar winds that transverse through space, carry electrical magnetic particles, that are crashing and being absorbed by our fragile shield.   Billions of watts and enormous electrical currents occur. "

Her eyes were witnessing stark intensification to his eyes.   She started to tremble, and interrupted him.   She gushed out, "Albert, you're scaring the hell out of me. "

"Good, " he said, then added, "That was my intention.   Data collected has proved that six of the eight time zones that were struck by disaster, occurred when the magnetic poles reversed polarity. Our very existence relies on a precarious balance of electromagnetic forces. "   He took her hand and stated solemnly, "The poles are about to reverse, As with any major catastrophe, several usually act in an odd sequence that can set the scenes of electromagnet storms that could devastate the earth, or most of it.   Laura, time is very previous, as precious as you are to me. "

Laura said seriously, "How much time do we have?"

He squeezed her hand, then patted it, and stated, "A week, two, six months, a year, impossible to predict, but it will occur within the year, or sooner.   Every minuscule electrical charge, or solar storm, or intense electrical storm at the precise moment of poles changing will directly affect the intensity of destruction.   I've determined where the annihilation might be of minimal saturation. "

She said gravely, "Damn it, Albert, I've just found you, and now you're saying our time is a burning fuse. "

He stated, "I'm sorry, I took this timely amorous moment, to deliver a speech, but I want you to know what is coming, as certain as tomorrow is another day."

There was terror in her eyes and she asked, "What can I do to assist you?"

He looked up again into her soft green eyes and stated, "Laura, I may have inadvertently led you astray as to my role. It is a business and personal scientific endeavor. No human being can prevent what's going to happen, but they could possibly reduce the destruction. Our goal is to intercept, accurately digest, and post the gravest of tragedies that have an effect on commodity exchanges. There, Laura, you have it all. Except this, my theory is that the powerful and instantaneous reversing of the magnetic poles caused the destruction of thousands of species, including the dinosaurs. This completely discounts the theory that hurling asteroids accounted for destruction of the great reptiles. My God," he said, "that's the longest I have talked in ten years," and laughed. That helped ease the tension of the conversation.

She said, "I have moderate instruments, if I am to be effective I will need several new tools."

He said excitedly, "That's what I like to hear, if we can buy it here, it's yours. One other matter, you are on the payroll, the same as Norma and Tom"

She said, "I can't. . . ."

He quickly intervened, "You are a professional and will be paid as such. I am the fortunate one to have a marine biologist on board."

She said, "You don't know how competent I am?"

Albert replied with eyes dancing, "You have a Master's Degree in your subject."

She replied, "That was four years ago."

He grinned and retaliated, "Knowledge is power and once grasped, you open the mystery that you still retain the techniques necessary to those mysteries." He leaned over, gave and received a smart kiss.

She said, "I don't want you to go."

He replied, "What the hell, I'm in charge of our group and I'd be delighted to stay." They exchanged glances, then they each rolled and wrapped their exciting arms around each other, snuggling and wearing smiles. Their hands competed for their partner's complacence, and discovering each other's contentment, their lips touched and amorous warmth flowed through their bodies. They lay back, his arm still under her neck.

Albert said, "Just pack your necessary items for a five day, not less than three day stay. Have you lived in close quarters before?"

She answered, "No," then added, "I won't just be your mistress."

He quickly replied, "I don't favor that word mistress, it implies female slavery to the wants of a male."

She laughed lightly and said, "Yes, that is the perception in this male oriented world. I want to contribute more than. . . ."

He interrupted her, "Honest relationships, yes, I don't want our affection to be tainted. I sincerely care about us, not what others might imagine."   He removed his arm gently that was cradling her head, captured her green eyes, and said, "You're a special change of menu from my perspective. You are a blessing and I am very fond of you, Laura. I can say, I love you, the depth I'm anticipating will blossom."

She said tenuously, "I love you too, Albert. Your springing in my life at this moment is like being tossed a live preserver in rough waters." Their eyes became misty, and they kissed softly, but lingeringly.

He said, "Can I get you anything?"

She replied, "No, I have everything. Do that finger massage again. The other one is jealous." She laughed.

He looked into her gracious eyes with pleasure and said, skillfully, "Counter clockwise is nice, too."

"Yes," she responded, and he fulfilled her request.

She stated, "Yesterday's paper had an article about an ozone hole in Antarctica. It was sixty percent destroyed by manmade gases and pollution."

He said, "You see, you're already contributing valuable information. I haven't read a newspaper in ten days."

She said, timidly, "That does feel good, you have the Midas touch."

He grinned and said, "You have two lovely Midases.   Do you know that word you used, 'Midas'?"

"Yes," she said, still watching with marvelous speculation.

He replied, "If you spell it backwards and put an "s" between "i" and "m", you spell the word sadism. Meaning, by dictionary terms, the infliction of pain on a love object, as a means of obtaining sexual release."

She said and smiled, in jest, but graciously, "You have the Albert's touch." He thanked her silently.

She stated, "Are you really serious about paying me?"

He quickly answered without looking up, or missing a gesture around her small bud. "Absolutely, your knowledge will be most valuable. For instance, one degree of temperature change in our seas, either direction, changes all the weather patterns in the world, causing turbulence and torrential rains. It could possibly be violent to some food crops, or droughts. That's right up your alley."

She still watched fascinated, almost hypnotically, as his busy finger every so often, would touch the end of her protuberance, with his thumb. She blinked, in unison and asked, as the tantalizing continued, "What is my salary?" more in jest, than just the supply of needed information.

He said, nonchalantly, "The same as Tom's and Norma's."

She asked tranquilly, "How much a month?"

He stated softly, avoiding her eyes, with room and board, over seven thousand.

She screamed and he lost control of his two principal tinted buttons, which went into intense gyrations. She screamed again in joy and asked in amazement, "Are you kidding me?"

He said, laughing, "you're certainly a lively. . . ."

She interrupted, excitedly, "a naked lady."

"Yes," he said, then added, "I'll say."

She said, "I'll do my best for you!"

He captured her green eyes, smiled, and stated, "You already have, Laura!"

She said shyly, "Not that, I mean my contribution."

He captured her eyes, grinned, and said nothing.

She said, "You men, you know what I mean."

"Yes," he said jokingly. She sat up straighter on the bed and placed her hands on his cheeks, and looked into his brown eyes, and sincerely asked, "I have to know, all this money, is it because of my permissiveness?"

He said boldly, "Good Lord, no. You'll earn your keep, there will be times, we will work into the wee hours. It's nonsense, that I would double jeopardy all involved. But let me say this, Laura, you do take me to the edge!"

She said, "I'm excited now, but tomorrow will be hectic."

He looked at his watch, then stated, "The equinox is dark, but it's morning already."

She said, to his appreciation, "My God." They snuggled, enjoying the quality, and slept contemporaneously.

Chapter 9

Laura let out a short gasp, and peeked under the covers at her partner, then smiled, as the grand memories flowed back. This had taken top billing to the trivial loss of her job. She looked at his watch, and then his contented face. She wanted to trace his facial features with her finger. She looked down at her breast, and fingered her nipple--it was not the same. She snuggled a little closer. He turned slightly towards her, his left hand fell gently, just above her left breast. She carefully moved his left hand down slightly. She felt mischievous, and selected his index finger, and slowly routed it to her nipple. She watched in fascination as her fingers were modules, his finger the receptor, to her sensuality which alerted her delight. She had never felt so amorous and fulfilled as a woman with another man. He opened his eyes, and the first thing he saw was his hand resting on her soft breast, and his finger romancing her fevered pink button.

He looked up and into her appreciative green eyes, and said, "My God, how long have I been stimulating you?" He smiled and started to pull his hand away.

She graciously caught his hand and returned it to its assigned task. She smiled and said, "It's soothing, did you have a nice rest?"

He stated, "Never better" and looked at his watch. He smiled and looked up into her eyes and stated, "Believe it or not, but my finger isn't tired."

She laughed softly. He used his other hand, and stroked his face several times, and spoke. "I need a shave. Do you have a razor, by chance?"

She replied, "Yes, Albert, I do. Do you want it now?"

"No, no," he replied, then added, "I want a fresh face when I face my colleagues this morning."

She asked, "Are you apprehensive?"

He quickly replied, "Good Lord, no. It's just the kidding I'm sure to receive, no matter how I condition my whereabouts all evening."

She asked, "Are you ashamed?"

He readily stated, "Nonsense, I would like to shout it to the world."

She queried, "What are you going to say to them?"

He said, "I feel like a teen-ager that raised hell, and now must qualify my deeds."

She laughed and stated, "Well, of course, if you would rather not say."

He said alertly, "Oh, oh, why is that women refer to that phrase, you know, you're indicating I'm hiding, or ashamed, or some damnable other motive."

She smiled and said, "Well, Albert, you said it yourself--motive."

He had to grin, and stated, "I'm going to tell them the truth, but first, I think I'll cancel the theory of celestial bodies."

She asked, "What if they solved it?"

He replied, "I don't know. I can't in fact, take it away, if they solved the puzzle."

"That's right." she said, "What will you do?"

He stated judiciously, "When in doubt, tell the truth, and I shall."

She smiled, and asked bilaterally, "And that is?"

He caught her off guard and she blinked. He planted a smart kiss on her unexpecting lips.

She said, tempting, "I like the unexpected. I'm so happy now. My mother answered a question that I asked as a hyped up teenager."

Being interested, he asked, "What was that?"

She replied, "I asked, when my Prince Charming would sweep me off my feet." At that time, her answer was ridiculously outrageous, until now. She had set the hook.

He incessantly asked, "Good God, you've really got a barn-door slammer. What did she tell you?"

Laura looked up, as to pay tribute to her mother and stated, "She told me life was a road with many forks, and down those roads are many bumps. She said the bumps aren't all bad, in fact, you'll find your bump in the road when you least expect it. That will be your sanctuary. Bless her, she was an oracle."   She paused, then asked, "Where do you hail from?"

He replied, my younger life was spent in Bay City, Michigan. About half of it was exploring Lake Huron, about 22,000 square miles of water and no islands. It was awash in violent storms. It was there that I learned the deep respect for power that such bodies possess. Did you know that Lake Superior is a little over 20 feet above sea level, in relation to Lake Huron?"

"No," she replied.

He looked and caught her eyes, and asked, "Am I that bump in the road?"

She convincingly curled her arms about him and stated, "I usually don't cuss, but you damn tootin' you are!"

"Oh, God," he said excitedly, and they hugged one another. They lay and fantasies saturated their charged bodies. Each perceived a different silhouette, but the nucleus was the same goal. He glanced at his watch, and asked, "Will you be all right for while this morning. I have to join Tom and Norma, then I can come back here, or you meet me at the White Tern, Slip Number ---, yes, No. 28."

She cried out, gleefully. She added, "I have to buy some instruments and pack some duds, it would facilitate our timing if I met you at your slip."

He said, "Great, do you have a checking account?"

"Yes," she answered, and added, "The Northern Bank."

He asked, "Do you have a deposit slip?"

She threw off her sheet and slipped on her robe. He began to dress. She returned with a deposit slip in her hand and relinquished it.

"Thanks. I'll deposit a month's salary in your account first, then I'll head home. Yes, and save your receipts for all necessary equipment. When can you wrap it up, and meet us at the slip?"

She said excitedly, "Heavens, I have to pack, purchase, and oh yes, I have to return where I worked and clean out my locker. Mmm, I should say about 11:36."

They both laughed.

He stated, "Pack light, and I would suggest you don't miss breakfast or lunch. We'll dine tonight on the high seas."

She blinked, and threw her arms around him and said thrillingly, "Just think--a banquet aboard the *Magnetic*."

They kissed quickly and he said, "I don't know about a banquet, you'll have to ask Tom about dinner. He's a crackerjack of a cook."

"Yes," she replied, "he was eloquent at Charlie's."

He finished dressing and checked his watch, and turned to face her, stating, "I bless your mother's wisdom. It seems inconceivable that one day later, I would have missed you, and you would still be looking for that bump in the road. I would have lost the woman that is so precious to me. God, it scares me. Having not met, we would be wandering in the crowds, filling our time with unimportant tasks to falsely facilitate our lonely lives. Lying to ourselves, how important, just to make it through the day and repeat the next day, week and month. God, what an appalling excuse for living. What a damnable waste of our uncharted lives."

Tears were beginning to peep out of her eyes, assiduously confirming the legitimacy of his statements. They embraced, and he felt her warm tears on his cheek. He gave her an extra squeeze and she reciprocated.

He said, almost in a whisper, "Imagine, over four billion people. God, I'm so blessed in finding you Laura. I'm skittish now and that puts me on the defensive."

She said, with moisture in her eyes, "You take special care, I'll be at your slip before noon, and yes, good luck with Tom and Norma."

"Thanks," he replied and added, "One more kiss to hold me."

They embraced, he kissed her salty sweet eyes and she rubbed her cheek.

He noted the gesture and stated, "Like our first meeting, a close shave." They smiled and Albert left. As he drove, leaving Diamond Avenue, he felt the colossal significance of his strong new love. Laura was a ponderously new challenge. He had to spread his protective net further using precious precision and caution, as he had more to savor than ever before.

Norma and Tom turned over simultaneously and woke each other, to their delight. Tom said, "Good morning, Miss Moon."

She smiled and replied, "Good morning, Mr. Spoone," and added, "Look at the time, it's breakfast time."

He grinned and stated, "Just like a woman, lust, and food, not necessarily in that order."

Norma kiddingly added, "You're not out of order." They laughed.

He asked, "Bacon or sausage, my queen?"

She smiled, and said, ruffling her hair, "The queen would prefer four long crisps."

He asked, "Is my queen refreshingly considering immersing in H_2O?"

She lifted her head and replied, "But, of course, you may scrub my back, if you wish."

He answered, "Permit me to pre-set our morning nectar, and I will join you in the enclosure."

They began to slip out of bed, both enjoying the flippant remarks. As Tom sat up, and then stood up, he was unaware that Norma had deboweled a pillow case. She snapped and accurately stung his bare buns. He cried out.

She gleefully shouted, "<u>Bon</u>-Jour Monsieur."

Rubbing his smitten area he stated, "Now, that's the same as back shooting."

"Yes," she fondly stated, and added, "<u>Butt</u>", and spelled it, then continued, "A little lower."

He had to laugh and responded, "I will rule the enclosure."

She replied, "Only because of your numbers."

He turned to face her direction, as he put on her robe, and asked, "What do you mean, by numbers?"

She giggled and said, "Silly, you and your staff!" She added, "You look cute in my robe, I've never noticed your knees before."

He playfully grimaced, and said, "You're going to get it."

She waved, and said, "I hope so, Lover." She bounded out of bed, and hurried to the shower. Tom prepared the ingredients for their morning coffee, and selected the temperature under the cool pot with care. He hung her robe up in the warm bathroom, opened the glass door, entered the tiny steaming modular, and quickly closed the door. His staff touched her several times as he positioned himself behind her.

She said, "I feel your staff is here with us."

He replied, happily, "You have two boobs with you!" He devilishly groped under her arms and covered them. She reached behind her and captured his.

He stated, "What a wonderful start of a fabulous day!"

She said, "Yes it is, and it's going to be gorgeous." They both unhanded their captures. She backhanded the wash cloth and soap, and asked, "Please."

He said, "You want me to bud, bud, abur?"

She exclaimed, "What?"

He said mindfully, "That's a rub-a-dub-dub, backwards."

She laughed and said, "God, oh conscious one, yes I was." They both laughed aloud as he soaped her back and felt each other's presence, enhanced with the warm vibrating pulsating sheets of water. They exchanged places and duties, with only a kiss intervening.

Albert had made the deposit in Laura's checking account. He braked on Downer Street, and pawed at his face and chin whiskers. The bristles were quite evident. He got out, bounced up the steps and slapped his key in the lock. The intrusion, initiated the metallic movements of the metal mechanism. He entered and was met with a net of perfume essences of morning characteristics. He rushed to the source, where high pitch spits…and hisses…of volcanism was empowering the hot element. He set the ferociously, perking coffeepot, on a less temperate plate. He found the dishrag, and mopped up the abandoned fluids. He could faintly hear a muffled suppression of water emitting from the bathroom. He smiled at the thought of walking in on them, announcing his presence. He discarded the thought. There was going to be a host of questions about his absence, and his plans for today. He retrieved a cup and its mate from the cabinet, and poured the steaming hot juice, its odors rapidly filling the room. It was as capacious, to that of burning spices of incense, to the olfaction. He blew to cool the hot liquid, but barely affected its temperature. In nervous anticipation, he spooned the coffee, and ejected more on the small surface. He dared, then touched his waiting lips, burning them slightly. He doused it down like a spoonful of medicine. He smiled. It was good medicine. Another spoon, and a satisfying contentment began to assimilate his body. "God," he thought, "This is what the captain ordered." He sat and nursed his black perfume, and thought of Laura.

Norma sprang toward the kitchen, her nose and bathrobe flaring. She saw Albert and said, "Hi, Cap, top of the morning to you."

He could see and smell the freshness of her body. Her hair, excitingly wet, gave the illusion of a "what the hell look." Her voice was crisp and spunky.

He said, "You look as fresh as a daisy, and twice as perky."

She replied slyly, "Well, you know the old saying."

He said, grinning, "No, I don't."

Tom was yelling from the bathroom. Albert said, "What's his problem?"

She said, fingering her hair, "Probably wants a bathrobe."

He said, "Pour yourself a cup, and I'll tend to Tom," then added, "By the way, your hair and Laura's are remarkably beautiful when engulfed with water."

She said candidly, "Oop's, Cap, Thank you for the compliment, but you just stepped barefoot in the hot, hot coals."

He smiled and said, "I know, it's like sticking your toe in the water before you dive in, sometimes it's better to ease in gently."

She smiled and fetched a cup of coffee. Tom was still shouting, as Albert entered the bathroom with Tom's bathrobe. He held it outside the shower and said loudly, "Tom!"

Tom's blue eyes brightened as he captured his robe and stated, "I'm preserving the limited amount of hot water."

Albert grinned and turned on the faucet in the sink, that had an "H" on the handle, and let it run over his fingers. Tom tied his sash and watched Albert's reaction to the availability of hot water. Albert shut it off and said, questioningly, "Gosh, that's peculiar, Tom, it's just lukewarm, if that."

Tom said, in deference, "Well, it takes a great deal more water for women."

They returned to the kitchen where Norma was sitting under a spacious cloud of blue smoke, wearing a smile. She said, "No, Tom, that's the Captain's chair."

"Oh," said Tom. She had poured a cup of nectar for him.

Albert stated, "We were just discussing how much more water it takes to bathe a female versus a male."

Tom attempted to look innocent, and shrugged his shoulders, and said weakly, "Small hot water tank."

She said, "Speaking of hot water, Cap, that looks to me like an eight hour shadow."

Albert straightened up. They both deliberately stared at him in anticipation. Albert divided his eyes between the two, then stated, "I am pleased to announce we will have the presence of another colleague who has accepted, and will join us on the *Magnetic*."

Tom and Norma's eyes widened, carefully tracked, and searched, each other's eyes, as if to gain inspiration. She quietly inhaled, and blew a cloud. Tom and Norma redirected their attention back to Albert. Albert continued, "A marine biologist is just what the doctor ordered!"

Norma said smartly, "If identity, as to our new associate is what I think, you should have ended your statement, "Just what the Captain ordered!"

Tom snickered. She smiled and covered her mouth. Albert knowingly, under speculation, continued, "Laura has a Master's Degree in her field, and her contribution will be as critical, as any of our special fields. I have one more announcement to make. The mystery of the two words, celestial bodies will be withdrawn."

That shocked both of them.

Tom blurted out, "Christ!" She blended in with, "Oh, God!" Albert saw the naked distention.

Tom barked out, "We've solved it, that is Norma solved your riddle. She looked apprehensively at Albert, and he captured that tense, eager look in her eyes.

Albert stated, "We have to face facts. There are only two staterooms aboard."

Tom stated, "The facts are you intended Miss Moon and me to share a stateroom?"

She waved her hand in the air and stated, "That's my Tom!"

Albert said in a monologue tempo, "Gracious, you really did solve it, what we now have is a debauchery." Three pairs of eyes acknowledged the fact.

Tom said sullenly, "I'm a poor bed partner, I get nervous at night, and toss and turn, and damn it, I'm not a misogynic!"

Albert said, "Good Lord, what does that word mean?"

Norma said, "Yeah, what the hell?"

Tom stated, with chin up, "In essence (disgust of women)."

Norma shrieked a laugh and Albert split a long grin. She said, speaking to Albert, "You said it yourself, it's tight living in those close quarters." She exhausted her last wall of smoke and killed her butt ruthlessly. Albert witnessed the destruction in epistasis.

Tom said, "I'll fix us a nice breakfast--eggs, bacon, steamed lemon rice and toast."

She reflected, "Mmmm. . ."

Albert said, "I will designate quarters after breakfast. A couple of hopeful "aye, ayes" were spoken. Albert stated, "I'll heat a pot of water to shave with and then I'll have a cool shower!" Albert searched for their eyes, but failing, got up, set to his task, and left for the bathroom.

Norma caught Tom's hand and gave it an affirmed hand squeeze and attempted to smile. He began to prepare their tasty breakfast. She could faintly hear coarse singing coming from the bathroom.

Tom caught her attention, and pointed with his spatula in the direction of the masked tones. The flavors of bacon and eggs, melted butter, toast and lemon, generated a rich and stimulating blending. The strong exciting odors of bacon and lemon were dominating warriors, with toast echoing and twinkling.

She got up, and the sweet cloud of fragrances was more distinct. She spun around, spoke loud and exulted, "Jesus, I'm a lucky woman."

Tom's blue eyes matched the succulence of his dishes. He smiled, and said, "You're right, you know!" They shared smiles and she broke and went to alert Albert. The off-tune singing had stopped. She crooked her right index finger, and commissioned it to strike the panel door soundly.

He opened, thinking he might be King "X". She stated, "The chef says come and get it, or whatever."

He said, "Your Tom is not only the finest of cuisine cooks, but has the swiftness of a cat."

She asked, "Male or female feline."

He said, "White flag, for now, let's eat. Those aromas are captivating."

He offered his arm and she took it. They busied themselves with the sumptuous breakfast.

The crew of three, soon to be four, bathrobed, lightly laughing, and enjoying one of life's many pleasures, were unaware that the missing crew member was about to place her life in jeopardy.

Norma napkined her lips, got up and planted an unexpected kiss on Tom's forehead. He looked up in admiration, and asked, "What was that for?"

She replied, "For the lovely breakfast, and for double slicing of my buttered toast."

Albert conveyed their affection increasing, his adrenaline increasing for the want of Laura. He sipped his coffee, and used his toast as a plate cleaner, and popped it into his mouth. He stated smartly, "Exquisite!" Then added, "Your grub is a belly buster. I'll have to increase my exercising."

She commented, "There are creative ways to maintain your waistline." Tom winked at Albert and was finishing his few morsels.

She said, "Well, Cap, now is as good a time as any. What's your decision?"

Tom joined in, "Let's see, are we under a renegade Captain, a pirate of sorts, or an officer in charge, of the proud *Magnetic*."

Albert hastily stated, "Good Lord, don't use that word pirate within hearing range of Ruth Dander, it will remind her of the . . . ."

Tom smartly stated, "The sea shine."

Albert replied dauntlessly, "Yes, yes."

She related to Tom, "You really lambasted our Cap, you are refreshingly surprising me."

Albert chimed in, "Amen, to the latter part of your statement."

Tom said, kiddingly, "I have a superb recipe for lamb basted mousse."

Expounding laughter except from Tom, who said, "It's true--eggs, paprika and cayenne pepper, gelatin, and milk." They kept laughing and Tom stopped his oration in disgust.

Albert still smiling said, "I'm sorry, Tom, but it was funny, and you were so serious."

Norma tried to keep her eyes away from Tom's face for fear of laughing again.

Tom replied, "Preparation of food is a serious undertaking, like any sophisticated, scientific agenda."

Albert replied, straight faced, "I sincerely apologize Tom, mundanely."

She spoke up, "Ditto, on my assumption."

"Well," said Albert, then added, "It's going to be. . ." changing his pitch of voice, he stated, "Do you know the name of our craft, the *Magnetic* contains two special words, besides Laura's last name?"

She said, "Oh, God!"

Tom said nothing and Albert continued. "Using the same letters and redirecting their order, the two words that reveals are..."Magic Net." He had them stilled and attentive, both were transmitting signals, between them, starring the key word magic. Albert continued, "Net has several definite meanings, but listen to what they are." His voice was almost a whisper, he went on, "(1) ... An entrapping situation... (2) final... (3) . . .yield . . . (4) . . .the essence. . . ." He changed his pitch again and stated, "Put to work, in a sentence, it would read -- the essence of an entrapping situation final, yields, magic." Not a sound, just breathing could be heard.

Tom and Norma, kept that sentence busy in their minds. Albert broke the stillness and stated, "Each man, and woman, will have a shipmate, a mate, and will be quartered as such."

"Jesus," Tom shouted, and wrapped his arms about Norma, who in turn was kissing and hugging. Albert enjoyed their antics and Tom said, "The Cap and Laura will bunk together, and you, my dear, are my designated bed partner."

She smiled, and said, "You could have played it less male oriented!"

Tom asked skillfully, "And the female description would be. . . ."

She sipped her last swallow of coffee and stated, "A covenant roommate." Albert and Tom grinned, and flashed male signals, they cared little for the dialogue, only the outcome. She continued, "I'm going to mention this very item to Laura, and I might add, she'll hail it as peremptorily important over designated bed partners." The men laughed heartily, but realizing, and knowing her description, was the hesitation. She said smartly, "You men!"

Tom said crisply, "You women, and your floral commentary of female words."

She asked carefully, "Such as...?"

Tom looked at Albert, smiled, and stated, "Your outspoken female lascivious words, such as sleeping with and affairs, are as casual as good morning."

His bright blue eyes claiming victory, Albert said, "I'm meeting Laura with her goods at the

White Tern at our slip at 11:56 a.m. I would appreciate you both being there upon her arrival."

Tom said, "I'll have to do grocery shopping for perishables. How many days stores, Cap?"

Albert replied "Three minimum, five maximum."

She asked, "Are you guys through with your male levitation? If so, my answer to my male colleagues, is that your 24 hour hunt and pursual of sexual gratification, has brought forth to bare, the modification of four letter words."

Tom stated excitedly, "See there, she said it herself, male levitation, means only one thing to a man, and you stated, brought to bare, spelled b-a-r-e."

Albert said, "Good Lord."

She had to laugh and replied, "Jesus, you're good, Tom." They all experienced pleasurable sport.

Tom said, "Cap, I'll need the Briarwood for transportation of food stuffs, and gear."

Albert leaned back and stated, "Yes, of course, I'll Yellow it to the slip, take my bag with you." Tom nodded.

She said, "I'm looking forward to seeing and meeting the Big Ruth Dander, I just might ask about her experience on seashine, one gal to another."

Albert swallowed and sternly stated, "Good God, don't you dare bring that unpleasant situation into focus again!"

Tom snickered and flashed a smile of appreciation and intimidation, and asked, "By the way, Cap, how did you compensate for ripping her dress off?"

Albert stared through Tom and stated, "I didn't rip her dress off, just her blouse. It was the albatross, and half-eaten sandwich, that were the instigators, and of course, Miss Dander."

She said, enjoying the conversation, "How high were her swells on her sea chest."

He echoed, "Good Lord."

Tom and Norma were having a good time. Albert continued, "Both of you know how to spike the last nail!"

Tom spoke, "Your crew, including Laura, will maintain you commensurately."

Albert nodded in delight and said, "Laura is receiving the same salary, I thought you would want to know."

She clapped and followed with, "Good for you, Cap, I'll try to refrain from discussing the sea shine with Ruth, although I feel like a traitor."

Albert nodded and smiled in agreement. "I'm going to the pay phone and call a cab."

Tom asked, "Can I drive you?"

"No," said Albert, "I'll have a few places to go. You and Norma have to load your gear and purchases in the jeep. I'll pack first, in fact, now," and turned to leave. He stopped, and asked, "Do either of you scuba?" They both gestured a negative response.

Norma spoke up, "Your Laura will likely fill the bill, I believe it's mandatory for a Master's in Marine Biology."

Albert asked Norma, "What size of underwater gear would be appropriate?"

She replied, "I haven't the slightest, in rubber, but dress size probably a 12.   Does the number 12 have any specific significance other than size?"

Tom chimed in, "Not unless you are a fat lady,"   Tom quickly received a vindictive stare from Norma.

Albert said, "Oh, oh, gotta pack."

Laura arrived at her destination, Room 106. No one was at her old desk.   She moved to the anteroom, to her locker and removed a rather large plastic bag from her purse. She had decided earlier to leave some of her mementos. The main reason for her return was to retrieve some of the books used for her Master's degree.   She necktied her sack, left the small cubicle, and took a left to the nucleus of information and the cataloging room.   It was a small room and was divided into three time zones.   The latter was one year, or more.   The first zone was thirty days and the second zone was three months.   All arrivals and departures, cargo, name of vessel and registry were listed.   One side of the wall was the antiquated and typewritten; the opposite wall was cubed and contained all the latest video tapes.   She proceeded to the wall where all land and sea charts were kept.   She took copies of the shore lines, known fathoms off Alaska, and their perimeters. She sequenced the maps, rolled them, and placed them in the plastic. Half of the rolls extended out of the bag.

Arnold Pulse, a cousin of Devlen Dawson, her former boss, had returned from Nature's call. He noticed the door open, stepped behind it and was in a position to monitor the intruder through the thin, but long opening. He saw a woman stuff long rolls into a heavy plastic bag. He took note of her hair color, face, and stature. He also noticed her ease, and familiarity of movement as she turned in his direction. He scooted quickly to the locker room, peeking out, and watched her rear. He moved swiftly to his new desk, dialed, and waited for the connection. He heard, "Yeah, Dawson here."

Arnold excitingly related what he has just witnessed, including her statistics.

His caller shouted, "Jesus Christ, what did she take?"

Arnold replied, "How the hell do I know what it was, the sack was loaded."

Dawson asked angrily, "Where the hell were you?"

He answered, "I was in the can."

Dawson replied sternly, "And you left the office unlocked, . . . well?"

Arnold said disgustingly, "Damn it, I had to go. Who do you think she is? You told me to keep a low profile, or I would have encountered her."

Dawson barked back, "Yeah, yeah, listen, don't, and I mean do not, tell anyone, even your wife about what went on, you got it. . . .?"

Arnold muttered, "Yes."

Devlen deliberately slammed the receiver down, and sarcastically piped, "Son of a bitch." He quickly decided to handle this situation himself; besides he had to maintain his usefulness. "Damn that bitch," he thought. No one would pull the rug out, only pad it. He definitely decided to get to the office, and delete any possible inappropriate records. He would also retrieve Miss Laura McAgneti's address, now!"

Chapter 10

Tom thrust his shopping cart forward. Norma was finding it difficult to keep pace and stated, "Slow down, Tom, or you'll crash that basket on wheels."

He replied in a stately manner, "I have many aisles to visit." She lunged forward and grabbed the backside of his belt, and pulled him to a stop. The balding store manager witnessed the episode, and recognized the couple who had devastated aisles six and seven. Now she was taking indecent liberties as to the placement of her hands. Two couples, and an old man attested to her fondling. The old man grinned in appreciation.

Tom said, "Stop that!"

She replied, "I did, just that, Jesus, when you wrap your hands around that plastic bar you must believe it's a throttle."

Tom genuinely sated, "We all shop differently, some callously, some seductively," then smiled and added, "You can be my sea anchor."

She snapped, "Are you calling me an anchor?"

He lowered his voice and said, "I could have said something nasty."

She looked into his blue eyes, and stated, "Such as?"

He replied, "I could have said sea bag." Emphasizing the word, 'could,' then added, "A sea anchor is a canvas they throw overboard to retard drifting."

Her eyes sparkled and she stated, "One more crack, Tom, and you will be in that canvas, retarded."

Tom said, "We're getting close to King X time."

She perkily stated, "You bet, and don't you forget, you don't have a King X, remember?"

He thought that one over and said, "All right, you manhandle--oops, you take over guidance of this basket."

She smiled ardently and said, "That's my boy," and she did just that.

The balding store manager nervously stayed one aisle away, playing peek-a-boo with Tom and Norma. He registered a relaxed sigh of relief when they left and his store was still intact. He quickly popped two extra tums into his awaiting mouth.

They transferred their goods to the Briarwood. She gave him a mischievous smile, and said, "Thomas."

He alertly spoke up, "Whoa, you just called me Thomas, meaning only one thing, you want a favor."

She quickly said, "Your male tie is showing, Tom."

He asked, "Well, what do you want me to do?"

She said smartly, "Let's find a store, where we can buy a cap for the Cap."

He glowed and said, "That's a dandy idea, and yes, my male tie is flying, in unison with my mouth."

She locked in a smile at his honesty, and stated, "That's right, you know."

He laughed and said, "Norma, I have an amazing play on words in regard to our first names."

She asked, "And what may that be, exalted one?"

He said, "The equinox of Norma and Tom is motor man."

She laughed and said, "Just another male anomalous reference."

He grinned and remarked, "No, there isn't any other logical letter combination."

She pointed to a store and exclaimed, "There's a store, pull over and we'll get our Captain his due."

She asked, "Do they have captain's hats with covers?"

He grinned and said, "I was going to let you shop, but that question alerted me to your being a mad hatter."

She smiled coyly and replied, "I thought that precise phase would send up the Jolly Roger." They exited the jeep and were about to enter the store when she asked, "What size?"

He said, confidently, "Ask for a seven and one-eighth."

She said, "Okay, Tom, let's get with it."

"Yeah," he said, and added, "Make it fast, my perishables are just that."

Norma, smiling graciously, made the purchase and insisted that they gift wrap it. Tom looked nervously about, and at his watch, with wild visions of his milk, cottage cheese, and eggs encrusted with decaying matter, or of living organisms."

She touched him and he jumped, and said, "What's got you so jumpy, let's go."   As they pulled away, Norma was still mentally admiring the fancy gold relief and shining bill. It was Albert's kind of reverence. The Briarwood was stuffed with bags of groceries and emergency equipment, bedding and three and one-half pieces of luggage that didn't match, and a captain's hat. The smell was a mix of spices, new pots, pans, and utensils. Tom had also purchased two coffee pots, one 16 cup electric, and one eight cupper, that used the old perking device. His keen mind also included hot pads and toothpicks. It was like going through his mind how Albert described the galley, with a microwave oven (Whirlpool), the six cubic foot refrigerator and freezer, full size princess stove, with a rotisserie. He even thought to buy one fishing rod, complete, to catch and prepare fish dinners. His repartee included just two electrical conveniences, excluding the coffee pot. He brought the big chrome toaster, four holder, from the kitchen on Downer Street, and had purchased a hand held, electric mixer.

They were on Ocean Dock Road, and she pointed and laughed at a weather beaten sign needing a couple coats of protection. They wheeled a sharp left, and the secretion of the sea was belching benevolently, violently, with its aromatic scents, imbued perfumes. The familiar and intense screaming of sea birds, motors, and horns all orchestrated in the rhythm and accompanied by the clapping waves. Seas command and holds hostage up to 72 percent of our exposed earth. This justifies the authority of rule by conquest. More of man's previous yellow gold lies in the depths of the brine, than its counterparts above sea level.

There was no visible road abutment to the main dock that spread out like a rectangular spider web. The White Tern was a greying culprit, resulting from the onslaught of the deadly ultra-radiations of the sun, and the icy crystals of long winters of Anchorage, Alaska. Half the state nestles within the confines of the municipality--where salmon spawn in Ship Creek during the summer, near downtown, and its gleaming high rises, and shopping centers that mince and mingle with muskeg (peat marshes). The smells of the city with the sea as salt, and the fine restaurants of oriental cuisine with lustful odors of Japanese, Chinese, Korean and Thai blend in the marriages that add pepper to the spicey city.

Tom spat out, "Damn, I wish I could back up to that dock."

Norma stated, "You stay here, I'm going to introduce myself to Ruth Dander."

He said, as she bailed out of the jeep, "Won't do any good."

She countered, as she slammed the door, "We'll see!" She walked into Ruth's office, using the door from the dock access. Ruth was apparently unaware of her presence. Her chubbiness filled and strained the old oak office chair. Her back was to Norma. Ruth was bent over, literally beating forcefully upon her old typewriter keys, using the hectic indignation of the moment as her whipping post. Norma called out timidly, "Hello there."

The battering stopped, and the mighty oak, holding the endowed Ruth Dander spun a dazzling 180°, using her flat heels to stop the spinning. She barked out, "What can I do for you?"

Norma approached slowly and stated, "I'm in the employ of Mr. Albert Dryfuss. We're here to store goods on the *Magnetic*."

Ruth sized her up and stated, "So you have my permission as long as you don't defile my waters, or harbor."

Norma said, "May I call you Ruth, my name is Norma." She nodded, and Norma spread a smile. "We have oodles of goods, could we back our jeep up to your dock walk?"

Ruth leaned back and the chair creaked in defiance. She crossed her pudgy legs and stated, "Deary, if you back in westward to the sign that reads 'No loitering,' keep the tail of your jeep perpendicular to the sign, 'No smoking beyond this point,' I'll give you permission to cuddle up to my dock."

Norma smiled, and genuinely enjoyed her gift of the language and replied, "I thank you, and I'm sure Albert will thank you."

Ruth stated, "Albert probably will, he's a gent." Norma noticed the ashtray on her desk laden with butts, pointed, and asked, "Just between us. . .why do you have a no smoking sign on the docks?"

Ruth briskly retaliated, "Two very good environmental reasons: (1) a careless match, or cigarette discarded, could devastate any of the boats in my slip with gasoline and canvass that were left indiscriminately. The whole batch would be put to the torch, singeing my very existence."

"That certainly makes very good sense," Norma replied. "What's the other reason?"

Ruth replied, "I don't tolerate defiling my waters." Ruth kicked and spun around, braked, and began punishing her reprieved typewriter.

Norma relayed the good news to Tom, who said, "Albert did say there were all kinds of signs, forbidding everything except breathing." Norma related to him how he was to maneuver the jeep with reference to every sign, and ended with, "You had better not chart other courses, or you will face Ruth Dander and her dander."

Tom asked, as he was backing up, "What kind of gal is she?"

Norma thought about that question and stated, "Compared to what" then added, "I know, my relation as to how high her jibs are flying."

He smiled and said, "Yes," then went on. "Christ, look at all those signs, what was the perpendicular sign?"

She nudged him with a playful elbow and stated, "No smoking beyond his point, Jesus, look out Tom, don't scratch her walk dock, or you'll have your answer, in person, quicker than you think."

"She must be some sea walrus, she certainly has you and the Cap buffaloed."

She put her hand on his arm to get his attention and said sharply, "Hold it, Tom, what the hell do a walrus and a buffalo have in common?"

Tom said, off handedly, "Oh, three things," to which she subjectively asked, "What's that and don't include being mammals or females."

He turned a little and flashed his blue eyes, and stated, "First the walrus and the cape buffalo love water, second, one has horns and the walrus has tusks. But, both bone implements are used in defending themselves, and third, they love to hump."

She listened attentively and placed a smacker on his lips, and said, "God, you're good."

He replied, smiling, "I know," and added laughing, "but you are too."

They both slipped out of the Briarwood and opened the tailgate. A refreshing breeze tinseled her hair. She looked up and assimilated the sea smells and sounds. She lowered her eyes, giggled, pointed and said, "Look at that sign, Tom."

He followed her gesture with his eyes and read the sign aloud, "Pollution allowed." The enhanced whiteness of bird crap had covered the No. Tom laughed and stated, "Miss Moon, the one unescapable adventure of my life, will soon be revealed, a mystery for you and me."

She said, cautiously, "Watch where you are stepping, Tom." The bursting cries of sea birds filled the air as they scattered. Tom and Norma had a light burden of goods as they approached slip No. 28. She exclaimed, "God, look at that! When I get this load on board, I'm going to pinch myself, this is a luxurious craft."

He said, "Let's go aboard, and I'll pinch you. I wonder where our Cap is?"

She replied, "You may be a wizard on most subjects, but you flunk time."

Tom looked at his watch and exclaimed, "Oh," then added, "I would liked to have spent more time on aisle ten."

She flashed him a domesticated smile. They climbed aboard, through the swing aft transom gate.

He said excitedly, "Let's see the galley!"

She nodded in jubilation. They oh'ed and ah'ed, as they moved through the salon, midstateroom, and then Tom's pleasure castle, a full electric galley. They unloaded their cargo to the cabinets containing the twin stainless steel sinks. Tom outstretched his arms, and anxiously sermonized his voice. "This is my nucleus, my hub, I will create. . . ."

She smartly interrupted him, "Dr. Frankenstein, I presume?"

"No, no, that latter portion of your statement is incorrect. The statement, I presume, was the highlight of a splendid old cinema, when Spencer Tracy in Darkest Africa met and stated, "Dr. Livingstone, I presume?"

She marveled at his precise accuracy, and sent him a warm smile. She said, "Tom, it's my turn for conviviality." She turned and went through the door, and screamed. He rushed in and crashed into her. Both caught each other but fell to the floor.

He said, "What. . .?"

She pointed and yelled back, "Look, look at my temple, my exalted. . . "

He intervened and said, "Yeah, it's sea blue," then got excited and stated, "The best of all sacred towers, our Olympus's join one another, with just a swinging door.

She put her hand to her mouth and said, "Dreams do come true," and she fondled the full tub enclosure. She said, "Look, Tom, two portholes in my temple."

He slipped his arms about her and kissed her, then said, still holding tight, "Remember, Norma, there's a Laura who loves to embrace the water too."

She said, "I know, us gals have ways of sharing the good things together."

He looked shocked, and exclaimed, "Are you and Laura going to bathe together?"

She emitted a long laugh and stated, "Silly, of course not physically, but timely."

He said, "We had better unload the rest of our cargo. My perishables are gasping for cooler air."

"Okay," she said, then trailed off, "I'll peak at my swimmerette later."

They retraced their steps many times, enjoying the movement of goods. They carried the final parcels, and he moved the jeep to a parking lot of sort beside the White Tern office. Norma returned, and couldn't resist peeking into the forward stateroom where Albert and Laura would cocoon. It was beautiful, even a private head. She wasn't the least bit jealous, but was didactically happy. She had found her love, Tom, on a grand yacht, a sumptuous salary, and a year of adventure.

She heard Tom calling out, "Where's my roustabout?"

He opened the door leading into the forward stateroom. She turned around smiling and said, "Just checking the stateroom, this has three port lights." He looked around and stated, "This forward suite is sensuous, but being with you is far more compelling. Our decor is the same."

She smiled and said, "Honest, I'm not apprehensive."

He took her hand and turned her face gently to mirror her eyes and stated, "We have the finest stability of harmony in our charming midstateroom. It is the smoothest ride on board."

She smiled at the thought and said, "We had better get the beds sheeted."

He backed off, and said, "No way, and you know why. I have many chores in the galley, stuff and duff, to do. They'll be here any time. Did you mention to Ruth about leaving the jeep?" She indicated a no and he went on, "Well, we'll leave that academic subject to Albert."

Albert arrived at the White Tern, paid his driver, and glanced at his watch--11:50 a.m. He turned, decided to skirt the main building housing Ruth and wait on the walk dock for Laura. He failed to make it to the corner when he heard Ruth's shrill voice, "Are you attempting to ship off without so much as a bon voyage?" He stopped, he had just spotted the Briarwood and turned and walked back to where Ruth was glistening in the sun. He stated, "Nonsense, I was checking out the parking lot to see if my colleagues had arrived."

She smiled and replied, "Yes, I met one of your crew. Nice looker. How long are you shipping out, more than a day?"

He looked at her, but not in her eyes and stated, "What has that knowledge of inference, consequentially of importance."

She barked, "I run a tight ship, no cars parked over 24 hours without proper registry."

He paused, and asked, "What is the charge for proper papers?"

She replied, "A fiver a day, will keep your jeep from being quarantined."

He responded, "How much for a week?"

She replied, "I like a gent with business brawn, $25 a week, a sailor leave's bargain."

He spoke, "You are most gracious," and passed the green.

She said, as she accepted the bills, "Your jeep will have to fend for itself for two days. You might want to run up a Jolly Roger on its antenna in defiance."

He laughed and thanked her for her consideration.

She said, as he turned to go, "May your sea chest stay dry."

He replied, "Same to you, Ruth," and suddenly realized the female rendition, including the sea shine incident. He muttered, "Good Lord."

A yellow cab pulled up, stopped, and Laura got out with a smile and a suitcase, two large boxes, and a long roll of maps. He moved to her quickly.

She said, "Hi, everything set?"

He smiled and replied, "It is now!" They gathered their goods, and he said, "Let's scoot around the building to the docks."

She said, "It's such a gorgeous day."

He replied, "Yes, the sea fowl would also agree." They both inhaled the perfume, where the sea was clapping in approval to have reached the unassailable land.

They approached his slip, and she exclaimed, "Oh, Albert, it's a beautiful yacht."

He said, "Just wait until you see the comfort, exciting interior and colors." Albert got in and she transferred her cargo to him, and he said, as he extended his hand, "Welcome aboard!" He wanted to take her in his arms right then.

Tom moved aft and met Albert and Laura in the salon. Tom smiled, and threw up both hands, and stated, "Isn't this grand?"

She said, "It's an infinitum."

Tom asked, "What's that word?"

She smiled and replied, "Without end or limit," then asked, "Where's Norma?

Tom replied, "Domesticating the beds."

She smiled and said, "I'll leave these maps on the table, Okay?"

Albert nodded affirmatively, and said a little nervously, "I'll show you our forward stateroom."

Tom grinned, and stated, "You have permission to pass through my galley."

They apprehensively looked at each other, trying to capture hidden messages. She remarked on how wonderfully the galley was laid out. Tom stayed in the salon. They were about to enter their stateroom, but Norma stepped through and said, "Hi, Cap, Hi, Laura. Just getting things shipshape."

Laura emitted a third "hi."

Norma asked, "Where's Tom?"

Albert said, "We left him in the salon."

Norma said, "That's good news. That galley is like a magnet and Tom's the iron filings. I'll see you two, I'm about to gather my filings."

They laughed, and he pointed, indicating for her go to first, into their boudoir. He laid her suitcase on the bed next to his and pointed starboard. She opened the clothes closet. He said, "We'll have to share."

She smiled and replied, "Sharing is what life is all about."

He wanted to grab and squeeze her, but instead he looked down at the spacious bed, with designer pillows. He said modestly, "You may have your choice of sleeping accommodations, starboard or port?"

She looked and stated, "No, no, this is your cruiser," she insisted.

He replied, "It is by right that I condescend happily. I insist, Laura."

She smiled refreshingly, and smartly stated, "I'm a starboard girl."

He smiled, and said glibly, "Excellent, why don't you unpack, get settled. I want to check the engines." He pointed to the doors leading to the private head, and access to the bathtub. She moved toward him and they met, kissed and hugged.

She said, "This is a grand home!"

He excitedly exclaimed, "I'm so happy you approve" and pursued another taste of the last demur meeting.

She said, "Those maps in the salon are Alaska's maritime shoreline depth charts, sea gates and some important sea mounts."

He smiled and replied, "Your keen cognition of the marine mysteries matches your loveliness." He enjoyed romancing the English language. He winked at her and left. He thought about what he had just done, and marveled at his swift remembrance of his youthful life. He hadn't winked sensuously for over 20 years. He entered the salon where Tom and Norma were debating.

She had just stated, "You men force us to wear garments that all women would gladly revolt against, and that's not the only item."

Tom smiled, and looked idealistically at Albert, then back to her, and asked, "What was the first item?"

She looked long at Tom, and exclaimed, "You men sure love to rattle your virile. The first garment is panty hose. The dreaded sadistic female steel chastity belt."

Albert alerted two pair of eyes when he muttered, "Good Lord." Tom's laugh was infectiously higher pitched and fell on deaf ears, except for Norma.

She dramatically flashed her eyes and both men took note of the ardor. Tom wilted. A slow but perfidious smile swept across her face and she stated in a low spooky voice, "By God, I'm going to put one of those thigh snappers on you, Tom, and by God, you're going to wear that elastic cocoon for 16 hours. Then we will discuss the auspicious laughter of your exclusiveness."

Albert could feel the pain emulating from Tom's face.

Tom nervously said, "Come on, Norma, I didn't invent that device."

She cunningly countered,"Yes, Tom, it is a d-e-v-i-c-e."

Albert retreated but added as he left, "Tom, let me know the day of the elastic." He patted Tom on the shoulder as he left.

She waited till Tom tried to catch her eyes. She looked into his blues and stated, "Yes, Tom, and the other set of accessories that are fluffs of flamboyancy."

He heroically stated sweetly, "Norma, do you know by jettison the letters of your first name only one word precariously remains?"

She obviously knew of his intention, but was curious, as to his unnerving ability to snatch the situation, and flip it, like his favorite spatula. She flippantly said, "Okay, oh, elusive one, but we're not by a slingshot finished with my substratums and your autism. Well, speak up, Tom!"

He stated with a bit of hesitancy, "Your first name is Norma, a highly regarded estate, a mansion and the soothing words, spelling it out, m-a-n-o-r."

She thought that over and pulled his face to her, and laid a welcomed kiss on his lips. His blue eyes danced to her orchestration, his eyes quickly and metabolically changed. Then, he said, "Jesus, you are sincere about me donning women's duds?"

She snapped back, "You can bet your Jolly Roger on it!"

He sadly lamented, "The Cap already has heard your flamboyant apprehensiveness."

She gloated, "That next to last word, flamboyant, did it, that male egotistical word, containing the word boy. For better than no other choice, we'll call this a scientific experiment."

He looked into her bright eyes and evangelical smile.

She spoke tartly, "I King X you Tom, on the elastic garment."

He said, stonily, "You just King X'd me!"

She smiled and replied, "I know, and another thing Tom, thank you for that lovely scrambled first name. You also have but one Webster's Dictionary work for your first name."

He said, dejectedly, "Go ahead, shoot me down, I'm in flames and crashing anyhow."

She felt foolishly reluctant, but continued with her submission. Your name, letters transposed becomes…and spelled it out, m-o-t, meaning pithy."

He said, bluely, "I never knew."

She reached over with her hand and lifted his chin up to capture his blue eyes and stated, "There is another meaning for m-o-t."

He sighed and relentlessly said, "What is it?"

She cheerfully said, "Witty sayings."

He perked up, and harnessed a wide smile and spoke, "Really."

She flashed a concerned smile, "You can bet your next King's X on it, but the King's X comes back to me, you know?" She grinned, "I know, but only after the King's X deed is completed and the 16 hours of agonizing gartner snapping elastic discomfort is completed." She added smartly, "By the way, Tom, you have your choice." She waited, then added, "Tan or beige?"

Albert made his way to the lower helm. He had noticed some very bulky 9" x 12" envelopes taped to the helmsman chair. The book like catalog included packs on navigation group, engine start-up, maintenance, and every specification of the 3888 Bayliner. He removed them and sat in the helmsman chair, a little awkwardly, and reached down and adjusted it for his frame. An attentive smile was on his face. He quickly removed the contents and reviewed all of the essentials that would prevail in place before starting his diesels, and have a trial run. The more he read, the more fascinated he became. He suddenly stopped, a vital thought flashed in his busy mind. He would ask Norma before they untied.

Tom checked the freezer and refrigerator and went topside for some fresh air, and to hopefully change his attitude with regard to the squeezing corset. He presumed to walk past Albert without so much as a word. Albert felt his presence, looked up and saw Tom moving precariously.

Albert said, "Tom, Tom." Tom stopped, regained some of his levity, and relied,"I was going topside for a breath of air."

Albert realized that Tom had been solemnly sanctioned, or had misspoken."

Albert said, "Never mind, Tom, "Ill ask Norma."

He said, "Yeah, Cap," and continued to aft. He then hand railed to the bow pulpit. He inhaled and assimilated the undisciplined screeching piercing cries of sea birds.

Norma moved swiftly, looking for Tom. As she approached, Albert said, "Norma, just the person I want to see."

She asked, still on the move, "Not now, Cap, have you seen Tom?" He replied blankly, "top side."

She picked up her gait and Albert thought that Tom was in bilge waters.

Norma saw Tom at the bow pulpit. The fresh light breeze filtered through her hair, and whisked her skirt, in an unpredictable rhythm. She lightly came behind and circled him with her warmth.

He grinned and they kissed. They held one another amid the high shrills of the busy birds. She asked, "Tom, may I ask you about King's X?"

He replied, "What is it you want to know?"

She said, still holding one another, "Can you King X someone, that's degrading, you know?"

He said, "No, nothing that is injurious, or degrading would be recognized."

She smiled and said, "Well, you're off the hook, Tom."

He said, "I wish I could be off the hook, but you King X'd me, fair and square. An exercise in frivolous futility. I know this is a man's world, we rarely perceive what it would be like to be a female in a male oriented world."

She felt indignant, and said, "Look, Tom, I'll withdraw my King's X and let's get back to our world."

He stated, "You've nailed the complexness in our world. No, I am going through with your exercise, then maybe I'll feel more lamentable in your cause, to fulfill, our pleasures."

She said, "Okay, Tom, but let me set your straight on one issue. We gals get our kicks, just like you guys, but more sensual." She winked at him and that made him grin and perked him up. She said, "Come on, Tom, let's get cracking."

He asked, "When are you going to familiarize me with the panty pull ons?"

She said graciously, "You certainly have a way with words." They made their way back to where Albert was reading his perspectives in the lower helm.

They both greeted him with a smile, and a "hi" which amazed him. How are grave controversial perplexities resolved so promptly? Albert asked Norma, "Did you secure the necessary Coast Guard and Marine emergency paraphernalia?"

She smiled and said, "You bet, Cap, right down to the life preservers and flare guns."

He acknowledged, "Wonderful. Tom, are all stores aboard?"

He replied, "You bet. I'm finished here. We are about to ship out."

Laura had finished unpacking and looked about when she heard the diesels revving up.

Devlen Dawson, with his ruby red buckle that strained to corset his belly at bay, had obtained Laura's address, and turned off Cranberry onto Diamond Street. He drove his car, like he fed his body, with preponderance. He sighed, then braked. His loose pink flesh bounced, following his every motion, like an echo. He slammed the doors of his Sable L.S. wagon. It rocked, then remained at attention. He hobbled up the steps and aimed his pudgy index finger on the button of the alerting fixture. He looked about, quickly then emitted a little gas. He was becoming irritated. He had thought to reconcile and be cordial, how sorry he was, that she lost her position. Then he would devise a word game, and have her tell of her presence at certain times. That was his syllogism, a scheme, that had proved to be one of his better working attributes. He punished the doorbell. The drapes were drawn, and he finally concluded that she was not inside. He looked about with a snitcher's eye, then checked her mailbox. It was empty. He swore, and slipped and scooted behind the wheel of his Sable. He disgustedly gunned his car, spun his tires, and left black rubber granules imbedded in the asphalt.

Chapter 11

Tom weighed the ties. Albert for the first time, felt the awed sensations of glamorous power in his twin 210 H.P. diesels. With engines engaged in reverse, and the low ebb of the throttle, he maneuvered the *Magnetic* from the slip. He headed east by northeast, utilizing the wise sea gate to the end of the long inlet, then port, north by northwest. He had his radio on and had set the engines for less than one-third. He checked all the instruments, pressures, and temperatures--voltmeter, tachometer, and all gages were indicating normalcies. He glanced at his Ritchie compass. He had neglected to have scratch pads and a ledger at his side. He would document his course, temperatures, wind speed, swells and times. He thought he might add a constancy of pertinence of the day, a mix of log, and a diary. He had to verify this and evaluate the dual controls from the flybridge. This was as good an opportunity as any.

Laura was moving aft and entered the lower helm, on her way topside. She said with a bright smile, "Hi, Captain, everything shipshape?"

He returned a generous smile and replied, "You can bet your sea mile on it."

She stated, "The engines are so quiet, and I'm amazed at the ease this 38 footer slices and cuts the water."

He replied, "I, too, am impressed. Come with me to the flybridge." The sea breeze was casually increasing its low level of intrusion. The faint salt odors, powder puffed their noses.

She fluffed her hair and said, "I love the sea, with all its grandeur, and possession of power. I wanted to understand some of its character. I enjoyed all my classes. Most were on the spot training off Seattle's Port of Friday. They spend hundreds of billions to study space and its entities. But the real adventure is my longing to understand the mysteries of the sea."

He smiled, and said, "We are mutually mated, Laura. The seas play a vital role and blend in a rhythm, with our galaxy. Each and every sector is connected by magnetic lines of force, a massive display of electrical current, absorbing, changing polarity, and discharging and assimilation of electrical charges. Your aqueous is the custodian, of all life on earth. Laura, please take control of the wheel. Feel this nine-ton razor scalpel the defense of the seas."

She smiled ardently and replied, "I would love to."

He briefly took note of the compass reading, and stated, "Hold it steady on that heading." He pointed to the compass, and she nodded accordingly. The fly bridge also projected a majestic radar wing, with recessed lighting, two weatherproof speakers, and a 7" remote control spotlight. An 8' antenna gave the illusion of a clean swept mammoth sting ray. He moved behind her, and kissed her neck, to her enjoyment. He said just above a whisper, "I'm going below, mate, to check on the dual controls. I'll be right back!"

She nodded, and replied, "See ya mate!"

He liked that, coming from Laura. He hastened down to the lower helm, and verified his instrument indications, then returned to the fly bridge.

Norma found the half case of yellow scratch pads, and deployed them in various areas. She became sidetracked, and interested in the frequency intensity of the implementation of the radio magnetic antenna. She had researched the effects of radio frequency waves 6 to 20 megacycles, bombarding the human brain over a period of time. One megacycle equals one million cycles per second. The continuous aiming dislodges the brain's normal functioning, which then becomes permanently damaged. Albert had instructed Bayliner to mount his special antenna on the aft loop, in the flybridge. The lead-ins, and metering devices were located next to the CD player and stereo system. This module is located directly behind the helmsman's chair in the lower helm. It was exquisitely installed and was powerful for its size. She reached over and turned on the stereo system.

Albert and Laura were surprised at the pleasant music. Tom in the galley, kept tune with his pen, as he began to write down and plan their first dinner on the high seas. Norma dialed the modulator until she came up with the cycles of station K-O-O-L. They were primarily playing old big band hit songs of the 40's, 50's and 60's. They had just announced the next tune, a rendition played by Lex Baxter (Drinking the Lonely Wine), a sumptuous favorite with the blending of orchestration and the harmonizing of the singers.

Laura said, "There has never been a more beautiful song, that has the intimacy of instrumental and vocalistic romancing one another. Albert, I want very much for that rendition to be our song."

He smiled as the beautiful music flowed through the *Magnetic*. They listened, *When the clouds roll by, I'll come to you, wherever you may be, I'll be true, sipping my lonely wine*. Albert and Laura slipped their arms about each other. He whispered, "I would be delighted to share that wonder with you."

Tom left the galley and met Norma in the lower helm. She asked Tom, "Do you know if there are any folding chairs on board?"

He replied, "I don't know for sure, but I doubt it. Let's check with Cap and Laura."

She nodded, and they hesitated long enough to brush their lips. They said nothing, their eyes said it all. Greetings were exchanged by all when they met at the fly bridge.

Laura spoke First, "Did you hear that lovely old piece by Les Baxter?"

Tom replied, "Yes, but there is one thing that disturbs me about that piece."

Albert looked at Tom as did Laura. Albert asked, "What?"

Tom's blue eyes danced and he replied, "It didn't last long enough, it is beautiful." They smiled in agreement. Laura looked at Albert and said, "We have a song now. It's been a favorite of mine for years."

Norma said, "That's great, next time we get to port, you and I will find a cassette, or recording, with both our songs." Laura flashed an Okay, you bet sign.

Albert said, "The wind is picking up a little bit, but the forecast still remains moderate."

Norma stated, "That's a powerful and nifty radio electromagnetic unit."

Albert said, "I'm hoping we can use that instrument with our other measuring devices and blend them into vital information."

Tom said, "Yes, just a little blending in cooking creates surreptitiously, in determining a tasty plate. His counterparts smiled affectionately and reflected one of his favorite subjects. They were cruising in Cook Inlet south by southwest with land to the starboard, and port. Just ahead the open sea was the only prominence.

Albert stated, "I want each of you to take the wheel, and familiarize yourself with the feel of the *Magnetic*."

Tom and Norma smiled and welcomed the gesture of taking it literally, hands on. Albert looked to the sky, the sun had just set, and gestured with his arm, pointing up and stated, "The magnetosphere is the area around outer earth that exerts a stronger field than the influence on solar or galaxy off the earth's surface. The Van Allen belts are highly charged electrical and magnetic charted particles, that are trapped in its magnetic field." He lowered his arm, but held their attention, so he continued, "These increased energies react to the presence of electrified particles, and induce one another. Upon colliding they produce a more violent build-up of electromagnetic properties. Satellite measurements have proven that energies from power lines are similarly amplified, high over our earth, a phenomenon known as Power Line Harmonic Resonance (PLHR). Radio and micro-wave energies, also resonate in the magnetosphere. This added amplified energy interacts with the electrified magnetized particles in the Van Allen belts, producing heat, and fallout of charged particles."

Norma stated, "God, I know this is our field, but it scares the pants right off me."

Laura said coldly, "My pants too."

Tom elected not to reflect on the previous statements.

Albert said, "Let me say a bit more, then we'll drop the nebulous of this subject." He continued, "We can't evade what is true and proven, but nurse the reality. Tests have proven that this massive ever-continuing amount of (PLHR) over America has created, and rearranged the duct currents that arrange and build gorges for our air streams to follow, like a canyon for a river. Continuation of releasing of (PLHR) in our atmosphere will result in electrical magnetic pollution of our ozone layer and above... and perhaps hasten the destruction of life as we know it. Our electric magnetic bond within our galaxy is held in principle, that <u>all</u> principles remain stable. Now today, as in the past, the earth's pulsing magnetic field combines with the solar winds to induce large currents in the Van Allen Belts." All was quiet as they assimilated the terrorizing statements.

Laura asked, "Albert, what schedule have you set up for manning the wheel?"

That brought a sharp reaction to all but Laura. They tried to capture each other's eyes, but finding no revelation, moved to another pair of questionables.

Albert broke the dilemma, slightly flustered, and stated, "Thank you Laura, for reminding me. I have yet to determine the shifts."

Norma asked boldly, "Laura, how can you say or accept that word 'manning' in all consciousness. That obsolete, renegade male stigmatic word, is another crucifixion on womanhood. The definition of that word is a complement of men.   The continued use of such, and you and I will have more tub time!"

Tom said, in defense, "We couldn't say womanizing the wheel or femaling the wheel."

Norma responded to Laura, "See what a male dominated world this is?"

Albert said, "Gracious!"

Norma was just starting to get warmed up, "And another male oriented item, the helmsman's chair."

Albert muttered, "Good Lord!"

Laura laughed refreshingly, Tom looked at Albert, and stated, "She's right, you know."

Albert menially said, "I didn't put the male tag on these descriptive items."

Norma said, "Another debauchery. Lower helm, why not lower herm?"

Tom said, "She's got you there, Cap."

Laura said amusingly, "I never even thought about all these male questionables."

Norma spoke up again, "And another dandy, seaman!"

Tom laughed, and said impassively, "Not fair, Norma, they rendered that lingo change long ago. The word is actually a feminized word, sea fairer. Not spelled in context, but refers to the pronunciation of fairer of the sexes."

Laura said, giggling, "You're right you know."

Even Albert had to grin broadly.

Norma said reluctantly, "I'll withdraw that one stigma and replace it with man overboard."

Albert said, "Good Lord." They all thought that over.

Norma shored up her descriptive, "Have any of you ever heard someone cry, 'woman overboard', not even in the movies?"

They were all amused at her compilation. Their craft and compass needle were on a heading of south by southwest in Cook Inlet. A sea mark on the starboard side was aptly tapering into the sea. The flocks of seabirds were still prominent, and the screaming still filled the changing light. The twilight would soon be upon them. The cordiality of the warm bright sun was fanning its last gentle rays. The broken sea began taking on a darker, more formative agenda. The women began to feel and address the cooling breezes. They used body language, rather than verbal conjecture, to convey their discomfort.

Albert picked up on it first and stated, "Let's go below, and I'll switch on the running lights.

The gals flashed a low profile secret message between them, congratulating each other on their gestures. The intercom radio was playing a rendition of "Oh, Look at you Now," with Peggy Lee singing. *So you are the guy that laughs at Blue Diamond rings, it's one of those things, oh, look at me now.* The gals were doing what intrigues all men, moving their bodies in a sensualistic fashion, in rhythm, to the perky tune.

Tom yelled, "Ouch, what was that for?"

Norma said unruffled, "I forgot to pinch myself, and remembered your cute buns were available."

Albert had a smirk across his face and Laura giggled in enjoyment.

Tom said hollowly, "Oh!" Grins were all around, except for Tom acknowledging a tweak or tinge of embarrassment. Albert sat in the helmsman chair. The others seated themselves in the salon which had three gorgeous decorative settees without arms, placed to form and opening U. The salon is one step up from the lower helm, which is directly across from the helmsman's chair, with its swinging foot rest. The chair was adjustable to any conceivable position. The arms were generously padded. Four impressive chrome hand throttles, to the right side of the positioned chair, the stately stainless steel wheel nd in front, and directly between the handsome chair and instrumentation, non-glaring, illuminated panel. The large Ritchie, flush mounted compass is located just to port of the extensive lighted display. An unprecedented 360^0 tinted windows enlightened the spectacular view. The generous main salon contained colorful lounge seating for more than eight. A large crowned oval framed glass, adjustable table, complemented the settees, and added to the elegance of luxury. Plush, colorful carpets gave variance to the richness to which this craft was inspired. Albert throttled down. The quilted headliner romanced with the beautiful bright colored billowing lined festive drapes. Two speakers encompassed the regal main salon. A wet bar, ice maker, stereo JVC control gave further evidence of the ultimate in motor yachts. Monitoring systems, and magnetic circuit breaks were just another testament of how this flotilla was crafted. Luxury and safety were ultimately matched. A dead heat, for performance and beauty, a far cry from the svelte boats of the past. The sunlight was slowly being swallowed in this part of the world, and the cooler winds kissed the waves, as they wandered in their own endless journeys. Albert asked Tom to turn on the zone control heat.

Tom remarked, "You bet, Cap, have you decided who the wheel person will be, especially after eight bells?"

Norma smiled contentedly at Tom's careful selection of words. The gals both loved these male jostling tactics for self-fulfillment.

Albert asked Tom, "Do you have dinner planned?"

The female spectators hung on every word.

Tom replied with self-reliance, "Yes, do you wish me to start now?" Albert was unassertive and looked across and up at the women, only to receive stares, awaiting his judgment. He said, "Four bells at sit down, with be contiguous."

Norma whispered in Laura's ear; she blinked and whispered back. They both had a definite, dialectical, jaunty smile, like a contented ceremony of affections. Then the women, still languishing in their inert smiles, focused their attention to their counterparts.

Tom answered Albert, "That would be an affirmative."

Laura asked, "What are you preparing for our first meal on the *Magnetic's* maiden voyage?"

Albert was proud of Laura's take with his vessel.

Tom smiled and replied, "A Tom's surprise." That news brought smiles of anticipation.

Albert was restlessly shifting in his chair, wrestling with the complexity of the assignment of duty at the wheel, the first night out. He was in charge and Tom, as well as Albert, wanted to be with their shipmates. He pondered how to impartially, yet in an unbiased manner, decide this quandary. The soft music bounced with the throb of the engines, Linda Rhonstead was singing the 1940 Academy Award song, "When You Wish upon a Star."

Laura was enjoying her new friends, and was building a more solid relationship, with the passing of every minute. She decided to take a chancy, but logical tactic. She wore a small smile, stepped down, and whispered in Albert's ear.

Tom and Norma looked on questionably, but intensely. Albert relinquished a pent up sight and a refreshing smile surfaced. His eyes cried out and delivered a thanks, well done to Laura. Laura stepped up, and sat on the plush lounge, next to Norma. Laura whispered in Norma's ear, and both had amusing grins. Norma patted Laura lightly on her knee several times.

Tom felt completely out of the apparent going-ons. He was unable to decipher the hieroglyphics that lay hidden in their eye contact. The speakers flowed with a Big Band repertoire, Les Brown's orchestra, with Tex on the sax, and Susanna McCorkall Singing, *How Much Do I love You, I'll tell you no lies, how deep is the ocean, how high is the sky. . . .* The girls tapped their feet and moved in other interesting ways in unison, dueting unconsciously.

Tom said in unpreferred solitude, "My Princess stove is waiting for me."

The girls waved, and Albert crooked his index finger, a come here. He did and Albert gestured for him to stoop over, and then whispered in Tom's ear. Tom straightened up with a worldly fresh face, smiled gratefully, and topped it off with a smart finger salute from his forehead, and made his way to the Princess stove.

Laura asked Norma, "Are you going to help Tom with dinner, or do you want me to help?"

Norma smiled and replied, "Thanks, but no, my Tom loves his soignee. I'll slip in a little later and see how he's doing. Did you see the amplitude on that radio magnetic receptor?"

Laura replied, "No, I'm sorry, I'll have to brush up on its system."

Norma said, "It's quite computerized, the reliance is on simplicity."

Laura replied, "Great, what is your attribute in your field of electromagnetic forces?"

Norma replied, "The magnetic deviation, and magnetic moments, which in simpler terms, the product of the distance between the poles."

Albert was busy writing on the large scratch pads, and glanced occasionally at the illuminated panel.

Laura asked, "Does Albert still hold fast with chain reaction, when the poles change their polarization?"

Norma replied, seriously, "Yes, it gives me goose bumps, when I am reminded of that part, and maybe all of us. . . ." They grasped and squeezed hands and both shivered, avoiding each other's eyes. Norma said, in low controlled voice, "Laura, we are both so blessed, we now have to hold every precious moment as if it is the finality."

Laura said, "Yes, you're so right."

Norma nodded, digesting the fated conversation and said, "I'm going to check on my Tom," and she slipped out of the lounge and headed for the galley, just a little shaken.

Laura got up and stepped behind Albert, and delicately kissed his neck. The charming music sentimentally billowed its rhythm throughout the many speakers.

Tom began preparing dinner. The perky beat of the lively tune had Tom bouncing and bobbing. The emergence of mellow and sweet odors were beginning to socialize with each other, enhancing their own hardy redolence. The mood of the music changed with the Canadian Spitfire Band finishing "Sentimental Journey." Glen Miller's smooth rendition of Moonlight Cocktail, greeted on all board with soulfulness. Tom was concentrating on his cooking skills, and music, and had no idea Norma was right behind him. She touched him affectionately, but he jumped, and they embraced. She released her prey and they kissed again, Norma exerting more pressure in her hold on Tom. She asked, "Can I help?" He smiled and said, "That's a new wrinkle."

Laura noticed it first but said nothing, and had no realization that the rhythms of the sea would be intensifying. The *Magnetic* had just begun discriminately dipping and falling as she cut through the nervous waters.

Norma asked, "What's my favorite chef preparing?"

He smiled in reverence and stated, "I'll tell you, but keep our secret a surprise."

She was off balance and a slight roll upset her equilibrium. Tom caught her just as she nodded. She attempted to smile with a thank you, but didn't pull it off.

He asked, "Are you all right?"

She nodded and said with a greater amount of zest, "What's for dinner?"

He stated softly, "Rock Cornish Hen's Bombay."

She said, "Just the description sparkles my cravings."

He stated, "If you are really sincere, you may prepare the salad."

She smiled and replied, "I will use the spectrum lying between blue and yellow. With your permission, I will dice sprigs of radish and celery for topping."

He grinned, "Don't forget to add sliced olives, the heavy can." They both grinned in remembrance. Tom flashed her a benevolent smile of approval.

He spoke, more to jog his memory, than conversation, "Soy sauce, ginger, wine, sage, fresh mushrooms, parsley, and green seedless grapes for garnishment, to complement the four cornish game hens." The cooking perfumes were strengthening, enhancing the aura in the galley, and traces of permeating into every cranny and enveloping lightly throughout the entire craft. Tom was simmering the combined giblets and the matching ingredients.

He said, checking his watch, "Maiden basting time, in thirty minutes. The galley's placement is but a scant few feet, dead center midships. The leeing of the *Magnetic* exerts approximately equal degrees of rolling with respect to bow, of aft. But the pitch, abrupt rise and fall alternatively precipitously is best absorbed and contained at midships. A simile would be equivalent to galloping a horse and not exerting pressure in the stirrups to cushion your bouncing buns.

Albert was conscious of the heaving, rolling the rise and fall like a twisting seesaw. He increased his speed and that would compensate for a degree or so, off the roll. Normally a craft smaller or larger would have a faint fluttering during this contemporaneous display of riding up and down, and rolling from side to side in the churning waters. The constant yawing, and heaving, twisting, motions that do not chime with our innards, also sway, dip, and lift, but not conceptual with being a clumsy land-lubber. The slow roll to port and starboard, and plunging and rising each maneuver waiting in sequence is called the temperance, as most seafarers are aware, is the displacement that determines the disfigurement of the angry seas.

Albert attempted a repartee for Laura's sake, saying "The news of our maiden voyage has been

suspiciously leaked to Oceanus, God of the Sea."
Albert slipped a confident smile on his face, then
asked Laura, "Please take the wheel. I've set the
trim tabs. I want to check on a bilge pump in the
forward stateroom."

She asked with concern, "Is anything
wrong?"

He picked up his smile a little and stated,
"Oh no, it's just the indicator light on the automatic
bilge pump is still flickering."

She sat down and adjusted the helmsman
chair and glanced at the compass heading. Albert
moved forward and entered the galley.

Norma spoke up and asked, "Cap, what's
going on out there?"

He put on a hearty smile and said, on the
move, "The sea has its own way of christening all
new crafts on her territories, excuse me."

Tom thought Albert was going to the head,
and would have a chat on his way back. Albert
looked in the stowage compartment under the bed
and retrieved a flashlight. He removed the panel
that gave access to the bilge, and bilge pump. He
jockeyed the flashlight in many positions, using the
light to refract. He watched for reflector light to
bounce back from the surface of the water, there
was none. The automatic pumps were working. He
arose, and still clinging to the flashlight moved into

the galley. Tom looked down and gestured, and said, "Lighting a problem?"

Albert replied, "No, no everything is fine." He titled his head upward and added, "There certainly is a deep channel of charming and fragrant odor." Albert looked at Norma, and stated, "Tom rarely shares his passionate cuisine with anyone."

Tom smiled and replied, "He's right, you know!"

She said, "We enjoyed the big bands on Station K-O-O-L Anchorage. I know you turned it off to monitor weather and ship messages. Any news on the weather?"

Albert replied, "You heard, as I, not so much as a tinkle. We probably are in a spurious squall, in which case it will soon pass. The redolence escaping from that oven is assuring that our dinner will be on time." Tom nodded.

Albert moved aft, retracing his path back to the lower helm. Norma asked Tom, "Why does the Cap believe we'll be out of these churning waters soon?"

Tom folded his arms and replied, "For one thing, no news on the air, means it's a small disturbance, and that word spurious means, as you probably know, in context, is illegitimate birth, false and other related words--just another mystery of the sea." He laughed and added, "It's weird and spooky, the weather all around us is ideal, romancing the waters, except in this isolated moving storm."

She sliced a radish in two, and placed his half on his lips. He opened up, she smiled and tapped it in. She took a bite out of her half and asked, while chewing noisily, "Do you know if these baby wind storms ever develop into big time dandies?"

He replied, "If the temperature difference increases, that affects the pressure fronts, it could readily escalate into a violent storm." Tom said, still chewing, "You forgot to salt the radish, and by the way, by juggling letters of the word salt, spells last."

Silence taffeyed out. Finally she said, "What do you want Swami, a pat on the back?"

He grinned and said, "No, another halved radish with salt. . . .please!" She obliged and asked, "One dash or two, oh shrewd one?"

He held up two fingers. She repeated her first performance and then she tasted his salty lips. He chewed, spinning the radish in his mouth, and crunching it, emitting flashing pops. He opened the oven and forked the searing rack out with a small ladle and carefully basted the lightly tanned hens with their own juices that were flowing profusely through their stretched skins. He closed the busy baking chamber.

She said excitedly, "Let's give Cap his cap at dinner tonight!"

He said, " A splendid idea, and why not spruce it up with two lighted candles?"

She asked, "Do you have four holders and candles?"

He laughed and said, "Norma, Norma, remember, I was with you when you got the candles and stands?"

She replied, "Oh God, why make a big deal out of a pinch of forgetfulness, Tom, Tom, Tom."

He said, "I didn't mean it to. . . "

She interrupted, "I know, Tom, this time I apologize. These unfamiliar motions truly make me nervous. I would like one candle, for each of us, it's more intimate."

He said, "All right, but we must be careful. The worst horror of a mariner is bioluminescence!" She looked startled and asked, "What the hell is that word?" He answered, "It is hell, a fire at sea."

The *Magnetic* was cutting less turbulent waters and was smoothing to a gracious cruise. He asked, "Will you set the table in the lounge, I'm sure Laura will give us a hand." He thought about what he had said, and the previous discussion. He followed through and stated, "She can assist." She laughed and said, "Tom, on board, that hand is like a glove, they go together. Let's put it on the shelf, and call it a neutral King's X, Okay?"

He grinned, and said, "You bet. I never heard of neutral King's X, but it sounds logical." Tom gestured to where the utensils were. She gathered them up, and he handed her napkins, following with a quick kiss.

She said, shaking her head, and smiling graciously, "Tom, you are a marvel."

He replied, "And you Norma are marvelous."

She left the happy busy galley with her goods.

Albert noted the calming, and asked Laura to dial to K-O-O-L for romantic music. Doris Day was part way through her vocal of "It's Magic," with Les Brown's orchestra. Laura saw Norma with the place settings and asked to help.

Norma smiled and nodded in the direction of the galley. Laura stepped into the galley and emitted, "Mmmm, that smells luscious."

He pointed to the four candles with holders and said, "Please don't light them."

She replied with her, "Okay."

He said, "Wait up a second, Okay, open up." He popped a pitted olive in her mouth. She rolled it around inside her mouth and repeated her "Mmmm" and left.

Albert saw Laura carrying the decorations and stated, "Tom certainly knows how to set a spread."

Norma remarked, "I share in part of this festive wax ceremony."

He said, "I should have known, you are the key to his coil."

She gave him a double look, then decided it was a right smart compliment. She searched her mind for a male frivolous meaning and found none. She still wondered . . . .Albert repeated the song title, "It's Magic!"

Chapter 12

Both women returned to the galley with smiles. Tom had just turned the oven off, and was dividing up the salad onto four plastic plates. He pointed at the colorful bowls, salt, pepper and butter dish. He said, "Laura, Norma and I have a little surprise for Albert."

Laura searched both their eyes, and Tom was waiting for Laura to pick up the lead.

She did, and stated, "Yes, we'll be making the presentation after dinner. It will be from the three of us."

Laura said, "You guys have been so kind, I've neglected to buy any gifts." Norma smiled with feminine tenderness and stated, "No, Laura, this is really a Captain's due."

Laura asked, "What is it?"

With a big smile, Norma said smartly, glancing at Tom, "A captain's cap. . . with all that wiggly raised gold leaf, and shining bill!"

Laura was excited and said, "Oh, let me pay my share!"

Both said no. . . Tom said, "There are many marine secrets that you had in your lovely computer, that's where the real value lies."

Norma grinned and stated, "My Tom has a way with words, most of the time, and this is one of his worthy culminations."

He briskly stated, "Hurry up, mates, my hens won't last much longer."

They picked up their pace and carried their goods to the large glass table in the lounge. Judy Garland was the reminiscence of some old favorites--"Somewhere over the Rainbow." Norma said to Laura, "Judy was a classy gal." Laura nodded in tribute.

Norma said, after they had placed their goods, "Let's both go back to the galley, you know, you can be my screen, on the way back. Laura glanced at Albert and smiled. He didn't capture it, he was doodling on his scratch pads in deep concentration. Laura replied, "You can bet your magnetic lines of force on it."

Norma laughed and retaliated, "You can waffle your dorsals on that too."

They followed one another to the galley, swinging their hips slightly with the music. Tom had placed four large oval steel platters in the hot oven.   The music. . ."*Where Troubles Melt like Lemon Drops, Away upon the Chimney Tops, That's where I'll find you.*" He gave Laura four champagne glasses for shipment to the elegant setting. Norma hurriedly slipped into their cabin, and had to hustle to catch up with Laura and the ribboned box at her right side. She placed the hat box behind one of the many pillows on the settee's lounge seat. Norma used the lounge ice maker, and half-filled the silvery ice bucket.

Laura brought the chilled bottle of bubbly, and wiggled it down through the glistening cubes. The chef, in accordance, served the main course. The hot juicy hens were expelling bouquets of mouth watering redolence. Laura said, "Come on, Albert sit on the end, next to me."

He did and his face and eyes enjoyed what was before him. Norma looked about at each and asked, "May I?"

They nodded, and she used her lighter to flame the waxy wicks. Albert had set the controls. The four of them in a hesitant, but close enough to call, unison, touched glasses, filled with effervescent of bursting bubbles. The clinking, complimented the four smiling faces.

Albert smacked his lips and stood up and said, "As with maritime practice, the Captain has his say on his maiden voyage, at his first dinner at sea."

They all looked up to him with happy faces, enjoying every word.

He added excitedly, "Let's eat! They laughed and quickly began to savor their tasty hens, Bombay. They relished each morsel. They ate, and took turns complimenting Tom on his exquisite spread.

Tom had also prepared, garnished carrots, and a small saucer of licorice sticks, that were in the shape of a ship's wheel. The glowing dancing flames of the four candles played subdued flashes of eminence on all of their spellbound faces. Laura pulled the wine bottle from its cradle, and held it up and was about to speak and pour. . . .

There came a clamorous resounding loud crashing. The *Magnetic* lurched hellishly, like a cataclysmic star board, in a resonance of screams, bodies flying in the air with expressions of wide horrifying eyes. Steel plates and uneaten hulls of hens and burning candles were air borne. Arms and legs were flaying in grotesqueness, as screams wailed, new cries welled, as a second loud and powerful object slammed the midships of the *Magnetic*. Albert was still in midair and headed head first, down the walkway between the instrument panel and lounge, screaming till his tonsils were begging for mercy "FIRE....." Laura was sitting next to Albert and Tom was thrown slightly starboard, and was slammed upward into the pleated ceiling headliner, still air borne, and struck the front of Laura's body upon the ceiling. Both fell as the *Magnetic* shook like a bone in an angry dog's mouth. Tom yelled agonizingly as he landed head first with his legs flaying against the windows. Laura's lamenting scream was locked in her throat, as she fell head first, striking Tom's crotch with his legs spread wide, and mercifully kicking at the windows. He screamed with the initial pain, then the falling weight of Laura. His torment turned to wailing moans. Tom was still upside down and Laura was on top of him in the same position. Her dress covered her in the darkness. On the way down Laura's flaying hand had found its way into Tom's shirt and tore off his buttons. Frantically trying to free her hand, however, she was not conscious to know where it was. Tom was having difficulty swallowing air. He was trying to breathe and spit out blood from inside his mouth. There were long strands or locks of hair in his mouth, mixing with his blood.

Albert had been seated at the aisle, port side, and facing aft. The first forceful collision sent him up and down head first into lower helm like an unwrapped ball. His head struck as his momentum and frame sent him into the walkway between the wheel and the lounge. The object that took the brunt of Albert's fall was a hand rail attached to the starboard enclosure that adjoined the large instrument panel. The left side of his forehead tore into the handrail and blood spurted from his body being banged back and forth in the walkway. Then, as the *Magnetic* dove, it pitched Albert's body tumbling down and careening, until he finally lay in a wilted heap on the teak parquet floor in the galley! The only movements were a slow throbbing in his stomach and the bright flow of blood from his left forehead that filled his left eye and was trying to reach the parquet deck.

Norma had been sitting on the opposite side She emitted one long shrilling scream and was also air borne and hurled, and drilled upon the landing. Her head grazed, gashing and squeezing past the foot rest, hit and wedged up between the large steel cylinder that supported the all purpose chair. The steel shaft was the recipient of Norma's bleeding head. The thrust of her head had peeled the skin from her cheek bone, and locked her head in a pinned position under the chair. The blow had knocked her out and shut off her wailing scream. She felt no pain now as her skin was torn and her left ear ripped. Blood quickly boiled out as she lay shoeless, with her head in a gruesome position and began to reach the 50 oz. polyester, scotch guarded carpet. Tom was still gagging on Laura's hair and bleeding mouth. He couldn't understand being upside down with a bulky weight on top of him, jammed against the slanted windows. All he really wanted was a little more air. He spit again, trying to get that matting out of his confused mouth. He desperately tried to inhale, but more strands sunk deeper into his gasping throat. Laura was dizzy, bewildered, and totally confused. She was bent, and sore, and everything was dark. Breathing was approaching a panic, hysteria and something was pulling her hair. She began thrashing with her right hand, in frustration, her fingers grabbed hair and flesh, but not her own. She kicked in panic, but her shoes just struck and slid on the hard slippery surface of glass, hurting her toes. She swore.

Tom heard a muffled "Jesus Christ." He was gagging and had no will or oxygen to scream or swear. She thought desperately, "I'm going to . . . She tried, and with both legs and both arms in one burst of strength, thrust forward. Three out of four succeeded, one arm snagged in weight, and clothes. Unfortunately for Laura, she had inadvertently, frantically attempted, to flay and push off just as an intense abnormal violent dip of a wave had heaved and smashed against the cutting edge of the *Magnetic*, only to fall in a deep absence of water. A displacement that followed a swell of the bow of the *Magnetic* had risen and now fell, with a thunderous hollow crash. Laura had pushed off of Tom, then she was suspended upside down, seemingly forever, as the craft fell, unsupportive of its liquid highway. Like a plane hitting an air pocket, all seemed to be held in mid-air, only the straps told you otherwise. Her feet struck the pleated head liner again. The radio had been emitting quality music all during this nightmare. K-O-O-L was playing -- *You Got to Accentuate the Positive, Eliminate the Negative, Latch on to the Affirmative -- But Don't Mess with Mr. in Between.*

The sudden fall sent everything not tied down into flying missiles. Again steel platters, wooden holders were all cascading, colliding with glass, and porcelain splintering upon impact. The horrendous fall of half of the nine tons shook Norma like a rag doll, with its head held tight. It further lodged her head and the rip in her ear, peeled and tore deeper, increasing the size of her laceration, and the amount of blood occurring. She was unconscious, and looked so ridiculous, her body being shaken like a rug, the end that was free to snap. She was unaware of the two lighted candles that had fallen on her dress were slowly burning a black hole, slowly smoldering, then a flame burst, and the fire began to eat, and gobble the cotton cloth that covered her.

Tom finally got a welcome breath of air. He was relieved when the matted strands disappeared, and the pressing weight, that had hovered over his inverted body. The weightlessness had turned his suspended somersault a little cock-eyed and he looked up at the ceiling, and caught sight of a woman's body above him, heading, head first at him. He screamed, and she screamed and careened off his legs, and struck the ridge of the settee. She kicked her feet like a diver just before impact to no avail. Tom was kicking upwards in defiance. Her body lay partly on Tom and on the settee, and wedged against the window.

The second candle that defied going out had rolled out of the metal plate and come to a stop on the deck covering the polyester carpet.

Albert's listless body was slammed and rolled until his head struck, with a thud, on the lower counter, reopening the slash that was just beginning to congeal. The music continued, *Don't*

Mess with Mister In Between" . . . Except for the music and light weeping, moans from Tom, and Laura, the most noticeable sound was the gulping and gurgling of the wine that was rolling rushing out of the narrow neck of the bottle.

The *Magnetic* was rolling slower, and bobbing less, regaining some respectability. Laura's diving body impacted and broke Tom's legs apart. Her head glanced off Tom's groin again, both screaming, one because of anticipation, the other because of the pain. They both captured each other's horrified eyes for just a second. Her body weight carried her to a slow somersault, and she then fell to lower helm in a heap. Tears were frantically flowing, as she thought this nightmare would never end. She lay silently in a curl. A smarting pain was becoming unbearable on her right thumb and index finger. It seemed it took a long time for her mind to react, and to draw her painful hand for inspection. She wanted to see what was happening to her. She smelled them first, then saw the pink to black burned skin. She muttered, "God Almighty." She raised her head and twisted it as far as possible. She thought she saw the outline of two shoes way off. She pushed up with all the strength she had left, upward and twisting her head to see what was causing the shifting light and shadows. She screamed. . . . Cloth was burning, but it wasn't hers. Norma lay still, her head cruxed in between the medal shaft, and the helmsman chair. Laura attempted to raise her body with her left arm, and screamed and let her body drop back to the deck in agony. Her left arm, or shoulder, must be broken or dislocated. She gritted her teeth, and tasted blood. Her tongue was smarting. Some of her hair was in her mouth. She spat to dislodge it. She was spitting up blood, and the hair matting with the blood, still hung in the side of the her mouth from between her lips. She drew strength from within, and groping with her right hand, caught the rolling champagne bottle. She flinched as her burned thumb made the first contact with the elusive container. She got control of it and slid closer, pushing with her painful feet, and right elbow. She hoped there was still some liquid left.

Tom was still groaning in agonizing pain. He was still in a vee position, as if he was trying to touch his toes and had relaxed. He had no idea what had happened, or which way was up. He was also unaware how close he was to the aisle walk through. He grimaced and let his body chart its own course, using the least strength. He painfully did a slow forward roll, and fell to the lounge seat. He saw flashes as he moved toward the dancing and flickering lights. As he righted his vision, with reality, he briefly saw the inevitable. He was unable to stop his momentum.

The radio was now emitting music by Franky Carl, *The rumors are flying, that you got me sighing.* . . . Laura saw a black shadow to her upper left, and again screamed compassionately as his hurdling body fell from the lounge to lower helm just across and down, striking her back, and outstretched legs.

Tom's scream was a wailing of "No - no....!" as he fell crashing on her legs.

He uttered, "Jesus," then lay on the carpeted deck of the lower helm. Her body had taken the full brunt of his weight.

She cried, "Tom, ... please. . . my arm."

He made one more painful move, and lifted his weight off her. He was now sitting on the floor among the scattered platters, plates and glass. He was trying desperately to focus his eyes. He saw two outreached shoeless, hosed, feet and legs belong to his love. There were moving bursts of yellow and blue, and the little flames magnetized, their size on the reflective cabinet, that held the entertainment complex.

Laura lay, face down, with searing pain, and thoughts. She had never in her life had been held as a hostage to the bowels of hell for so long. Still in a tunnel of horror, and the throbbing pain in her left shoulder, she knew something was important. She tried to raise and push off with her right hand, looked and saw the two burned fingers. She cried out in desperation -- "Fire -- fire" -- those dreaded words, drove solace to Tom's mind, and was beginning to shake the cobwebs as he realized the closest fire was Norma's dress. He reached for the bottle of wine, and fumbled it, then secured a firmer grip, and emptied it on the curling flames. He shook the bottle and patted the rest with his hand as it spit back in defiance then quieted. He then saw the ugly position that Norma was in. He ducked his head down, looked up and saw the grisly picture. Her head up, and jammed, under and beyond the metal shaft of the support for the helmsman's chair.

He uttered, "Mother of God," in horror. The second unattended fire was just beyond Norma's outstretched body. He grabbed a pillow that was partly stained with creamy thousand island dressing. He could smell it's odor as he beat the little fire with agonizing blows, taking some of his frustration out on the burning hole in the carpet. Part of his mind told him to stop, the other, if he did so he would have to determine if his love was dead or alive. He threw the pillow down, and for a moment not wanting to act, listened to the music of Franky Carl, *I'm not denying . . .I'm falling in love with you. . . .*

Laura had heard the fluffing noise from Tom's frustrating swats. She thought to try one more time to raise herself at least to a sitting position. She looked around, and saw that her legs were caught and hooked in the passageway and realized the Albert was nowhere in the area. She cried out desperately to Tom, "Where's Albert?"

Tom was yanked back to reality and said stonily, "I don't know, I was just going to find out."

She asked, "How's Norma?"

He said, close to tears, "I don't know, I was just going to find out." He dragged himself closer to Norma, and peered under the seat, and wailed, "Oh, Jesus." She was so limp.

He brushed some broken debris, out of his way, and didn't realize he had cut his palm, and the blood was oozing between his fingers. He scooted on closer, to facilitate her position. He snuggled up as close as his sore groin and knees would permit. He had to straddle her, to get close, to where her head was jammed. He squeezed his knees together against Norma's waist for balance. He ducked down and up, let his hands grope slowly, till both of his hands were on each side of her face. Appalling thoughts raced through his mind--she's already dead--her neck is broken...." He said in as calm a voice as he dared, "Laura can you help me?"

She asked, "What can I do, I only have the use of one arm?" The chilling news stung his conscious mind. Laura said, in a sinuous, singsong voice, "Where's Albert? Why doesn't he help us?"

Tom said, "First things first, the fires are out. Help me free Norma. Her head is jammed underneath this chair. When I say go, or when you're ready, pull Norma away from the chair."

She said, methodically, "I only have the use of one arm."

He thought that over and replied, "Use your good arm, but first put her feet together, Okay."

She said, "Okay, give me time."

He said, "Okay, then we'll look for Albert."

She said, "Yes," and groaned as she positioned herself to fulfill her task. She tried to line up Norma's legs without touching her two burned fingers. It worked. She lunged and shoved her hand under Norma's ankles, obviously in severe pain. She wailed, "Oh, oh," . . .She said nervously. "I've got her, I'm going to try and pull her."

He said, "Good pull, Laura -- pull. . . ."

She did an inch at a time. He had to spread his knees outward to let her body pass through, while holding her head like a cradle. Laura slowly dragged Norma's body, until her head was clear of the steel cylinder support.

Tom looked about and reached across, and snatched a wayward pillow and placed it under her head, with great care. He clearly saw a torn ear and the blood in her matted hair. He then focused on her chest first, not wanting to take the final step. He knew, as he saw movement, his eyes burning to see, a trivial rise, or fall of her chest. He shouted, "She's alive, God damn it, she's alive." He glanced to see what Laura was doing. She was looking queerly, and examining her right burned thumb, and index finger.

She said calmly, "Thank God, where's Albert?"

He said, "I'm going to look for him." He looked gruesome with his shirt all torn in front, caked blood on his face and lips. He used the handy sturdy helmsman chair to right himself. He had no idea what had happened, but he was damn sore and tender. His muscles had been stretched and bruise. He looked down at Norma, then Laura, and realized that he couldn't put it off any longer. He was apprehensive of what he might find. He foolishly looked up to see if any windows had been smashed, then suddenly realized there would have been a cool breeze if they had. Enough of that surmising, he thought, just use good sense. Then a cloudy thought crept into his mind, this all made no sense. He moved, using the implements at hand to steady himself. Christ, his privates and groin sure hurt. He used his right hand to steady himself on the instrument panel as he moved forward to look for Albert. He slipped his hand down from the tilted panel to grasp the one and only hand rail, and his eyes captured splotches of dried blood. He stopped and examined it closer. No doubt it was blood, and no debris was this far starboard, it was all toward aft. His mind raced back. They were all sitting side by side eating when . . . his mind flashed over the horrible events, some blurred, some rational. He looked down and thought he spotted shoes on the galley floor. He turned and looked back, not knowing why Laura had her back to him, still sitting on the floor. Thank God, their eyes had not met. A few steps more and he was in his galley. He wondered how Albert was thrown this far and in this direction. Then his mind began imprinting images of Norma's head, driven up, and under the chair. Both of them had been on the outside aisle, that's why he and Laura had ended up on top of one another. He said, "Jesus," as he dropped to one knee in pain, and then settle on both knees. Albert's left face was flat to the deck. He went for Albert's wrist, and held his breath. He muttered, "God, Jesus, I can feel his pulse. He positioned himself over Albert for better balance and turned his head upward. The nasty gash was prevalent on his left forehead. Clotted blood was also in his hair and he was unconscious. He stared at the placid face, only then did he notice his shirt was buttonless and ripped to shreds in the front. He wondered how that happened. The music streamed through the shaken craft, *Dreams -- while the smoke rings rise in the air -- You'll find your share of melodies there -- Dreams, are never as bad as they seem -- so -- dream -- dream -- dream.* Tom caught himself again, just listening to the soothing music. He blinked, and got back on track. He removed what was left of his shirt and balled it up, and what appeared to be reverence, slipped it under the Cap's head. He silently thanked someone, or something, that the sea claimed no lives on the *Magnetic*. He was not a religious man, his choice was close to pragmatism. He hurried back with the good news. As he told her, tears were already swelling in anticipation, regardless, if favorable or otherwise. The salt laced drops flowed evenly down her cheeks and moved swiftly down the lateral wall of the nose, then streaming on her mouth. She tasted the happy saline fluid through her quivering lips. She looked up and Tom dreaded what most men deplore, to try and capture the revelation in women's eyes, that is emotionally, crying. He turned his head slightly.

She asked, "Where is he?" as the leaking continued.

He started to bend down and help lift her up. She said, with sobs less frequent, "No Tom, remember my left shoulder. Where's your shirt?"

He then went for her waist, and asked, "Can you lift that pillow on the way up?"

She glanced down, and said, "I think so."

He said, "Isn't the music wonderful?"

She had stopped crying, and she said, "What is the name of that song, God I must be crazy even thinking about it."

He replied, "No, just like me, and the rest of humanity, your human reflexes are returning. By the way, that melody is familiar, 'It's in the Mood Backwards,'" Ray Anthony's orchestra' called -- 'Drive In' . . . ." They worked together and she stood up, with Tom shirtless, his arm around her waist.

He said, "Albert's in my galley." They made their way, and Laura saw Albert stretched out, with a ball of cloth under his head. He helped her to drop slowly on to her bruised knees. He moved to the other side, and went down on one knee. He said, "I'll lift Albert's head and you scoot the pillow under it. Okay?"

She blankly said, "Okay."

Tom snatched his balled up shirt and sent it flying across the teak parquet. She managed a slight smile, and "Thank you, Tom."

He didn't respond to her small gesture, but said, "For now, let's stay by our mates until they recover. There's always the possibility of hysteria. When they come to, they will have our arms and eyes to comfort them. Okay?"

He asked, as he went to the cabinet drawer, "Will your arm be Okay till our loved ones are with us?"

She replied, "Yes, we're lucky to be alive." Her swollen eyes attested to the wonder of escaping death. He wet two dishtowels, wrung them out, and gave one to Laura. He opened a small plastic bottle of Vitamin E, 1000 I.U. Soft Jels, took his paring knife, and punched a hole in one end. He knelt down next to Laura who was holding Albert's head. He said, "Give me your right hand." He squeezed the oblong capsule. The thick clear liquid began to slowly transfer over to the burns on Laura's thumb and finger. She said, regaining a perk in her voice, "God that feels good. You are a wonder!"

Tom smiled and said, "No, no, but this Vitamin E is. It is the finest enhancer for repairing damaged cells and its denseness coats the nerve ends from the air. That's why you have instant relief. I'm taking another capsule and will put it on Norma's cheek bone, and torn ear. I'll fix another one, and you can recoat your finger and put some on your's and Albert's lips."

She was swathing him again, and replied, "Yes, please do."

Tom was anxious to get back to Norma. He fixed her a flex jel and said, "I'm going, Laura, hollar if you need me, Okay?"

She smiled, emotionally beseeched, and said, "You're right, you know."

They both captured a stark, personal smile. She briskly said, "Get back to Norma, and make sure our running lights are on, Okay?"

He emitted a small laugh. "Okay, thanks for putting me back on track." He knelt down, shoveling garbage out of his way and began removing the crusting blood around Norma's ear. He used the other end of the wet towel to wipe her eyes. He pricked a hole in the Gelcap and applied it liberally to the torn flesh, and skin on her cheek bone. The running lights were on. He had time for contemplation as he scanned the mess of half eaten hen carcasses, pickles, olives, and an array of broken dinnerware. He thought of the champagne, the precious liquid that had remained in the bottle. Then he dashed that as madness, why hadn't he familiarized himself as to the location of the fire extinguishers?

A licorice stick was next to his knee. He popped it into his mouth without hesitation. After this unholy calamity a few likely germs were not adversaries. The music unending was Steve Lawrence and Edie Gormet, a duel rendition, *Ain't got a barrel of monkeys . . . traveling along . . . . singing a song . . . side by side.* The *Magnetic* although crippled, cruised through the pitch black night and relentless nervous sea with no one at its helm, heading -- "

Chapter 13

For more than one hour, no one on board the *Magnetic*, had so much as glanced at the lawrence, the digital depth sounder, or the two Richie compasses. Both magnetic needles pointing the direction of the course that bore south by southeast. The craft was approximately 200 miles out of Anchorage, Alaska. Dead reckoning from this present course was one of the larger island, Kodiak. A smaller island, due eastward out of the sea, is less than 30 miles from Kodiak. It is solely named, Atognak. The ferry sailing lanes connect Kodiak and Seward, Alaska, then westward to the Aleutian group. A slight loop is the sea gate, which includes the Gulf of Alaska. The Bering Sea lies north of the tail (Aleutian Islands). Two small islands lie south of Kodiak Island. Further Northwest likes the Beaufort Sea. The Bering Strait is a division of make believe in the salty brine, in defiance between the U.S.S.R. and the U.S.A. The winds were again picking up, and the choppy waters were magically becoming waves. Whitecaps would follow, with further intensity of the currents, or weather displacements. Port Graham lay ahead port side with Kamishak Bay to the starboard. Atognak Island is heavily forested and is a national forest, called the Chogach. It has jagged reefs, rising sharply, some like razors that appear, then disappear. The sea is the matador, choosing which it will reveal and what time, in an endless, deadly game, in which the house is the restless sea. Sea marks and oil derricks are clearly chartered on maps, radar and light beacons, point out the most treacherous hidden adversaries, a warning to all crafts that move on these waters to be aware. The *Magnetic* is churning through these waters, at 12 knots, with no one at the wheel.

Laura almost screamed in joy when Albert popped open his brown, bewildered eyes, and asked, "What happened?" She touched his cheek with her little finger, and smiled down at him and replied, "We don't know!"

He asked, "Is everyone all right?" She could hardly wait to reply, "Yes, yes, but Norma is still out, I think,."

He began to move, then stopped and stated, "Gracious, I don't relish getting up."

She said in a cheerful voice, "Then don't, Albert, Tom can handle the situation."

He asked, "Is he at the wheel, are we still moving?"

She replied, "I don't think so, I really don't know, Norma has a torn ear, and bruised face."

He said, "Good Lord, are you all right?"

She answered lightly, "Just my left arm and shoulder."

He reached up and gently grasped her right wrist and pulled her hand down to examine her burned thumb and index finger. He muttered, "Jesus, I'm going to get up."

She said, "You have a nasty gash on your forehead. Be careful when you move."

He asked, in a low tone, "How the hell did I get down here?"

Laura jokingly said, "All by yourself, Albert," then quickly added, "I don't know, Albert, but we are so lucky to be alive!"

He fumbled and reached up and pulled a drawer half-way out. He used it to raise himself and pulled his bulk up, unsettled and swaying, but hanging onto the counter top. He looked down at Laura on the floor, her puffy eyes looking up in compassion and said, "Come on, I'm going to lift you and put you in your berth." He then noticed all the scattered paraphernalia on the galley deck. He shook his head, and then was sorry he did. They made their way, holding, hobbling, and so very grateful to be enable to embrace each other. He helped her lay down and double pillowed her head. He looked down on the floor and found another pillow and placed it carefully under her arm. He sat on the side of the bed and kissed her, as if it was for the last time. He said, in irony, "I've never felt so vulnerable in my entire life--I'll be back as quickly as I can."

She smiled, and gallantly said, "I'll be all right. Let's hope Norma is. I have this spooky feeling. . . ." She added, "Do you have an automatic pilot?"

He said, "No, I wanted to chart and have hands on. Why didn't someone shut the engines off or idle them. This is my fault. I nearly killed us all."

She said in support, "The *Magnetic* would have rendered us like a waif. The intensity of the waves would surely have carried us, and put us at the mercy of the sea."

He thought about what she had said, and added, "I have to go, Laura."

She gestured with a hang touch smile. He was shocked by the disarray, moving to the lounge and lower helm. He saw Tom hunched over Norma, his back to him. He noticed she was shoeless, and lying still. He tried to evade stepping, or kicking, on the cluttered floor, as it was impractical. Tom heard him. Albert reached across to reduce the throttle, and disengage the engines to neutral.

Tom said, "Hi, Cap. Glad to see you up and about. How's Laura doing?"

"I haven't had time to check on her arm, but she seems remarkably fit. How's Norma doing?"

Tom looked into his exciting eyes and sated, "She's alive, that's all that matters now."

Albert said, "That is a nasty rip on her ear."

Tom said sullenly, "I hope her neck is all right, her head was driven up and lodged under the helmsman chair."

Albert replied with anxiety, "Good Lord, can I get you anything, maybe a shirt?"

Tom looked down at his hairy chest and replied, "Yes, and a blanket for Norma." Albert left noisily. Tom heard him approach again and stood up, and took a shirt from his outstretched hand.

Albert stated, "I'm not going to radio the Coast Guard as long as the *Magnetic* is seaworthy for emergency medical attention. If Norma doesn't respond, I'll get cracking. Actually, we should be somewhere close to Port Graham, or Kodiak."

Norma moaned slightly, and moved her right arm. Both men peered intensely with anticipation. She opened her eyes and looked up into two excited, but grateful faces, and asked, "What happened?"

Tom covered his hand with hers, and gave it a pleasing squeeze.

Albert spoke first, "We really don't know."

Tom chimed in, "Everything is fine now, can you move your neck?"

She cried, "Oh, oh..." as she slowly turned her head. She asked, "Where's Laura, is she all right?"

Albert stated, "Like you, she's alive, but may have a busted shoulder or arm. Now that you are with us once again, I'm going back to her. Welcome back, Miss Moon, you had us biting our lips."

She looked up at Tom and stated, "Tom, it looks like you took that to heart. Are you all right?"

Albert spoke up, with a little grin, "You had better wash your face, you look terrible."

Tom replied, "Yes, I will, and I am all right, but my groin is still a bit sore. Laura did a head first on it!"

Albert and Norma both cried out, "What!"

Tom said, "It's not that, I was upside down, and Laura was above me, also upside down, I think, well, anyhow, she was air borne, and she came down head first and hit my groin with her head." Albert said, "Good Lord!" and Norma added, "Jesus!"

Tom added, "Twice. We got jammed on top ridge of the settee and up against the tapering glass windows."

She said, "Christ. . .where did you land, Cap?"

He replied, "I woke up on the floor in the galley."

She said, "Good God, how the hell we all made it, I'll never know, but I do know one thing, I feel like hell, in more places than one!"

Tom said, "You were on fire, too."

She looked down at her dress and swore, "Jesus Christ," then added, "What else happened?"

Albert replied, "We don't know yet."

She sighed, and stated diligently, "You guys help me up." They shuffled about and carefully lifted her to a standing position, but not without oh's and ah's.... She started to reach for her left ear, and Tom captured her hand and spoke,
"Don't, don't, Norma, it's torn." She stately said, "I want a mirror."

Tom said in a low pitch, "Your cheekbone lost a little skin."

She looked down and said, "God, what a mess."

Albert said, "Never mind about these incidentals. Tom, why don't you take her to your cabin?"

Norma said, spiritedly, "Cap, you said Tom's cabin."

Albert said, "Yes I did, didn't I?"

Tom nodded, then he added, to your joint cabin," and smiled. She snappily said, "Thank's Cap."

Albert said, "I'm about to find out just where the hell we are, and then tend to Laura."

Tom put his arm round her waist and they went forward. Albert heard Norma ask Tom, "Are your body parts all right?" He also heard Tom reply, "I don't know, I haven't had time to see." He heard a high pitched, "Oh."

Albert went to the fly bridge and turned on the remote search light, and made a few passes. He was certain the *Magnetic* was seaworthy, and just made out faint lights on the port side. He checked the raytheon radar, which he had installed at additional expense. The depth radar indicated over 300 feet of sea beneath them. Albert did not give his course or identity to the Coast Guard upon debarking, so it was wise to remain silent. Port Graham was less than an hour away. He would tie up and get medical help for Laura and Norma. He went below to the lower helm and pulled the charts from the overhead storage compartment, until he found the one he wanted. He engaged the engines from neutral to slow. He looked at the needle, and made a course correction, that set his course for Port Graham.

Their cabin was in a minor disarray. Norma had laid down, with help, then she sat up in pain, and said, "Damn."

Tom said, "Can I do something?"

She replied, "Yes, Tom, find some house slippers. I have to go to the head."

He smiled and said, "Okay." He found them and put them on her feet, then added, "want a lift?"

She replied, "No, Tom, I can make it. Please do me a favor while I'm gone?"

He happily replied, "What can I do for you?"

She said, "For both our sakes, wash that blood off around your lips." He grinned and said, "Okay," then added, "Do me a little favor, use the private head in the forward cabin and say hi to Laura. She was indispensable in getting you out from down under."

She smiled, still sitting on the bed and said, "Just another tiny request."

He said, "Okay, what is it?"

She said, coyly, "Check out your body parts!"

He smiled, and helped her up. He asked, "Say hello to Laura for me. I'll do the second request first, then I'll go to my galley, wash up, straighten a few things, then I'll meet you back here. Okay?"

She said, "That sounds like a good one, but what's with these Okay's. You rarely use those remarks."

He said, "Laura and I locked in on using Okay's, to keep our sanity, and get you out of that dreadful headlock Okay."

She hesitated and tried to decide the best course to take, just smile. . .or. . .to say, and she did "Okay," and left.

Tom checked himself out, saw his face, and noted how grotesque he looked. The blood had covered tips of hundreds of his bristles on his unshaven face around his mouth, and bloated lips.

He decided, then and there, to hell with everything. He was going to wash and shave now!

Norma tapped lightly and didn't wait for a response and walked in, and said with a pleasant little perk in her voice, "Hi, mate, how you doing?"

Laura smiled, "Pretty good, except my left wing. I'm so happy that you are up and about."

Norma said with special significance, "I want to thank you for literally straightening me out, with a busted wing and all." She nodded, and stated, "Laura, we are all so lucky to be alive. Do you know what happened?"

Laura replied, "No, just the tremendous crash, and I was hit headlong and scrambled. I must have blanked out immediately." She smiled and continued, "You sure were, and on fire too!"

Norma asked, "Look, can I use your head?"

Laura changed her tone and responded like a command, "Norma, please don't ever say that again. This is not my cabin, or my head, all of us have a run on the *Magnetic*."

Norma turned, smiled, and said, "Good girl, I was just checking. You'll be fine," Norma heard her say as she closed the door. "I should have known . . .s-i-s. " Norma wasn't sure about that last word.

Albert returned with a smile and asked, "How're you doing?"

Laura replied, with the same and said, "Fine, now."

He stated, "An hour away, and we'll dock and get a doctor for you and Norma."

Laura replied, "I really don't think it is broken. A couple of aspirins would help me now."

Albert went to his drawer and took out four of the white pills. He started to go in the head, and Laura alertly stated, "Norma's using it!"

He said, "Oh," went to the galley, drew a glass of water and returned to his cabin, as Norma exited the head.

She spoke first, "Hi, Cap," and saw the glass in his hand. He moved over, and handed Laura the two aspirins. Norma asked," Can you spare two more for a shipmate?" He left the glass in Laura's right hand, a little unsteady, with the testy two fingers. Norma said, "Thanks, Cap."

He moved back to help Laura wash her pills down.

Norma said, when Laura finished, "There's plenty of water, may I?"

He handed her the partially filled glass. She popped the two, and drowned them and handed the glass back to Albert. He retrieved two more from his catch, and finished his drink. He smiled thinly and tried for a moment, to capture either one of their thoughts. . . .He then began to examine Laura. He said, "She's right about it being more flexible, and it even looks better."

Norma had mirrored her ear and asked, "What's that sticky soft stuff in my cut?"

Albert looked bewildered and Laura said brightly, "Your Tom put blobs of Vitamin E on it to seal the exposed nerves and helps repair the cells. My thumb and finger are so much better."

He said, "Tom is still full of surprises. He has more answers than questions."

Norma said, "It doesn't hurt very much." All was quiet for a moment, then Norma jolted them with an excited voice, "Laura. . .Laura. . . this cabin has access, just a sliding door, and we're in business. Let's you and I soak! Albert can get a towel, make a sling, and you and I will soak, soak, soak. . . ."

Laura grinned and said loudly, "Done!"

He said, caught between the two excited gals, "I'll take the wheel."

Norma said, "Don't worry, Cap! I won't let her arm or shoulder soak, but the rest of her can."

He said, "Good Lord."

Laura said with a snicker, "That sounds funny."

Norma said, looking at Albert, "I know, it gets them every time. See you, Cap. You and yours are about to rise and soak."

Laura crooked her little finger and he bent down, and they exchanged sensualities. Albert fixed a sling, and heard Norma say on his way out, "Two of them burned and that little one still has that feminine magic." He heard them both giggle as he closed the door.

Albert went in search of Tom and found him in the other head, toweling his face. Albert said, "Tom, you still amaze me at every turn."

Tom folded the towel on the rack and replied, "How's that, Cap?"

Albert shook his head, grinned and said, "Once again, you're one jump ahead of the devil himself."

Tom said, "I don't get it."

Albert laughed and said, "You just did, another five minutes, and there wouldn't be one drop of hot water."

Tom thought, then said, with enthusiasm, "I know, Norma's going to soak to christen the tub."

Albert said, "Well, Tom, you got it half right!"

Tom thought about that last statement, then said, "But she said."

Albert asked, "What?"

Tom replied, "Never mind, I'm cleaned up and I'm going to scrub the lounge area."

Norma, wearing a bathrobe came bustling in.

Albert said, whimsically, "Come on in."

She challenged his statement and did just that, closed the door, sealing her dare.

Albert said, "Good Lord."

She said, "That'll teach you, Cap, I just wanted a kiss before my soak, and remind Tom of something."

Albert said, "I'll go."

She said, "No." She squeezed past Albert and placed a smart kiss on Tom's clean fresh lips. She said, "Now, Tom, I've just one thing to say before Laura and I will pleasure."

Albert said, "Good heavens."

She added, "Tom, remember Cap!" and she pointed to Albert. She left as quickly as she came.

Albert was bewildered and said, "What was that all about?"

Tom said, with substance, "She just wanted a kiss."

Albert said, "That was nice of her to think of me."

Tom replied, "Yes it was, wasn't it?"

Tom said, "Cap, you could use a shave yourself, . . . you know. . . Laura."

He quickly reacted and said, "Yes, of course, I'll heat a kettle, then I'll correct our course if need be."

Tom hurried out to the lounge and looked for Albert's new hat. He found it and the ribbon still bound, but the bow was crushed and splattered with seasoned dinner. He put a pillow over it and took it to his stateroom, then retraced his steps. He picked up all the wood, steel, oval platters, and stacked them, then began piling the refuse on them, until he had an armful. Down to the galley and back until he had accounted for all the particulars of their interrupted dinner.

Albert did his navigation with zeal for he loved it, and he was on course. He asked Tom on one of his many trips, if he needed another mate, but Tom declined. Tom had found the broom and was sweeping everything left into a pile.

Albert said, "Tom, that was an elegant spread, the flavor was strikingly wonderful. That cut on my forehead is better, thanks to your Vitamin E."

Tom said, "I have burned many a finger during my training in cooking, and those soft tabs are an instant fire extinguisher for pain. You'll be fit as a fiddle, Cap."

Albert stated, "I'm steering for Port Graham, and we'll have the women tended to."

Tom replied, "Good, I knew you wouldn't let them or me down. How soon?" and Albert replied that it would be less than an hour.

Tom asked, "When they inquire as to what happened, what's your strategy?"

Albert pondered, then said, "I"m going with the truth, that we had a collision."

Tom asked, "And when they ask, who and where. . . ."

Albert said, "I believe I'll use the lingo of Ruth Dander, and tell them I was struck in the bow, and midships, by a hit and run pirate, approximately 150 miles out of Anchorage, with my running lights on."

Tom said, "It plays for me, Cap, your Laura has a hard head, hair and all."

Albert smiled and said, "She's smart you know. She landed on an old softy."

Tom grimaced in remembrance and said, precariously, "I hope you're wrong on that score, Cap!"

The humming and laughter could be heard easily in the galley, and faintly in the lower helm. Albert and Tom would grin when hearing the high shrill joyful giggles that would transcend from the busy tub. Albert had set the course from departure, using the port ridge, of the shipping lanes, although these sea lanes are far from being congested. He still pondered, what dastardly captain, would commit such a malicious act. The immensity of the collision, and that bowel swell after the second blow at midships gave every indication, that a large seagoing ship was indeed, the debauchery. There were any number of reasonable possibilities, for the blatant impact, but the audacity of not rendering assistance, or radioing the Coast Guard to fix the latitude and longitude of the collision was irreprehensible. He thought of the probability of ever finding the culprit, and the renegade ship that used the *Magnetic* as a battering ram. These waters, according to this charts, had safe sea gates and lanes. Outside of these safe waters were the pinnacles and shoals, both port and starboard. Dark hidden nests of boiling, churning, suctorial action of the splashing dark and white waters lay in wait. They slapped and licked the sharp rock pinnacles that lay hidden at will, of the gestures, of wild tides and currents with the many surges of brine, redirecting its ferment, agitation and producing spidery tentacles of notorious undertows. Its dramatic vacuum, sucking and spewing the scum and foam, and drowning all that dared ride on its violent highway, where rocks braced for the next crushing wave. The Cook Inlet was in a generous mood, the winds were moderate, the visibility was clear, it was possible to penetrate the darkness. The inlet was like a gentle rocking cradle, with the *Magnetic* one of its babies.

Norma was all smiles, and said gaily, "Isn't this just the greatest, I waited so long to soak, soaking with a mate who also enjoys one of the finer pleasures, makes it even more special."

Laura spoke, "This is scrumptious, not worth flying without a plane, but one terrific follow-up."

Norma reluctantly said, "Laura...a tiny sticking point, I didn't want to play the fool, that's why I kept silent in asking Tom, or Albert. . . ."

Laura asked, "Can I help give your mind a rest, please. . . a little lower to the left. . . oh, yes. . . "

Norma said, "Just between us gals, the Cap, Albert told me, and all of us, that we have only a 200 gallon tank of water. Now, I've done some estimation in regard, with thirst, galley use, heads, washing, and the men shaving, there's little water left for one tub. . . ."

Laura laughed refreshingly, and finally said, "I'm not laughing at you, Norma, but the dilemma presents you with a very serious situation."

Norma said, a little perturbed, "This predicament is shared by you, too!"

Laura grinned with elation and said, "The best of news mate, you have a whole ocean to soak in."

Norma stated, annoyed, "Yeah, that's what Tom said. I remember his pert words exactly. . . The sea is nice to soak in too. . . . "

Laura laughed again and said, brightly, "Your Tom was making a play on words in your behalf." She splashed the water, and added, "This is sea water, there are two water systems on board."

Norma stopped washing and swore, "Christ, you must think I'm an idiot, it never dawned on me. Hot damn, you're right, we can never run out of soaking substances."

Laura laughed and said smiling, "I've never heard the sea described so elegantly."

Norma perkily remarked, "You have now mate, how's the wing doing?"

Laura replied, "It's sore, but better, thanks to your substance."

Norma said, jokingly, "Let's whistle to that childhood favorite song in Snow White, *Whistle while you soak!*" This brought a snicker from Laura.

Laura retaliated with, "You're an Oak's with me, Miss Moon!"

Norma sincerely said, "You're Okay with me too!"

Laura giggled and said lightly, "Your Oak's is spelled O-A-K-S, that's the only word; mixing up letters of your favorite word, Soak."

Albert and Tom heard the snickering clearly. Both men failed, by position, to capture each other's eyes, but both were thinking identical thoughts. They both recognized that all of them came close to being killed, and how fortunate to have their mates and hear their joyful laughter, although they were in pain.

Port side lights of Port Graham were becoming more prominent. The flashing cardinal buoys with flashing bursts of light, were bobbing like deep water guides. Albert looked at his watch, and spoke out, "Good Lord, it's one a.m. in the morning. Tom was on his knees, with a brush, scrubbing their dinner and components off the carpet.

Tom said, "Yeah, it's been one hell of a night. Say, Cap, set me straight, before dinner, all that whispering. Whose idea was it to just anchor so that we could all be with our mates. He smiled, "It was Laura's."

Tom said, "Well, I'll be damned, she's a real cracker jack."

Albert said proudly, "To me, she's the prize in the cracker jack!"

Tom said, "I have a suggestion...seeing the hour, and we're all sapped, why don't we just tie up, or anchor, and seek medical aid at a respectable time?"

Albert thought that over and replied, "That makes good sense."

Tom added, "We're just doing what Laura suggested. . . . ."

Albert perked up and said heartily, "You're right, you know!"

They grinned at each other rather foolishly, in anticipation of being with their own.

Tom said, "Another suggestion . . .whisker those whiskers, Cap."

Albert remarked, "Good Lord, I forgot my heated pot!"

Tom smiled and stated, "I shut it down."

Albert responded, "And that's what I'm going to do, shut us down." He maneuvered, then asked Tom to go forward, and release the anchor in the recessed fore deck, then move aft in repetition.

Norma had just finished drying Laura and was toweling herself when both looked into each other's eyes as the engines stopped. Norma, just one armed, Laura with a bathrobe, and slipped the temporary sling over her head and under her arm.

Norma was robed and Laura stated, "That was a grand soak, Miss Moon.

She replied, "My pleasure, Miss McAgneti."

They both heard a tap at their cabin door and heard, "Albert here."

Norma smiled and gestured, and they both hollered in unison, "Mate's here."

Norma opened the door, and smiles held their reflections. He told them they had anchored, would get some rest, and would seek medical assistance at a more appropriate hour. Norma said, "Good night, you two."

Albert said, as a matter of fact, "It's nearly two a.m. in the morning."

Laura said brightly, "Thanks for the morning soak" and Norma replied, "Morning or night, its oaks for the soaks."

The *Magnetic* was tethered and bobbed sleepily in the darkness, romancing with its lapping, caressing waves.

Chapter 14

Devlen Dawson had learned the hard way, never trust anyone completely, except yourself. Another lesson, well taught, conformed to neat transactions. Never leave a trail, regardless of the minuscule of misdirected information. These two grueling roles have preserved his abatement and solidified his integrity. His character was perceived as clean as a freshly powdered, and diapered baby's butt. He had put out the word on Laura, as to her whereabouts, being unemployed by the government. Harbors and Immigration presented him with access on all arrivals and departures of all nationalities, and their cargos.

Soon after he began working for the government, he was approached within his first year with a lucrative proposition. Simply put, just look the other way on some mixed cargoes. Six months later he approached with his original contract, and stated he wanted a small percentage of all the action. A deal was struck, but he would only occasionally be asked to alter shipping invoices. His judgment to remain at that level, was a wise decision. Unknown to Dawson, the nucleus, he was working for had decided that any further increase in doing business would not be tolerated. New sources would be found, or management supplanted, and the cycle would continue. Dawson had no identification of these men, and wanted to keep it anonymous. He however, being a cautious and greedy man, wanted to supplement his covetousness.

Forty years ago two freighters, one a sister ship, slid down after the christening, one was of the latest in technology for that area--an era of building Portuguese cargo ships. Abused, and horrendously overworked, each change of owner wanted to quickly recapture the purchase price. Fundamental repairs were ignored to make the mighty buck. Years of debilitating pace had taken its toll. Six years ago they both became orphans, and were up for bid, or headed for scrap yards. Dawson got wind of these great old hulks and was advised by a little known man of the advantages of having two sister ships roaming the seas. He made no inspection of the freighters. He was advised, by the same man, to bid just above the scrap price. His bid was the winning one of the three submitted. Dawson became their new owner. He scrambled two registries, one Greek and the other Portuguese, and set up a dummy company. His new friend told him the minimum it would take to make both of them seaworthy, and knew of one captain who would work well in his operation. One of his old rusting freighters was named *Scarab*, the sister hulk was named *Baracs*. The crews were made up Asians, Filipinos, and South Africans. Wages were deplorable, as more jobs were lost, thousands scrambled for the few positions that became open, at hapless compensation. This atmosphere, and inter-capital-ized on monopolization had in essence, its crews comprised of the scum of the seafaring lot. These old freighters were journeyers for five years, wandering from port to port, not so much as raising an eye brow in all the seas, except the two poles. This enterprise had elevated Dawson into unprecedented wealth, far above his estimation at the time of purchase.

The crowded harbors ride a host of these old derelicts, flying little known flags of registry, of past owners. The cost, just to break even, in hope of higher freight costs, some day, kept hundreds of these from being scrapped. With the scrap price of steel trending downward, these coffins of the sea lingered, and further debilitated their seaworthiness. Dawson had made necessary repairs on his old vessels, and instantaneously updated when practicality was applied. The nucleus that Dawson was collaborating with, to his knowledge, had no privy to Dawson's two phantom ships. In his pretentious view of the harbor from his home, he had at the expense of the government, permission to own and operate an elaborate radio transmitter from his home, thus vacillating its conveyance to contact his ships, anywhere in the world. Less than three hours after his unsuccessful attempt to contact Laura, word was sent, that she had cabbed, and met a man, and had disembarked from the slips of White Tern. He asked, and after further snooping, had uncovered the craft's name, heading, and that it sported a peculiar antenna at midships. Dawson's third rule was generous retribution for services rendered, and he always paid his informers. Strictly on a hunch he contacted his closest ship, as to its bearings, and discovered it was latitude 154°, longitude of 58°, approximately 415 nautical miles from Anchorage, sailing north by northwest at ten knots, and entering the Strait of Shelikof. Dawson coded all messages to his captains. He would alert the radio operator, then have the captain summoned when the message was sent. In the message was a word--one of three colors, which indicated the transmission at random. Simplicity was contagious to his mounting success. The three colors were the implication as to importance, and monies paid, if information was useful--red, white and blue. The latter was prognosis of the highest award minimum of five figures. The coded message blue, in essence, ordered an accident at sea, giving the heading, type of craft, name, and a strange looking antenna at midships.

The so-called friend who had tipped off Dawson about the two old freighters, had another reason than just money, for relating his information. He also wanted to be captain of one of the old ships, and he was successful on both counts. His name was Dork Ninkeki, of Malayan descent, and he was proud of the fact that he was still able to speak his native tongue, Malayo. He grew up on the south tip of Borneo in the city of Banjarmasin. He was captain of the *Baracs*. High frequency (HFUI) ultra installation was installed and low frequency (L.F.) was state of the art, and had proved to be a wise investment. (L.F.) for radar under water, had saved his ass more than once, into harbors that normally would not complement the draft of this antiquated dinosaur of the sea. Its bilge pumps, seemed to be incessantly purging back to the sea what the ocean had penetrated. It mattered not, how decrepit it appeared, Dork had the command, and was salient to his crew. He laid out his rules, and if they disobeyed, they knew the consequences and that they would receive with unrestrained pleasure. He was not a cruel man. When he was born, crying and flaying his arms and legs, his brown skin, was drawn taught over his skull. A scowl was thus his birthmark. The tense skin then never became elastic. His squint, black eyes, gave no quarter, with his persistent frown, his judgmental attribute was, obey my orders, and I'll protect your ass. He respected and garnered mariners, for he knew that the sea is an ever changing highway. The violence, or lack of change in his premier scowling, turned out to be a blessing. His adversaries were caught off guard by his blasphemic facial indifference. He was but 5'5", and lean muscled, in his middle 50's. Captain Dork had few scruples--another reason, beside his devout desire to own a captain's coat, with papers in the lapel pocket, to match his magnificence. Through diligence and crafty perception, a loin cloth was finally traded for a captain's coat at age 41. He labored on scores of scaling rust buckets at every station aboard with the exception of two--a cook, or captainship. Six long years with two of his decks, anchored in sea bed. He managed to be in command of one of the wonders of the liquid highways. The old plate with the larger rivets, had oxidized nearly one sixth of its original plate thickness. The keel was laid in 1936. He was the only one fool enough to accept the task of sailing over 6,000 miles, and after accepting his first command, he later found, his complement was less than one-half a crew. The further dilemma, after disembarkation was the realization that only one spoke his language, English or Malayo and that man died, the second day at sea. The second burial was the fourth day. He found himself in the shadowy intestines of a living hell, with the sea the Grim Reaper, antagonistically waiting to capture him, and have his ship to use as a headstone. He delivered his ship and cargo and received a modest token. He learned two years later he was the last to be in command as they had it towed to the scrap yard four months after he steamed into Singapore harbor. Scuttlebutt of his deeds, as captain on his maiden voyage, challenging the sea with that derelict and only four crewmen was a masterful feat. He quickly accepted his second command and a substantial increase in pay. He had a seaworthy ship and full crew. His only regret was that time slipped by so fast, and he was constantly being moved about to different ships. Just when he was beginning to learn the secrets and the pulse of his command, it was sold or the owner wanted him on another vessel. He had worked with Dawson, as well as with other owners, who respected the way he did business, nice and clean. His second dream, was to own and captain, his own ship. He always received an additional bonus, when his orders were completed as instructed, and when Dawson was admixed with the cargo. He took it on his own, for these reasons, to make that call about the two ships up for sale. He made the stipulation that he was the captain of one, and had a commitment of four years as such.

Out from Anchorage, Alaska is a swath of sea water known as the Cook Inlet. The basin is the commercial heart of Alaska, towered by North America's highest peak, Mount McKinley which rises 20,320 feet above sea level. The Alaskan and Aleutian ranges protect it from the ravages of the Siberian and the high arctic storms. The Cook Inlet itself carries marine warmer waters to the heart of the basin, where the warmer waters are captured, and directed to the shores. Cook Inlet tides have a maximum range of 34 feet at Anchorage. One exception was when tides reached 39 feet. The scores of rivers that drain off the high mountains deposit minerals and silt. The Knik Arm is most vulnerable, having to be dredged at various times.

There are a number of oil rigs rising off the top of the sea at the southern end of the inlet. Their platforms look like defiant steel monsters out of the waters. Above this, dominating all below, is the towering Augustine volcanic mountain. Its interior depths is like a momentous time capsule. The thrashed molten rock, hibernating, is just waiting to vent the earth's crust, one more time. It previously inverted its womb of hot lava in 1976. Normal temperatures range at winter 10° to 25° and summer temperatures 60° to 70°. Captain Dork knew of the two mysterious waves of the Cook Inlet. The first is called Tsunami, which is taken from the Japanese word (Tsu) meaning harbor and (Nami) meaning great wave. Tsunamis, can speed across the Pacific Ocean up to 600 mph. The waves are just a few feet high, but can be up to 100 miles from crest to crest. They cannot be felt if on a ship, or seen by air. But once they approach the shores, the shallow waves become taller and taller, concluding in a two foot wave. Traveling at 500 mph, it would become a voluminous teeming wave of 100 feet traveling at 30 mph as it neared the shores. These terrorizing waves have the capability of repeating every 15 to 30 minutes.

The second mystery wave is called the Seiche, it is a long rhythmic wave in a close or partly closed body of water, like Cook Inlet. This specifically is caused by earthquakes or atmospheric pressure. The water only moves up and down and can be active for a few minutes or up to several hours. The highest wave of this type was measured in Alaska at 1740 feet. This took place on July 9, 1958, and was due to a devastating earthquake. Few men of the sea, and indeed, fewer captains have lived to tell this true tale. But the knowledge of compounding, unleashed, uncontrollable power, is mind boggling. Each path of the sea has its own fingerprint, its containment, and all the variances from winds, rain and storms of every magnitude, dictating its torment or solace. Captain Dork never tired of reading the true stories of the sea. His romance with the ocean became more profound with each secret, only to find the facts compromising.

The exception to all his cargo ports of call was Singapore, located on the extreme south tip of the Malay Peninsula and separated by the Straits of Jo Hore. Sir Stamford Raffles founded a settlement in 1819 which is now the Port of Singapore. The Japanese overran this city in World War II in just 15 days. Reinforcements were promised, but never arrived.

Devlen Dawson was smuggling everything-- antiques, silks, gold and precious stones. He was smart enough to stay away from the drug market, and that was one of the reasons he was successful for so long. He made it a point to know his commanders and to establish a significant power of understanding with both captains, a complement of monomania, and compliance with his rigorous and uncompromising details. This included his persistence of drug free cargos, as well as crewmen. He made a modest income. As assurance became bonds, the expansion of his illegal cargo spectacularly expanded. Smuggling of choice, was the preference, and became the normality. Dawson could literally and fastidiously select what cargo to caress, through the medium of his two loyal and stringent captains. The captains had to discipline every soul on board. They would inflict excruciating pain and cause intense suffering, using their discretion, to have a drug free ship. Captain Dork had heard of a sadistic ritual that was sometimes performed many years ago on the Sulu Islands.

He was just a sea hand then when his slow moving freighter anchored in the Sulu Sea. Sulu was the largest of an archipelago of 272 islands, northeast from Borneo to the Philippines. They were riding high to take on a cargo of rice, indigo and hemp. This loading would require four days to accomplish. Dork was allowed shore time after his twelve hour shift, and he looked forward to it with anticipation. Islam, the primary religion of a semi-savage civilization, dropped the practice of slavery, piracy, and polygamy.

One of the more fascinating industries was pearl fishing. The oysters with the thick-shelled mother of pearl, were the wombs which produced

the alerment, a substance that forms in the inner layers of shells of nacreous mollusks, pearl oysters, and abalones. A pearl is formed when some irritating particle causes the oyster to cover it with a protective excretion, which hardens in time.
The Islamic religion, accepts an obligation to wage holy wars for worldwide conquest, and issues to conquered countries, the ultimatum "Embrace Islam, pay tribute, or die. Friday is their esteemed day. This ritual is not sacrilegious to their religion as it is performed on non-believers, but it did get morally efficient truths from tongues that wagged, never tell, never tell. The native court appointed police of the Moro tribe exclusively. They realistically performed, as they pleased, so that the result was incognizant, as to truth, only what they perceived, conforming to their judgment.

Dork obtained two small pouch bags in Hong Kong which were connected by strong strands of silk cord. This long necklace of silk rope hung around his brown throat, and had two pouches attached, which he kept concealed. Through many years and many countries, he had traded coins from scores of worldly and some not so worldly ports. He selected his coins to trade at night, using a flashlight, under the bleached covers of his berth. He wanted to see this practice, known as straight tongue, and selected a large silver high relief, Balboa, 1947, from the Republic of Panama. He polished it up with his sheet and put it in a hidden flap inside his waistline, for presentation, after his twelve hour shift. He knew it was chancy, but this was the spice of life, that maintained his mind and body.

He finished at six bells, cleaned and changed clothes, to seek the secret of truth. Darkness was spinning its covering as he exited from the old freighter. His lean brown body was clad in a short sleeved sailcloth shirt, and tan shorts. His height at 5'5" fit in well, but his skin was a tinge lighter, and that profound scowl on his tightly drawn face, separated him from any native they had ever seen. He, in fact, was being impressive and unimpressive, as not two people agreed to his perennial frown. His passion to some day command one of these great steel, heavy and clumsy ships, had him speaking in six different tongues, although not accurate in some of the associated words. The language was Spanish, that of romance with a regional dialect of Dutch/German, as the city of Sulu was built by the Spaniards.

It wasn't difficult to find police headquarters as this was Friday, their payday. He sauntered leisurely, weeding out the lowest escutcheon, then tried to capture the eyes which would allow him to watch interrogation of the tongue that never told, never told. He approached in a sideways manner so as not to evoke his presence as being hostile. His grim scowl had taught him to act and speak apologetically and be the humblest in order to gain their favor. He blundered through the language and indicated discretionary palaver, retrieved the silver coin, and made certain the others did not see the palm transaction. He waited with his scowl until the tall black and pure Dyak, removed his worn mahogany cadre stick from under his arm, and tapped him twice on his left shoulder with a grin. The lieutenant then indicated with his proud stick to sit. He did so, cross-legged, lowered his head and waited. . . .

Dawson captains had proved their worthiness by literally beating and shackling, if necessary, any crew member that was on, or delivered drugs. The bounty would then be thrown, as a gift for Oceanus, God of the Sea. Captains were given a percentage of the profit. It all worked like a precision Swiss watch. The captains were not given privileged information for two very good reasons. They, however, were informed as to the location in their bays. They protected the designated cargo of unknown contraband, like a relentless dog with a bone.

Captain Dork Ninkiki radioed a coded message that he had rammed the cruiser in question, striking it twice, but the craft apparently was seaworthy, as it stayed afloat, its running lights still faintly visible, adjacent to and paralleling the cropped swells. After unscrambling the message, Dawson swore, "God damn it." He then withdrew the charts for that area and decided to set up surveillance at Kodiak, Palmer and Port Graham. Laura was the one threat that could possibly unravel his elite venture. Dawson had people in nearly every village or town who welcomed his money. His network was like a mundane spider's web, woven like a net, from the south, Prince Rupert to Port Heiden and on to Fairbanks. He was never an ardent book reader, but he did manage to skirt through a few.

The crew aboard the *Magnetic* was excited and talkative. Tom had just prepared and placed breakfast before them in the lounge. It was one of his usual, substantial, and lip-smacking pleasures. The anchors had held fast, and the welcomed sleep, was like the sun emerging from behind a total eclipse. Their faces all attested to the fact that their lives had been spared, and not added to the endless food chain, to the aspiring fish, striving to stay alive in the treacherous fathoms in the sea.

Norma was half through her tasty vittles and was working on her second cup of hot, invigorating coffee. She asked prudently, "Is it all right to light up, . . .with everything that has happened, and this lovely breakfast, and. . . my soak? I want to top it off with a cigarette!"

Tom reached back and over and cracked the louvered window. They waited for Albert's response.

He crunched a piece of toast, with a wedge of crisp bacon on it, and said, "By all means,

Norma, I imagine it'll be like the frosting on the cake."

Under the table, and out of Albert's sight, Tom passed the ribboned, crushed package, to Norma's lap. She felt it, and inhaled a deep ball of smoke into her lungs, reclaiming most of it, and sent it slowly back, and spewed it from her mouth, with a passionate affection. She felt so attentive, she didn't hesitate, and startled Laura, when she briskly slipped the box to her lap. Tom, being the spectator, was watching the expressions, and captured the surprise in Laura's body language upon receiving the container.

Tom spoke up, "Albert. . . ." He looked up awaiting, and Tom continued, "The Captain has to have his due, congratulations from your celestial crew."

Laura retrieved the box from her lap, and held it out, wearing a strikingly beautiful smile. Albert was quite surprised, and readily slipped off the ribbon, opened the box with a gracious, "Good Lord!" His eyes sparkled with a twinkle of moisture, and he cordially plucked it out. It immediately made a statement as he put it on, and looked into their approving faces.

Tom smiled and said, "Cock it a little, Cap!"

Laura did just that, sitting next to him, and Albert grinned, with a commissioned cockiness.

Laura said, beaming, "You look so handsome and authoritative." They clapped. Albert stood up and stated, "Let me just say this to each of you. You are my family and with deepest respect, I thank you all."

They smiled and made small talk while the blue cloud of smoke snaked its way, found its path, and exited out of the lounge. Albert forked a generous bite of hash browns, and stabbed his bright yellow egg, transferred it to his busy mouth and said, "Mmmm. As soon as I finish we'll weigh anchor and tie up at Port Graham. I'm going to get in touch with someone and we'll have you gals checked out."

Laura said, "You too, Albert!"

Tom added, "She's right, you know!"

Albert grinned, using his half piece of toast as a soaking, swabbing mop, picking up very little but flavor, then popped it into his mouth. His jaws were rolling sideways with contentment, his eyes flashing with pride and anticipation. Albert said, "We're out of range of the Anchorage radio station K-O-O-L, or we would have had our concertos, so-so breakfast."

Tom grinned in appreciation. Norma drew a slow last drag, and held it, then expelled it quickly and graciously tamped her spent butt on her plate. Tom watched her, as did Laura. Norma sensed a qualm, a misgiving, and looked about to search their eyes, then looked down at her plate. She had an expression of regret and said, "Tom, it won't happen again. . . ." then quickly thought and added, "Okay."

Tom had to smile and repeated, "Okay."

Laura looked at Albert as he napkined and excused himself. He went directly to his cabin and mirrored himself, to check the compliment that majestically sat on his head. He smiled cockily, and finger saluted himself.

Tom cocked his head to one side, still chewing and said, "Laura, neither your shoulder, nor your arm is broken."

Norma smiled lightly and reveled, "Tell us, Oh Great One."

Tom replied, "No need to act vindictively, Norma, I've accepted your carelessness, with your beguiled butting. . . I know that Laura has swelling in the tendons, inflamed yes, but no broken bones or dislocations. . . .She landed on me, ... The Cap said I was an old softy." That brought giggles, then laughter. Tom said, with dignity, "My galley is waiting."

Norma rose as Tom did, and she captured his chin with a soft touch, and said, "My gully, I want a right smart smacker from you right now, . . .Okay?" They embraced. Tom gathered a load of dishes, and Norma gave him a little pat on the butt as he left.

Laura marvelled as to how quick a flippant and nervous remark could be quiescent. She said to Laura, "God, I'm lucky to have that man!"

Laura said, "Yes, you are, but why do you strive to walk that mystic line of acceptance, or dominance, and just maybe damnation?"

They looked into each other's eyes and Norma responded, "You can bet your tail..., as Tom so graciously put it, a little spice in our emotions!"

Albert started the engines and told the women they had about 20 minutes, before they would be dockside. They cleaned off the lounge table, and played pack horse, backing Tom, retreating to the galley in reverse.

Albert was dismayed at the thought of his own wisdom on docking at Fort Graham, instead of Homer, or Kodiak. Graham's population was less than 300, while Homer was over 4,000. One reason, which may or may not be valid, was maintaining secrecy. On the negative side, however, doctors and facilities were limited. He chose this path, well knowing that the first world's regional satellite communication network had been in place for many years, and that more than 26 small Alaskan towns and villages had been connected with this system of medical assistance. Many a baby was born with satellite communication, directed by a doctor, visually and verbally, taking and explaining each procedure, a talk through medical assistance.

Stan Marcus, age 54, with a dab of Abnaki (eater of raw flesh) Indian, received the radio message, craft unknown, but seeking medical assistance, will dock at 10:00 a.m. Two female patients, both mobile, and not critical. Stan was also one of three policemen and Chief of the Volunteer Fire Department. Dr. Ward Jenklin, a young 69, a thin weathered man, three times retired practitioner, and self-proclaimed authorized healer, was alerted. There was a one-room facility for first aid and minor treatment center, an outdated x-ray machine, and for all practical purposes, his medical bag. This generous, small room, connected with City Hall and the Fire Department. It was a masterful concoction of simplicity and very functional.

Due east of English Bay, lies Port Graham, located at the throat of the long inlet, which dissects its waters through approximately 13.2 miles to Windy Bay. Distances are nearly equal north to a small town called Seldovia. Many small villages are Russian named, as being the second relevance, of the Indians or Eskimos, which is one of the same. Coal was discovered in 1851 near Port Graham. The Russians tried fervently to develop this find, but it was an utter disaster and was abandoned in 1865.

Stan was born on Afognak Island, a mere 80 miles due south of Port Graham. He was in no hurry to make this his home. A worldly man in basics, his fondness for terra firm outweighed the lustfulness of the sea. In this grand country you have to have a marriage with the sea, at best, a romance, or you generally will find yourself heading south in search of a more moderate, less challenging environment. Stan walked to the docks. The diminishing harbor facilities were simplified, and modified, within a fraction of this size, and intensity, as of the 40's and 50's. It was a long safe cove, and the choice of many to retreat when the Cook Inlet or the Gulf of Alaska had violent storms, thus providing a safe harbor. After World War II modern highways connected Kenai, Homer, Seward and Anchorage. The precipitous Kenai Mountains on the south side of the Kenai Peninsula, cascading to the ocean floor, was a factor in preventing a major highway to Port Graham.

Dork felt a tapping on his shoulder and saw the stick retrieved under the black arm of the tall lieutenant. He ordered him to follow. Dork looked about, no one seemed concerned, and he nervously followed the towering officer. They went through a door, down a poorly lit and smelly hall. The tall lieutenant, whisked out his stick and tapped on the door. It swung open, and they both entered a stinking small room. Two bare bright bulbs with reflectors engulfed most of the room in a bright light. A man lay prone on a rough hewn high table called a strappado. His hands were pulled to each side and shackled, as were his ankles. He was stark naked, moving slightly in fear and anticipation. His eyes were wide in horror. The man, in his 30's, started to speak and was struck across his lips with the officer's cadre stick. His lips began to swell, and he tried to thrash his head from side to side, until he felt the point of the stick on his forehead, and saw the look in the lieutenant's eyes. The officer dismissed the guard, waited until he left, then locked the door. The tall black officer turned and spoke to Dork. The essence of his message, if words came from his mouth about witnessing this interrogation, were that he would be interned and face the fate of the tongue that never tells. The white eyes of the officer widened. He asked the shackled man a question, as he plucked a pair of rubber gloves from a box, and pulled them over his long black fingers. He snapped them loudly and each time the terrified man jumped. The rubber covered hand punched the stomach, just above his scrotum. The man cried out, thinking he was about to be cut. The lieutenant laughed, turned and spoke to Dork. "See, a non-believer. He thought I was going to mark him. I will ask one last time. He did ask the trembling nude man, "You did steal the pig!" The restless, quivering man shook his head fearlessly sideways, afraid to speak. He laid his stick on the top of the cabinet, that contained the implements adjacent to the confined prone man. He opened the top drawer, withdrew a thick and soiled, dirty towel that contained dried spittle and blood of scores of victims who had preceded him, in the tongues that wagged, never tell, never tell. He shook the stained towel, and cruelly laced it tightly around the man's mouth. He was in a state of hysteria. Not knowing, what was going to happen, his mind envisioned grotesque nightmares. He was unable to comprehend the purpose of the rubber gloves. The officer plucked a jar, and a straw from the same top drawer. He unscrewed the lid, used his index finger to wipe a smear of the pale lard, and covered the outside of one end of the straw about three inches. Dork was beginning to be excited. He moved to better position himself for what was about to happen. He laid the straw on the cabinet, with the greasy end projecting over the counter top and used his cadre stick to weight it down. He pulled open the drawer below it, and snatched out a small black, four ounce bottle, shook it, then set it down on the counter. He retrieved a small hand carved wood funnel that projected a long thin tube, so tiny it would slip into the opening of a straw. He grasped the straw, turned, leaned over his prisoner, and amid muffled screams, he forcefully jammed the greased end of the straw as far as possible into the enslaved man's urethra. Tears, thrashing, sweat, and moaning were the reactions of the pained, shocked and frightened soul.

Dork grabbed his groin, reflecting the pain that he was witnessing. His black eyes bugged out as he watched the performance of this macabre ritual. He felt, indifferent to the creature who lay so undignified, and the excitement had him mesmerized as to what was in that black bottle. He didn't realize that he had both hands clenched tightly in fists, and his nails dug through his sweaty palms, and blood was trickling down his wrists. The black officer was also sweating, not from fear, but with apprehension. He retrieved a small stained spoon from the drawer, laid it next to the bottle and funnel. He picked up his cadre stick smartly and the quickness of his movement made a swishing sound. His training had taught him artistry, for his cadre stick barely touched the nose of the shaking man. He looked into his horrified eyes, and tapped his stick lightly, indicating that he should settle down, or else. The trembling was reduced to wavering, and slight involuntary jerks. He laid his authoritative stick on the counter and looked into Dork's squinty black eyes, then removed the long cork. He tilted the bottle, filled about a third of the tarnished spoon with a white crystalline, thick liquid. The tall, thin, black lieutenant in salt stained kahkies grinned, exposing his perfect white, white teeth. He asked Dork to steady the straw in an upright stance. The man winced and groaned as Dork did what he was told. The officer placed the crude funnel in the straw, with Dork holding it steady. The officer then noticed Dork's blood on the straw which had been caused from clenching his palms. He then added the white liquid, filling one-third of the spoon. The prone prisoner saw the ceremony was not over and bit his tongue, his blood, seeping out of his mouth. Dork waited . .

.then chanced a glance at the grinning man in charge. As he poured the liquid in the funnel and grasped another straw, it took a minute for the solution to drain down to the end of the straw and move ever so slowly in the quivering man's uretha. The lieutenant smiled as he flayed the second straw in the air, further distressing the restrained and devastated man. Dork was holding the straw in a vertical position as the grinning officer leaned over, and using his gloved finger, coupled the second straw with the first, flashed his white teeth at Dork and inhaled, then surrounded the straw with his thick sweaty lips, and blew viciously. Dork saw the holocaust, screaming pain in the bound stretched out man's eyes, attempting to cry out, through the bloody gag. The pressure drove the liquid deep into the man's bladder. The lieutenant flashed his whites again at Dork and said, "Done," and he recklessly pulled the straw out of the man's penis, he lurched. He peeled off his gloves and with the two straws, dropped them in a small woven basket that was on the floor. He reached over and re-trieved his stick, and cruelly batted the side of the prisoner's nose, maintaining further control. He removed the man's gag, then asked, "Did you steal the pig?" The helpless man cried out despairingly, "Yes - yes - I did - yes. . . " The officer stated, "See... tongues that wag never tell, never tell, are non-believers, and justice will be served."

Dork was petrified by what he had just witnessed, man torturing man, by devious means, but nothing as compelling as just plain savagery and butchery. Dork looked up into the officer's accommodative sweaty face and asked, "May I ask about the white liquid in the black bottle." He flashed his glistening teeth, and in essence said no. . . Dork asked, "What does the white fluid do?" and pointed to the ravished naked man. The officer obliged and stated, "He may never pleasure again." The man on the table screamed upon hearing his impudence, and the terror in his throat was still expelling horrendous sounds. The cadre stick stung his neck maliciously. He poised for another blow and the wailing stopped.

Dork asked if he could buy 20 drams of the white thick juice. The black man touched Dork's shoulder lightly with the stick, and asked, "What have you to trade?" His eyes stared through Dork. Dork reached into his hidden cache on the inside of his broad waisted shorts, and retrieved another large coin. Before presenting it to him, Dork made certain, the side of the coin that showed a raging alligator with its powerful jaws open was face up. The coin was a 1956 four shilling, that was distributed as the coin of realm, in the smallest country in Africa, called Gambia. A former British Colony, located at the mouth of the Gambia River, it was surrounded by Senegal on the Atlantic coast. Peanuts and rice are the principal crops. The lieutenant's eyes owled as he respected such a creature. He turned it over, then back to the reptile side, and grinned at Dork. For a brief moment, the scowl faced little man had a brisk flair of hair rising on the back of his neck, impending danger, by this tall smirking black.

Dork indicated graciously he had no more coins, or loot on his person. His frown must have been convincing. The black bottle was retrieved, then held out for him to take. He did, with a bow, and thrust it into his pocket. He moved quickly out of the den of hell and gasped for a healing breath of air as he left the front of the station. The warm, soothing breeze filled his lungs. He scowled all the way back to the old freighter, and hid his precious black bottle.

Ten months later their freighter was anchored in the Arabian Sea at the port of Bombay, India. This was one of the finest natural harbors and the choice of ports of all India. Bombay City is located on an island, about eleven miles long and three and one-half miles wide. It is connected with the mainland and Salsette Island by causeways and cribs. The Magnificent Harbor is fourteen miles long and five miles wide. The only city that exceeds in enterprise is Calcutta. Dork, with permitted leave, and his black bottle tucked in his hidden waist pocket, moved through the busy narrow streets in search of an analytical chemist.

The language in India is broken down from a hundred to now fewer than ten dialects. Bengali, Telugu, Marathi and Gujaraty are the leading dialects spoken. He successfully found such an establishment, and using sign language, negotiated a price to break down the white liquid in the little black, long necked bottle. He had to pay for a small vial to leave them a sample and was told to come back in two days. His present home on the water had a five day portage. Dork was pleased with the commencement of solving the mysterious white liquid.

He worked hard that day and sweat was pouring off his face, splitting evenly on his two frown lines, then around his defined lips. His ship was taking on cotton, linseed, mustard, and tobacco. He would go ashore again at twilight and receive the results from the alchemist. He was excited, as he talked to the turbanned chemist. The thin dark-skinned Muslim insisted that he pay the remainder of the cost before receiving the breakdown of the liquid. He had changed some of his money into rupees to alleviate the haggling that always transpires when doing business in India. The little scrap of paper had the following components: $C_{17} H_{23} No._3$. Dork turned the little paper over. That was

the formula, and he became a little excited. This brought another man to settle the ruckus.

They found a familiar language and Dork explained, he also wanted the layman's ingredients. He passed the little paper to him and the head chemist told him -- shining blackberries, leaves and roots of a poisonous plant of the nightshade family. The secretion rendered is called atropine, a white crystalline alkaloid, and the plant is called belladonna.

Dork tried to free his scowl with delight. He now knew the secret and could replenish his supply. Dork asked one more question, placing two rupees in the pharmacist's palm. He asked, "What does it do?" The proprietor replied, "relaxes the muscles, but taken internally could be deadly. He crossed his fingers making an X, indicating poison, and shook his head. Dork was relentless in finding the elixir of truth. Another notch in his endless search of means to control and quantify man. He swore happily, although not able to discern his permanent scowl. Later as Captain, one of his crew, was unable to be restrained, and had been caught for the second time with a small catch of cocaine. Captain Dork then performed the ritual, the tongue that wagged, never tell, never tell. . .

Chapter 15

The U.S. Defense Department, in the 1940's, had built a major portion of the defense of Alaska. It is still being used. A huge communication network was designed to keep in contact with the furthest Alaskan outposts, whatever weather conditions--the code name was secret, but it is still called (White Alice). Forty-seven stations were equipped to receive and transmit signals, using a method of radio and relay. It was never used at that time on a large scale. Good telephone, and telegraphic communications are also maintained, no matter how turbulent the weather conditions become. Huge soup shaped, 60' high antennas, each weighing approximately 100 tons, are used to beam signals into the troposphere. This troposphere is a five mile layer that extends upward from the earth. Signals are scattered in the area. Just a tiny fraction of energy is used, and is sent out. It then arrives back at the receiving antennas, and amplified, until it's intelligible. Identical antennas spaced up to approximately 200 miles, receive and transmit if necessary. This area of the world has one of the finest communication systems, and is a comfort for all in Alaska.

Stan Marcus was on the dock and witnessed the moorage of the sleek Bayliner that bore the scars of two impact areas, both on the starboard side. But what really captured Stan's eye was the impressive dish antenna. He immediately assumed these were government people, they had to be, and must be damn important. They finished tying down, with Stan's assistance, and then introduced themselves. Albert looked about at the meager dock facilities, and stated, "I would like to diesel and water up. Where are the dockside inlets located?" Stan laughed and said, "Sorry, it's been quite a while since we had that luxury. You'll have to five gallon it from Mattie's. I would have him mix up a batch of rolsen, glyseol, anhydride, and some fibers, and have those fractured impacts reinforced."

Tom stated, "You're generously informed on synthetic derivatives."

Stan smiled and said, "Yup, I'd better be, I'm the fire chief, and I'm concerned about all combustible materials."

Norma said, "Well, we have quite a gathering, the captain, a fire chief and a swami." They all laughed.

Stan said, "I'm also one of three policeman. You might want to check under your hull for damage. We haven't a dry dock for that large a vessel. They were moving off the dock toward a large weathered structure. Stan said, "I just assumed by your call that you haven't had the opportunity to examine your lower structures."

Albert replied, "You're right as rain, Stan, and Laura was our best bet with her wet suit, but her arm and shoulder has received an extensive blow."

Stan said, "You'll be wanting to give me the particulars on just what happened and I'll relay it to the Coast Guard." They looked about nervously, except for Stan.

Albert immediately stated, "It's really fruitless to make a report, we were moving less than fifteen knots, with all running lights on, were in the middle of dinner, when the collision occurred. Two of us were knocked out, and the other two were in uncompromising positions so viewing was impossible. It was a hit and run, a cowardly pirate's act."

Stan shook his head in sympathy and said, "These waters are not as friendly as more intrusions are filing in each year. I'll send along your oral testimony."

Tom spoke up, "Stan, would you be breaking a confidence if you didn't notify the Coast Guard? The reason I'm asking, . . .for all of us .... we're damn mad this happened. It nearly cost us our lives. We are planning to scrape some paint smears off the bow, and have them analyzed and search for that renegade ourselves."

Albert quickly picked up the lead and stated, "Then we'll notify the property authorities. They might let their guard down, if they believe no one is looking for them."

Stan glanced at the faces of the two women, and saw no fidelity in their eyes. They entered the side door of this simple complex into Port Graham's one room hospital.

Dr. Ward Jenkin introduced himself, smiled, and selected Norma as his first patient. Dr. Ward asked the others to leave, with a wink, and stated, "I'd like to be alone with this gorgeous specie of a female." They smiled, and Tom heard Norma say as he left, "He's right, you know!" Laura heard it too and snickered. Albert grinned. The three of them, led by Stan, entered another room a bit larger that comprised the Police Department.

Laura asked Stan, "Does anyone own a wet suit, and have a knowledge of small crafts?"

Stan replied, "Yeah, you're in luck again, Wade Freedman, the owner of Mattie's all goods store, does quite a bit of diving and gets a kick out of retrieving mementos of World War II, and of course, any wrecks. The Navy had a bad habit of indiscriminately dumping overstocked inventory into Davey Jones Locker after the war was over."

Albert stated, "I really appreciate the complacency you have shown all of us."

Stan questioningly asked, "You don't have to answer my question, but what's with all that electromagnetic radio antenna fixed on your vessel. I presumed you were government connected?"

Albert appeared calm and stated, "No, no, I have some revelation of using this device for a fish finding instrument. It's a hobby of mine of sorts.

Stan frowned, further screwed up his face and stated, "That's a high frequency shaped antenna!"

Albert countered, "Yes, you're correct, but I'm attempting to integrate the high and low frequencies to increase the fold and scope of the searching areas."

Tom added, "We think it's plausible, but may not be adaptable...." He continued, "With your towering radio antennas, circumflexing all the Alaskan waters, we selected your home for our trials."

Stan thought that one over and concluded they could be what they claimed. They certainly had the answers. Stan then decided his action and stated, "If you prefer, I'm really not obligated to notify the Coast Guard, seeing that there were no eye witnesses."

Albert commented with a smile, "We thank you for your discretion."

Laura asked Stan, "Is there a cafe, or some place we could coffee up?"

Stan replied, I usually have a pot handy, but Mattie's has a little counter, that serves probably the finest cup of coffee you've ever tasted. That's the reason I don't have a pot on. No one can match his drink. But don't ask Wade what's in it. He has successfully retained his secret from the world."

Laura stated, "Tom makes a smart, savory pot of coffee too!"

Albert said, "Amen." Tom grinned.

Norma entered the room and stated, with a smile, "A stitch in times saves four. Imagine, Dr. Ward apologized for not having a non-colored thread complimenting my skin tone. He's so sweet. You're to go right in, Laura. He has such lovely hands."

Tom moved over to Norma and they embraced. He looked at her four stitches during the hug that followed. He said in admiration, "He is good, he used a knot stitch, very resilient."

Albert said, "Stan has informed of a counter where the finest coffee is prepared. I believe it would be pertinent that we all go together."

Tom said, "That's a Jolly Roger with me."

Norma said, "Aye, Cap," then added, "Stan, do you mind if I light up?" These pleasantries were undeniably sincere.

Stan replied, "Please do!" Just another engagement of words that brings new acquaintances closer to becoming friends. Norma offered Stan a cigarette, but he declined, with a "No, thanks," and added, "Being fire chief and all. . . ."

Albert said, "This must have been an elaborate port during the war."

Stan relied, "Yes, you could say that our inlet with its profoundness, made it a safe accompanying port, especially in bad weather. But our topographical location prevented us from becoming a larger entity. Since then the Navy has consolidated and moved to more temperate ports."

Norma stated, "I believe history may reveal, at least to you folks in Graham, that you may have been a blessing in disguise. If the rest of your men folk are as lambent as you and Dr. Ward."

The men all exchanged smiles, reflecting her eloquence. She was enjoying her smoking stick, and these three strapping men.

Stan asked, "If you like, while waiting, I can ring up Ted and see if he is available to diesel you up. . . .You'll have to set the terms yourself, you understand?" Albert nodded and said, "Yes, Stan, please call and I'll talk to him."

Stan dialed and told Ted the brief purpose of his call and to hold on, as he passed the headpiece to Albert. Albert asked if he would fuel, water and examine the visible structural midships and bow damage. Both men seemed pleased with each other's communication. Albert hung up and stated, "Stan, Ted seems to be a jack of all trades. We just consummated a mutual deal."

Stan said in earnest, "Yes, Ted is all that--skillful and he minds his own . . . He's also got his eye on a certain rifle at Mattie's and can use the extra . . . ."

Norma had just blown her last cloud, and was tapping her smoldering butt. Stan unconsciously watched her every step in killing the embers.

Tom, looking at Stan said, "I've never owned a pair of suspenders. I've seen plenty in my day, but that pair you have on, is as gentle as I have ever seen. Stan grinned and impressively snapped them one at a time to show his appreciation. That really got Norma's attention. She jumped up, stepped directly to the front of Stan, and just stopped short of touching bellies. Stan was taken by surprised and flushed immediately.

Albert muttered, "Oh, Oh."

Tom said in a hushed voice, "Jesus."

Norma stood defiantly and smiled. She boldly put her right thumb and index high on Stan's brightly brocaded suspender. Tom turned his head and shut his eyes. Albert just looked on flabbergasted.

Tom heard the snap. . ., and let out an "Oh, oh," and peeked. Stan was still standing at attention, his chin drawn in with ambivalence. Tom felt sorry for him, as he himself, had been the recipient of Norma's outrageousness. But he loved her valorousness ardently. Norma had her dander and femininity by the horns and was discriminately displaying it.

She said frankly, "God, you're right you know. These are fancy laced, bridled with handsomeness, they are the finest pants supporters I've ever seen!"

She added, "Tom, I insist you get a pair just like those of Stan's."

Albert stated, so Stan was certain to hear, "We are a spirited lot, and those adjusters are impressive."

Stan relaxed, and looked sheepish, for being benevolent so easily. He recovered and stated, "They have them at Mattie's," and he courageously placed his fingers under them and moderately fondled the straps, thus capturing some of his virility. The door opened, and Laura came in first, followed by Dr. Ward, both wearing smiles.

Tom stated as they entered, "No broken bones or dislocation."

Dr. Ward stated, "Is this your wife?" The room was silent, as the foursome thought of a judgmental response.

Tom confused, said, "No. . . but she's my future wife's bridesmaid."

Norma perked up, with brightened eyes, and looked on with appreciation. She remarked, louder than necessary, "You all heard it, as I, his proposal was as pretentious, as the falling of the seventh veil."

Tom said, "Jesus."

Albert said, "Good Lord," and Dr. Ward chimed in with "What a delight!" Stan just hung on to his fantasm.

Dr. Ward said, "You're right young man, just minor torn ligaments of the triceps and deltoid. She'll be fine. I would suggest, however, that one of you help her exercise the joint, two days from today. Ice would help occasionally."

Norma, all smiles and energy said, "I'll volunteer, and congratulations on two scores."

Tom asked, "Two scores....?"

Norma replied, "Yeah, you're apprehensiveness on Laura's prognosis, right as rain. ...the second score, the bridesmaid's ceremony has yet to materialize."

Albert said, enthusiastically, "She's right, you know."   Silence.

Then Tom proclaimed, "Norma, will you marry me?" and she replied, with freshness, "torn ear and all. . . Yes, Tom, yes. . ."   Tom approached her and they embraced, it was a magical moment, an affinity, a soulmate.

Stan experienced an incredible, vibrant sensation. Although just meeting them, their aura was ever present. Dr. Ward grinned, with remembrance of asking his wife to be, as long as they. . . , then a quick change to his lonesomeness.

Albert and Laura both moved in unison, to congratulate their counterparts. Each wishing, with fickleness, the romancing of this unique spectacle, wishing their partner would be as prudent as Tom, and scream their love to the world.

Albert gripped Tom's shoulder and said, "You never cease to amaze me, you lucky bastard, congratulations."

He turned to Norma and said, "You two are as well suited as two opposing poles or polarity, unable to function unless in tandem."

Stan said out loud, "That was beautiful."

Dr. Ward asked, in jest, "Which is positive and which is negative?"

Tom remarked, squeezing Norma, "It's of no consequence, we're operating on alternate currents."

Norma shouted proudly, "That's my Tom."

Albert, caught up in the zeal of excitement, yelled, "That is positively negative! . . .and added eloquently, "Laura, this amicableness has stirred me to ask for your hand in marriage!"

Laura screamed as did Norma, "Yes, Albert, yes."

Stan jumped up when the two women screamed and remarked, "Well, I'll be God damned."

Dr. Ward elbowed Stan kiddingly, and said, "By God, you're a lucky man."

Stan looked bewildered, then he smiled. Albert and Laura hugged and she winced a little when they wrapped their arms around each other.

Tom said, "I'm so happy about us all, I could. . . ."

Norma laughingly interrupted Tom and screamed, "Drinks on the House!" That brought laughter and further excitement. Dr. Ward moved toward Norma to officially wish her the best. Stan was watching his every move. As he bent to kiss her on the cheek, Norma dramatically lifted her hand up to his chin, gently guided his lips to hers, much to the surprise and delight of Dr. Ward. He said to her softly, "You've re-established my faith in humanity, may all your relevance be reconciled.: She then kissed him on the cheek. He moved over to congratulate Laura. Stan quickly moved to Norma's side, and proceeded to kiss her on the cheek. She must have felt his reluctance for she then kissed him smartly on the cheek. Dr. Ward caressed Laura on the cheek and said to her, "This is a tender romantic day, my heart goes out, to you, and yours, in resonance." Hand shakes were in order for the men. The excitement was still insulating and infusioned. Dr. Ward said, for all to hear, "I've never in my 69 years on this earth been a spectator, witnessing two couples who cried out with such affection and tenderness, reeling with reciprocity." Stan, caught up in the moment again stated, "Doc, that was a mouthful."

Norma said, "You're damn right it is."

Tom said to Albert, "We're the luckiest bastards alive."

Albert replied, "I hope we can stay that way, life is so precious for all of us now!"

Norma said to Laura, "We've had our last soak together."

Tom and Albert laughed at Norma's remarks, enjoying every word.

Laura raised her good arm and shouted, boastfully, "Bring on the preacher!"

Norma said with fire, "That a girl. . . ."

Albert asked, his two new friends, "Do you have a reverend in your domain?"

Dr. Ward smiled and replied, "Your command of the English language is refreshing. Just to set my sequence straight, my charge as a physician is my wedding gift, small as it is, in case I am unable to attend your festivities. I'm on call every day since I am the only doctor."

Tom said, excitedly, "We would like both our new friends at our double wedding. Jesus, I'm getting married, hot damn!"

Albert asked, "Will one or both of you, please take us to Mattie's, and we'll taste his renowned elixir."

Norma said, "Yeah, and Tom, don't forget to buy a pair of those pants supporters!"

Stan said, "Great, Doc and I will show you the way. I know I could use a lift with all this excitement."

Dr. Ward said, "Yes, and I'll find out where the Reverend is hiding. He, too, has a busy schedule ahead of him. Let's coffee up!"

Laura and Norma's eyes sent another coded message of contentment.

Norma giggled and said, "We'll both have soaking partners now!" Laura injected with, "Yes, but their response was a preference for showers -- remember?" Norma replied with a wink and said, "Oh, yes, I remember only too well, but remember, my sweet, what that little finger accomplished."

Laura smiled, but Norma felt her hesitancy.

Norma told the men to go ahead, that they would be there shortly. They kissed their husbands to be, and Norma got to it with, "What's wrong, Laura?"

She replied, "Do we have to constantly use our . . ." Norma spoke up, ". . .power!" Laura said, "Yes, I guess that's what I mean, but. . . ." Norma broke in again, "They've got the brawn and we have the incentive." She thought, then laughed, "Did I just call our femininity an incentive?"

Laura perked up and stated, "You did just that!"

Norma said, "Christ, don't you dare tell Tom, or anyone else, for that matter. This is the sort of situation I chastise Tom for."

Laura smiled and said, "I know."

Norma said with a cautious air, "Laura, I just thought of something, we have a tail twister. I want you to be my bridesmaid of honor. . . " Laura interrupted, "You've got to be mine too... then added, "Hmm--I see what you mean. If we have a double wedding, it will be impossible to do that."

Norma replied, "You can bet your butt on that."

Laura used her good arm, and flayed it excitedly, beamed and stated, "I know, we'll have three weddings, seeing's you, and Tom committed first, I'll be your bridesmaid, then you can be mine!"

Norma could hardly contain herself and spoke out, "Yes, yes, then we'll have a double wedding." They joyfully hugged one another with the expectations of their double, double wedding.

Laura asked with a doubtful look, "Will the guys go for it?"

Norma smiled broadly and stated, "That's one reason why they made gals tender and to keep them in line!" They both enjoyed that philosophy. Norma said, "I'm going to tell you the naked truth, bare my soul to you--my reasoning of adaptability toward men. I've never told this to anybody. Remember, we flirted a little when you mentioned about walking that mystic line, of acceptance of dominance, remember?"

Laura said, "Yes, I remember."

Norma continued, "I truly believe that all females, past, present and future, have a double task to our male counterparts."

Laura's good hand found hers and gave it a squeeze and received one back. Norma went on, "As to the incubators of life, our first task is obvious, whether it be a male or female in ovulation. Our second task, is to bring tranquility to our subverted males. What's your take on my thesis?"

Laura squeezed her hand and replied, "As a woman. . . "You can snap my garter any day." They laughed and left the one room police station in Port Graham. Both enjoyed the fact that they were women.

Stan held one of the doors open at Mattie's. Albert's and Tom's eyes immediately shifted over their head. There were hundreds of articles, hanging by rope and wire from the 12 foot ceiling, clearance for a seven footer. Ropes, whips, paraphernalia of every type for cars, horses, snowmobiles, clothes, hardware, food, candies, and fishing gear. This marvelous trading post was creatively stocked.

Albert said, "Good Lord, think of the inventory!"

Tom said, "Think of the dusting!"

The floor space seemed to be about 2000 feet, not counting the walls and ceilings. Doc and Stan strolled over to the six old bolted down stools at the little counter and sat down. Albert and Tom walked down one aisle, and Tom said emotionally, "Jesus, look at that!" It stood seven feet tall, glassed on three sides above the waist line. This antique booth held the face of a bearded, three dimensional, and turban bound mystic. The back-drop of this bold face was a circular spectacle, of stars and planets, used as a back drop. A miniature crane stood with a boom, and a double claw's pinchers that was to the left of trinkets mixed with large gum balls of many colors. There was an open shaft in the center of the prizes which connected to a shoot in front of the mechanical machine. There was a large button, painted long since and had worn off, two small wheel dials, one on each side of the button. The trinkets were mouth organs, ear rings, men's and women's rings, and two wrist watches with leather straps. There was a high relief medal cast, flowing framework from the large slot opening at the bottom to the ornate opening above, that read in cast letters, pay 25 cents." But using the dials, manipulation of the boom, and the mechanical claw used to pick up a prize and drop it into the open shaft. The button was pressed when the claws were positioned over the prize selection. Tom was taken with this mechanical challenge. The instructions were barely legible due to oxidation on the lower panel. A happy burly man moved toward them. He was wearing a full clean wrap around green apron over his chest, all the way to his kneecaps. He wore a bright plaid shirt, with the cuffs rolled up meticulously, on his muscular biceps. His eyes were a clear hazel, with swirls of red wavy hair spouting in an unorganized array. He thrust his big hand out and stated, "I'm Wade Freeman, propri-etor of Mattie's." They exchanged handshakes and first names. Tom bent closer to the Wizard Ma-chine and was peeking through the glass at the crane apparatus and its wizardry display. Tom fingered the machine and said, "That's a dandy!"

Wade replied, "Yeah, I picked it up when a Carny went busted over at Homer."

Albert said scanning, "I'm impressed, I've never seen so much merchandise displayed so cleverly. I expect you have everything including the sink."

Wade shook his head sideways and smiling, said, "No, not now, sold my last over a month a go."

Tom said, "That elegant pair of suspenders that Stan is sporting, I would like its twin, then I would like to tease my palate with some of your famous coffee."

Albert said "Amen" to that. "Would you be knowing where your reverend might be?"

Wade wiped his big hands on his apron, and said, "You'd be wanting Dell Kotnick. He dropped by earlier, and mentioned he was going to check on the Springers. . . . Want me to give him a ring?"

Albert stated, "Yes, please, you might mention to him, he has a double wedding to per- form, as soon as we get our license."

Stan had headed for the three men and overheard Albert's last words. He spoke out, "No problem, I'm also authorized to issue those happy papers."

Tom said, "It's nice to have friends, maybe you could nudge Dell--a little, maybe?"

The girls came through the double doors, each energetically shoving the doors wide open, like a strong breath of fresh air. They waved at Dr. Wade, and received one back. Both were oohing and aahing at the uniquely stuffed, and articulately displayed goods. They spotted the backs of Albert and Tom as they approached. They saw the antique mechanical game of chance. The men heard Norma say, "God, I haven't seen one of these in ages." The couples embraced with smart smackers. Norma looked through the glass at the mechanical wizard, and said glibly, "I might have known, you'd be romancing a gaming erector set."

Laura, grinning, was sincerely uninhibited now, since Norma had made privy her sanctuary. Albert said with dexterity, "Your Tom has been eyeing and caressing this mechanical teaser."

With a gesture of cock of the walk attitude, Tom stated diligently, his blue eyes flashing, "I dare say, I can take this little bugger!" Albert said, "Oh, Oh." Norma swung a right fist in the air and said, "Go for it!" Laura shrilled, "Take that mechanized ferret."

Dr. Ward and Stan slipped off their stools, and joined the noisy party. Tom was a little boy again, his tongue was moving nervously in his mouth, his palms were becoming wet. He wiped them on his shirt. He excitedly said to Albert, "Give me a quarter--please."

Albert dug in his pocket and passed the quarter to Tom's waiting hands. He had a little moment of hysteria, and swore, "Jesus," I almost dropped Washington."

Dr. Ward said, "The last person to pluck a prize from that cantankerous machine was a lad named Willy Dobbs, on his sixteenth birthday. He had a fine set of elusive hands." Stan spoke up,

"Yeah and those hands made him a papa three times!"

The girls snickered as Tom entered the quarter in the raised slot and there was a loud clunk. The machine vibrated, humming of gears, and brilliant lights came on inside the glass display, as well as the slow turning circular back drop, giving an illusion of a moving galaxy at night. Tom flinched as the lights came on.

Doc Ward shouted, "Let it warm up a little. . . ."

Norma gave Laura a friendly elbow, and spoke just for Laura's ears, "Wouldn't you know it, my Swami, attempting to outsmart another Swami." They giggled as Tom turned the little wheel, that started up the whole mechanism, spinning it to direct the positioning of the boom, with the double clam shell claws, hanging down from the tip of the boom. Tom quickly pressed the large paintless button, between the two wheels. The release mechanism tripped, and the clam shell jaws opened and came tumbling down. Tom touched the wheel to compensate. The open claw hit and Tom spun the other wheel, that closed the two hinged jaws, pincering on his prize selection, then back to the first wheel, to move the derrick in line with the open enclosure, to drop his prize in. His eyes were excited. He was excreting adrenalin, tints of moisture appeared on his brow. The spectators were tensing up.

Laura grabbed Norma's hand and squeezed hard. All of them were leaning closer, just a little more -- with anticipation. All their perfumes and body odors were mixing in the excitement. The boom was transporting in its jaws, Tom's prize, to place it in the open shoot. Suddenly the machine stopped humming, the lights went out, the spectrum stopped turning, the jaws opened and the trinket fell. . . . Tom bent over, his face pressed against the glass. He froze. As he watched his catch fall on the rim of the shoot opening, Tom then twisted his body in a pulsating fashion, even his butt was moving spasmodically. A performance reminiscent of the Huckle-Buck. All eyes were fixed on the prize that was teetering unsteadily on the edge of the trespass, except Norma's whose hormones began rising, watching Tom in those physical intense gyrations. It fell, and Tom, like a teenager at a high school contact sport, screamed, "Got ya. . . got ya." They all cheered and clapped.

Doc Wade shouted in rapture, "Well, I'll be damned." Stan vocalized loudly, "He's a you- bet-yah...!"

Norma, her body fluids mixing with her enthusiasm hollered, "That's my Tom," while Laura screamed with joy. Albert, said with a vibrant poignant voice, "Tom, you never cease to amaze me!"

They were all caught up in the frenzy. Simple madness over a two-bit wizard machine. Wade came rushing over, hearing all the commotion. Tom licked his lips, and thrust his hand in the cast iron entry box and grasped his prize. Wade moved around the circle and asked, "Everything Okay?" Doc Ward said, a little out of breath, "You just missed a marvel, the first shot and he plucked a prize!"

Stan said, "He sure did. . . ."

Wade said, "Let me shake the hand that is the best of all wizards."

Norma raised her voice for all to hear and stated, "The itinerary may be right, but the name isn't. He's my Tom, and he is a swami." She rushed to him, and planted a well done kiss and dropped her left hand casually, grabbing his torso and giving it a quick squeeze.

Wade shouted, "Drinks on the house." Norma said, to deaf ears, "That's what I said a while ago."

Tom slipped his award in his pocket and gave it a cocky tap. They all strolled over to the worn stools and sat at the counter except Wade. He recaptured his position, with a smile, "What'll it be folks, remember it's my treat!"

Tom sounded like a winner, "I'd like a cup of your java!" Wade looked and captured the eyes of each expressing their pleasure in Tom's choice. Five cups of Mattie's surprise and one for the proprietor!

Chapter 16

A large, ornate sign, with old fashioned flare lettering was home to the forward wall above the well used counter. It's bold letters read, "Coffee $1.00, refills $1.00."

Stan said, "These tasters have a way of romancing and tantalizing you all the way down." He gestured with his finger and stopped at his waistline, then ran his tongue over his upper lip. There was no pot in sight. Wade had exited through a swinging door adjoining a small pantry. He had to make two trips. The heavy mugged and steaming coffee sent aromas around, like a light spray of perfume.

Wade smiled and said, "Don't burn your lips. It's hot, hot."

Laura said, "Mmmm--it smells delicious."

Tom sipped too quickly and winced, after taking a sample. He remarked, "Jesus."

Albert blew and tempted his palate and responded, "Good Lord, this is a whisker puller."

Norma tasted it, and said, "What a romancing surge!"

Laura didn't respond, she was busy drinking it, hot or not. Then she said, "I thought chocolate was the habitude."

Doc raised his mug and stated, "To the newlyweds to be." Smiles sprang up with anticipation.

Doc added, "If it were I who held the formula for this arousing elixir I would have bottled it and be watching the moving grass skirts."

Albert smacked his lips and asked Wade, "By the way, when can we expect your spiritual leader?"

Wade supped, and replied, "On his way, maybe 20-25 minutes or less. I'm glad you folks are enjoying my potion."

Tom said nonchalantly, "I know what's in your alluring concoction."

The smile faded from Wade's face. Tom shifted on his stool, and looked about uneasily, then said, "Do you want me to write it down, whisper in your ear, or. . . ."

Wade looked perplexed, and in a stately manner said, "Are you serious?"

Albert joined in the conversation. "I'll tell you Wade, with a damn sure, if Tom says he knows, you can bet your store on it."

Tom felt whimsical and a little cocky as he walked behind the counter, and pointed to the swinging doorway. Stan downed the last drop, licked his lips, and spoke, "Ahh...I'm going back to the office and bring a back a couple of those shackled, leg-iron licenses." That received snickers and smiles. He stopped, then turned around and asked, "You folks, do you have proper I.D.?"

They nodded in unison. Tom put his arm around Wade's neck lightly, showing respect, and they went through the swinging door. Norma spotted an ashtray, then eagerly looked about for signs. Finding none, she smiled, and tapped one out, and lit up with a benevolent smile. Her eyes and contentment were those of a seductress as she secured the taste and blue aura of her white cigarette.

Laura said, "Your Tom seems to have done it again. Doesn't that scare you, just a teeny bit?"

Norma was feeling relaxed as she blew a soft meandering layer of smoke.

She replied, "Come, let's step over there." Laura got up and the two went into a huddle of sorts. Norma said, "I've got an ace up my sleeve." Laura stated, "What could possibly be a stalemate-- other than a King's X to counter his uncanny powers?"

Norma slipped that sacral smile on and responded, "Fact of the matter is, he's going to pick out his own nemesis, then by God, he's going to have to try it on. After he dons it, that will be that. If his mind wanders, and he brews up an incantation, I'll have him wear that ravaged elastic girdle of iron. . . .!"

Laura said, with hesitation, "Do you mean, you're going to put his body into a full body panty hose?"

Norma blew a cloud, and smartly replied, "You're damn tootin'!"

Laura giggled, then laughed, as she mentally pictured Tom in one of those high flyers, then added, "You know, knowing Tom with his mystical powers, I still feel sorry for him if he has to wear one of those pincher--creeping--wait a minute, Norma, for your own sake, you'd better be sure he has a jock strap underneath that strapping chastity belt."

Norma replied, "Jesus, I'm glad I told you, I must be slipping, fantasizing him, getting into that crotch puller. Thank you, Laura.. . ." She then added, "But I think he's run his leash, bless him." They exchanged good girl signals between them.

Tom and Wade came back through the swinging door and Tom recaptured his stool. Wade stated, "Tom and I made a pact--ah, Tom, go ahead. . . ."

Tom said, "In all fairness, I will give but one clue. I have given Wade my word not to reveal the ingredients of his fine drink."

Laura said with somberness in her voice, "Does that mean you'll never make such a pot, ever?"

Tom looked at Wade, then said, "If I do, it will be brewed in secrecy."

Norma said, "I'm relying on you to duplicate this luscious drink."

Albert spoke up, "What's the clue, Tom?"

Tom again looked at Wade, received a slight nod, and said, "Molasses and tongue!"   D o c stated quickly, "Is that it?"

Wade said, "Yep" and nodded. Doc added, "Tongue and molasses, yuk -- what a fowl taste."

Tom said, "These two words are just clues."

Norma ground out her burning stick, and asked Wade, "Do you have women's panty hose?"

Laura made certain not to catch Tom's eyes. Wade replied, "Just down from the Wizard on your left."

Norma said, "Come on, Tom."   They hooked arms and strolled. As they approached the Wizard, Tom put his index finger to his forehead, and gave a smart salute.

He then spoke to the mechanical wonder, "Hocus - Pocus - Got-ya, Got-ya. . ."   Norma frowned, she knew Tom was drawn and intimidated like a magnet to that two-bit taker, and Tom knew that she knew. Norma found the bin, with full panty hose, and dug through for sizes, and color. She selected two, and held them up for Tom to see. She asked, "Which do you like. . . .?" Tom pointed, indicating the light beige.

Albert struck a deal with Wade to scuba under the water line of the *Magnetic* tomorrow and examine its seaworthiness. Albert inquired about the firearm that Ted had his sights set on. Wade said inquiringly, "How did you know about that?"

Albert laughed, "No, no, Stan mentioned it to me. By the way, I wonder how Ted's doing?

Wade said, "You don't have to worry about that lad."

Doc spoke up, "I say not, I delivered that feisty ball of energy, he's a damn good kid."

Wade said smiling, "You'd never guess the type of rifle he picked out and has it on lay-away. My Dad passed away many years ago. He took it in on a swap--I don't recall what he traded. Well, anyhow, I can't figure why he wanted to buy that particular gun, when there's a batch of them with real fire power. No, he wanted that one, and I made him a special bargain. It's an under-over Model 1903-A3."

Albert said, "Lord, that's an early World War II issue."

Wade said, "I Know."

Albert stated, "They were right accurate, for a one shot, bolt action. But surely that rifle is inexpensive?"

Wade asked loudly, "Anyone for another cup?"

Albert said "Yes, please, and what about you, Laura?" She replied, "You bet, that's scrumptious." Doc said, "Me too!"

He left, then returned with fresh hot fill up's, the aromas welled up once again. Wade got a mug also and said, "You're right about the A-3 being reasonably priced. I cut it down to 38 bucks. But you know he has kids, and he's a good father. His family comes first. His family takes about all he makes, at least it seems that way."

Albert said, "I've never met him, but I liked what I heard on the phone, and he's quite prompt. Maybe I'll be in these waters again and need a jack of all trades." He withdrew his wallet and extracted two 20's and laid them on the counter, then added, "Give Ted the gun under some pretense, but please do not mention my intrusion. I'm going to pay his wage."

Tom and Norma captured their stools. Norma looked at their empty cups, and slid the panty hose across to Wade and held up her cup and asked, "Please." Tom said, "Second that." He took their cups in for another round trip.

Wade said, "He should be finished with your craft soon." He tapped the keys on the tall, ornate, and silver plated cash register. The foursome enjoyed the dainty clear, pleasant sounding ring, as the cash drawer sprang out. Ted came in and went directly to Wade, and told him the diesel topped out at 95 gallons. Albert presented himself, then shook hands. Stan came in with papers in hand, and following him, within a shadow, was the preacher, Dell Kotnick.

Albert asked Ted, "What is your charge?"

Ted looked about then stated, "It took me just over an hour is all."

Albert asked, "Will $20 cover it?

Ted smiled brightly and remarked, "Happy to oblige!"

Wade did some doodling on the scratch pad, circled the bottom figure, and turned it around, shoving it for Albert to read.

Stan was getting the information from the foursome on the licenses. Norma was joking around. Tom was nervous, realizing the step he was taking. Doc introduced Del Kotnick to all. Norma flashed a smile and a message to Laura and received a positive response.

Norma remarked, "I have a statement to make on behalf of both the bridesmaids. Entering into this tunnel of no return," -- Albert muttered, "Good Lord," Tom whirled on his stool nervously, attempting to accelerate the impossible -- time. She continued. "We insist, being quiescent, that we have three weddings!"

Tom swore, "Jesus," and fell off his spinning stool.

Albert remarked, "What!"

Laura said coyly, "Isn't that wonderful?"

Wade said excitedly, "Hot damn, that's a new one, you must have your ceremony here at Mattie's."

Doc shook his head in wonder and Stan said, "I'll have to get more license forms." Rev. Dell Kotnick held up his hand and said, "No, Stan, no need for more forms. The marriage license and copy of such is all that's needed. I'll perform one couple, then the other couple, then we'll have a renewal of vows, with both couples.

Doc said, "Well, I'll be damned."

Laura spoke up, "If it's all right with Norma this would be a grand chapel to have our wedding in--Mattie's Emporium."

Norma stated excitedly, "Done!" Tom got up from his fall and was brushing his pants, just to be busy.

Norma asked, "Okay with you, blue eyes?" Tom looked about apprehensively, and said lightly. "One request," it became very quiet, all waiting to hear what direction Tom was heading.

Norma asked with indulgence, "What is it, Tom?"

He replied, "I'd kind of like to get hitched next to the Wizard machine." They all laughed to Tom's disapproval. Norma marched up to Tom, and cradled his chin in her hand, and planked a smart kiss on his surprised lips, and said, "You can bet your new suspenders on it... " Tom's looked about, and inched out a nervous smile.

Laura asked "Albert, "How about you, Teddy Bear?"

Albert gulped. Norma winked at Laura, congratulating her on taking her first step with her philosophy towards the men in their lives. Albert moved over and whispered something in Tom's ear. Tom shook his head in a negative gesture and weakly smiled, and patted his pocket. Albert then went around the counter and whispered in Wade's ear. Doc, Stan and the Reverence looked suspiciously at each of them. Norma and Laura flashed messages of what was going on. Wade produced a smile, and gestured for Albert to follow.

Doc said to Laura with a grin, "Now be careful, don't excite that limb for two days--then your bridesmaid and you can exercise it, the way I explained."

She said, "I'll do just what you told me."

Norma got close to Tom and asked, "What was that all about, blue eyes?" Tom was trembling slightly and his eyes were darting in different directions. Norma realized his nervousness was sending distress signals with more prominence. She slipped her arms around his neck, and whispered in his ear, "Tom, I've got the shakes too, but they're inside me." He squeezed his life raft in reciprocation, and began to settle down and loosen up a bit, although his eyes weren't having any part of it.

He answered, "Uh, Albert doesn't have a ring!"

Norma swore, "Christ, I don't either, and I don't think Laura has one for Albert."

Tom said innocently, "But I got one for you, I took it from the Wizard." Norma looked up into his grateful blue shining eyes and broke down crying. Tom spoke up, "Jesus, what's the matter."

Laura and Reverend Dell moved to her side to comfort her. Laura was disturbed. She hadn't seen strong Norma break down, and wondered if Tom had antsy feet, or. . . Norma's ambient tears flowed down her cheeks. Dell pulled out a hanky, and handed it to Laura, and she put it to work dabbing and swabbing. Norma was crying in lurches now, letting it all out. The saline fluid was just about spent, and Norma looked up pathetically, eyes in disarray, peering into Laura's understanding and sensual face, and said crying, "I love him so!"

Albert came back, saw Norma, and glared at Tom, and asked, "What did you say to her?"

Blue eyes said weakly, "Just that I had a ring for her."

Norma was at the sniffling state. She shouted hysterically, "Tom said he took it from the Wizard."

Albert was dumbfounded. Wade was in a quandary. Doc shook his head constantly and Stan said, "Jesus Christ, what a day." Tom felt like a scoundrel and looked into their faces for solace, but none materialized. Laura remarked with reassurance, "Can't you see, she's just happy, then screwed up her face, and said defiantly, "What's a matter -- with -- all of you -- sniff - sniff. Don't you realize, this is a happy -- sniff -- occasion -- !"

They hugged each other, and their tears intimately washed each other's faces.

Doc spoke up soundly, "Let me tell all of you, I've lived longer than any man here, and -- I still -- can't stand to see a woman cry. . . . " He wiped his nose, and moved his hanky a little higher. Ted had witnessed this from the adjoining aisle. Although the father of three little ones, he was young compared to his counterparts. He felt uneasy. Crying was an elusive enemy that he had failed to quell with all his skills. Males usually grow up in wonder and maturity, but neither parent emulates what the hell to do when a woman cries. He also began to get nervous.

Wade broke the tension and shouted, "Folks, we are all gathered, and the Wizard is right down the aisle waiting. Let's get this party down the right path."

They got up, milling, and the foursome realized that time was no longer a luxury or relevant. Muscular Wade had scooted hurriedly and returned with a two foot plastic reel. Rolled inside was bright yellow sailcloth. He planted his number ten shoe on the trailing end, and gave it a slight kick, with his other matching ten. It rolled, while dispensing a bright yellow swath, until it struck the Wizard. Norma and Laura were overwhelmed by the display of ingenuity and hospitality. Laura said to Norma, "Remember--we talked about what a special gal -- Judy was. . .?"

Norma replied with a smile, "You bet, who would have thought this would be the grandest of weddings? We couldn't have selected such a highlight, if we had months of planning."

Young Ted walked quickly out of Mattie's. Albert and Tom were delighted, and Albert withdrew two small velvet ring cases and gave them both to Laura. He stated, "These statement rings are identical for Tom and me."

Laura passed one to Norma and remarked, "Thank you, sweetheart."

Norma asked with a question in her voice, "What if. . . ."

Albert interrupted,"I know mine fits, if Tom's finger doesn't concur, we can change his size, but not our commitments."

Rev. Dell Kotnick moved to his ceremonial position next to the Wizard, careful not to step on the yellow cloth. He spoke clearly, "Bring forth the first couple for congruous matrimony!"

Norma said in a stately manner, "Well, Blue Eyes, this is it, for better or for worse, down the yellow, to the preach and the Wizard."

Doc pulled out his old harmonica, and began a sweet rendition of the wedding march. Doc was good, he was wailing it, enhancing the enlightenment of the favorite exuberant lyric. Tom unprepared for music, took her arm, and began walking on unsteady legs, their feet romancing the yellow ribbon towards Dell, with the Wizard looking on. Laura and Albert walked in step behind them, then stopped behind their gender.

Stan ran, and just made it as they stopped, and thrust his arm between Norma and Tom. Norma was startled, then realizing, took the cluster of silk artificial flowers from Stan's nervous hands. The doors opened at Mattie's and Ted and his wife and three little ones came in, and quietly moved closer, where they could witness the triple wedding. One of the tots had a blue balloon on a string. Minister Dell Kotnick performed a soothing, but short ceremony. He suddenly stopped and asked, "Norma or Tom, before I conclude our emulation, would either, or both, care to romance their future mate, with a personal message of love?"

Tom, shifting his weight, his blue eyes barely containing the sockets, asked, "Are we married now?"

The minister stated softy, "No, I just have one precious sequence remaining.

Tom and Norma were holding damp hands as in a match of strength. Norma said, "I would like to say on this special occasion to my Tom, I'm losing my last name, Moon, and happily taking his name, Spoone. My new name, in the diction means, to make love. . . Tom . . .I pledge and will always honor, your name."

Fidgeting was Tom's foray. More nervous than ever, he said, "Jesus," all were touched by Norma's special promise. A tear escaped from Laura's eye, then the other reciprocated with sensuality, and tears cascaded down her cheeks. Her eyes were reflecting bright moisture fracturing of light, like a cluster of diamonds in the sunlight. Tom was wavering, and hesitantly stated, "Miss Moon, my love, has given me what I would promise, but never could deliver, to you -- thank you -- my love -- for the moon."

Norma looked up into his waiting blue eyes, feeling his overwhelming love and wanted to freeze this precious moment forever. She had to embrace him, relieving some of Tom's tension. The spectators, in suspended animation for just a minute, placed themselves on that beautiful sheath of yellow. The only exceptions were the three little ones looking on, with personification of innocence. After a warm kiss, Norma said for all to hear, "Our song will reign forever, as our love, "Shine on Harvest Moon," sniff, then "Oh, God."

Doc ran the scale, then played their sweet rolling song. Norma was about to explode with all of the unexpected but truly wonderful emulations. Tom patted the magnificent Wizard, and his wife to be, although he was still quivering.

Ted's wife began crying upon hearing the sacred words and became nervous. His kids looked up at him in wonder and guidance as to why their mother was crying. Ted said excitedly, "Oh, God," and took his trembling hand and captured his wife's, then gave it a squeeze.

Laura's tears welled up, and Albert's stomach was queasy. He was proud of them. Stan had his hanky about his face, and when he thought no one was looking, dabbed his eyes quickly, so moved was he by the proceedings. Wade's face was balanced and furrowed in benevolence. He was glad it was Tuesday, a slack day at Mattie's. The Reverend then completed his sequel with man and wife.

Norma was wearing the Wizard's captured ring and Tom's finger was adorned with a wide, plain wedding band. They emotionally kissed, their lips transmitting messages of devotion and sensualization. Tom was finally overcome by the tenseness and aggravation, that had him at bay. Possibly, maybe, because it was finally over, and he realized that he held his wife in his arms. All moved in for positive congratulations, hand shaking, and cheek kissing. Norma's and Tom's eyes said it all, emitting vibrant, exciting bursts of energy.

The deliverer stated, "Will the second chosen couple, step up to the enchantment of life?"

Doc picked up on the clue, and played another chorus of the familiar ceremony. Norma passed the little silk bouquet to Laura, with a warm hand pat.

Albert said, "That's our fascination, Laura." She looked up into his face with shining and tear stained eyes and said, "Yes, Albert, I want to thank you for getting involved in my world, now it's our world."

They kissed, then stepped up to the imaginary altar. Doc put his mouth organ away, realizing just one more performance. Midway through they exchanged rings. Reverend Dell stopped his ritual and asked for their special constancy. Laura received a Here's to You Honey, wink and she stated, "Our meeting was one of destiny, by a scant 24 hours we would never have met. He is my apostle of time."

Norma was flashing that-a-girl, but Laura was receiving. Albert stated, "My Laura is like the freshness of a vibrant star, hurdling through space, dispersing quadrillions of soft quantities of glittering light, and choosing me to be the only spectator of her loveliness."

Wade was struck by these solemn commitments, and saw moisture in Rev. Dell Kotnick's eyes. Albert and Laura also kissed. Ted's wife broke down and used Ted's shoulder as a blotter for her tears. Both of Ted's legs were anchored, tots' arms and legs each staked out their own. Their third boy, age five, started walking closer to the festivities, foregoing the indifferent, hushed and urgent cries from his Dad, "Billy, Billy."

Ted was an unmovable pillar wall, with his weeping wife and the two little ones, still clutching and wrapped about their Dad's legs. Billy made his way, stepping through, or around limbs until his blue balloon became lodged under Tom's right arm, which startled him for a moment. He moved his arm, and the blue rubber skinned object popped up a little higher, but was between him and the mechanical Wizard. Albert and Laura, having just heard the final magic words, hugged and squeezed each other.

Tom remarked, "Congratulations, Mr. and Mrs. Dryfuss." Norma moved in for hugs and warm wishes. The balloon was a trifle of a nuisance, and Tom brushed it aside and looked down, and discovered a little hand holding the end of the string. Billy's hazel eyes looked up into Tom's excited blues. Tom foolishly looked at the Wizard's face.

Norma came over spilling with happiness, looked down and said cheerfully, "Who's your friend?"

Tom leaned over and whispered in her ear, "I don't know." Norma laughed and said, "What are you whispering for?" Tom gestured excitedly and pointed to the little boy holding the balloon, who was still looking up.

Norma spoke up, "Tom, just ask his name."

Tom whispered again, "What if he's too young to talk?" Norma giggled, then laughed. She leaned over, placed her hands on her knees and asked, "Where are your parents?" The little fellow pulled on his balloon, using that hand, to point toward the door and said, "Over there." Norma attempted a ballerina stance, and succeeded long enough to see the pinned picture of fatherhood. Ted had given his wife his hanky and she waved it in response to Norma's waving. Tom turned to look, and his arm pressed on the balloon that was positioned and pinned with pressure, against the two-bit coin slot. There was a whopping bang! The girls screamed and Billy yelled, and started crying.

Albert looked about nervously. Doc came over to see and evaluate the situation, then attempted to console Billy.

Wade asked Stan, "What the hell happened?"

Billy was sobbing and pointing up at Tom, crying loudly, and said, "You broke my balloon." Tom's nerves were immediately in residence. His apprehensiveness was in drive position. He looked about in guiltlessness, and in defiance wailed, "No, I didn't do it, the Wizard did."

"He did it," Billy cried, even louder and still holding the limp string.

Rev. Dell announced, "Renewal of vows, as Doc pulled out his mouthpiece and tapped it a couple of times to get the juice out, and swung into a rendition of "Here Comes the Bride." Norma said briskly, "Come on Tom." Billy was down to sniffling with waves of higher and higher thrashing. Tom pointed to the little tyke, still holding one end of the drooping string. Norma grabbed Tom, and got him in step. Albert and Laura were already standing in place. They both glanced at Tom and Norma and all looked nervously to the ones adjacent to them. They listened to what the minister had to say. He spoke mainly on just one word--a word he said was relevant for all mankind. Norma sort of waved to get his attention. He stopped and Norma spoke up, "Would you please use another word more representative of the feminine gender?"

Tom coughed and Albert, somehow, contained himself. Laura smiled broadly. Preacher

Dell smiled and spoke, "Please negate the term mankind, and reinsert the word, consortium." This was a delight to Norma. Laura and Ted's wife grinned in appreciation. He continued, the word is "respect." Norma thought, "Oh, oh." He proclaimed that if that word was exercised between all beings, Utopia would be the consequence and not the elusive dream. He reiterated their names, then saying, what a wondrous exciting participation of contractual lives they had before them. The two women were both thinking that they had already survived the terrifying events and would be happy to ease off the gas pedal.

Dell held his arms in front of him and said, "Just one last worthy observation," and shouted, "Drinks on the house."

That brought excitement and smiles, and movement. Tom was still a bundle of nerves. That swallow-tailed lad was still standing next to the Wizard, staring up at him, and he finally had dropped the string. Norma grabbed Tom and swiped a grand kiss. Albert and Laura were also embracing. The foursome was stuffed with apprehension. Albert laid his hand on Tom's shoulder and he jumped.

Albert said, "Good Lord, man, you look terrible, give me a bill please for our reverend."

Doc had been observing the foursome with some interest. He keenly saw the anxiousness of Tom, and went to the other end of the store. Tom mechanically opened his wallet, and pulled a bill out. Albert plucked it from his moving fingers, matched the same bill, and thanked Dell Kotnick for a wonderful service. Upon shaking hands, he pressed the two Franklins into his palm. Ted and his wife, with the two youngest, moved to the counter area where Wade was awaiting their order. They ordered his fine coffee, and three strawberry sodas for his young'uns.

Ted was about to recapture Billy, when Doc approached, and handed him two balloons. Doc kept moving and tapped Tom on the shoulder. Tom winced, then turned around and to his surprise, Doc transferred a balloon to Tom's outstretched hand. Tom bowed several times with genuine appreciation, bent over and put the dangling string in Billy's hand.

Billy looked up picturesquely, and followed with an innocent smile that became alive with, "Thank you, Mister."

Tom smiled a little crooked, and said, "We're square now, partner!" They were all spectators to this event, except his kin and Wade.

Billy said, "Yeah, I guess so."

Laura asked, "What's a matter, little fella?" He looked up at them, rolling his words, and tugged on his balloon and spoke, "It's not blue..." and he walked away with the balloon following him. Tom looked contented as Billy shuffled to reclaim his family. Tom winked at the Wizard machine. Norma saw the wink, but interpreted it was for the cute little guy. She put her arm around him, and whispered in his ear, "Tom, my biological clock has stopped ticking." The thought of silhouetting his image and Norma's was irrelevant to Tom. He smiled warmly and said, "Gee-whiz, Shine on Harvest Moon."

Chapter 17

Wade and Tom teamed up, brewing the special coffee. They all had seconds, except for the three tots. Wade told Albert he would be at dockside, suited up at 1:00 p.m. tomorrow. Laura whispered in Norma's ear and produced a well done smile. They both asked their new husbands to purchase that vivid, sacred swath, of yellow sailcloth. Wade got his tape out, and Albert and Tom did the same with their wallets. Laura was feeling whimsical. She tucked her legs in and up, and spun on the low stool, gleefully laughing.

Albert said to her, "Careful, honey," then added, for Tom to hear, "Any ideas, spells, from magic land?" Norma was feeling full of hell too, and tried spinning faster.

Tom said, "I feel better now."

Albert said, "No, not you, Laura."

Laura spoke up, "What did you say, Teddy Bear. . ..?"

Tom said, "What was that, Huggy Bear?"

Albert paid no-never-mind, and asked Wade, "Is there a cafe where there are takeouts?"

Wade said, "You bet, Wong's Dragon Inn. It is but a block up the street, can't miss it."

Albert remarked to Tom, almost shouting, "How do Chinese takeouts grab you?" The women were still whirling and giggling, and having fun.

Wade, upon hearing Albert's conversation, said loudly, "Wong hasn't Chinese on his menu, his specialty is spaghetti."

Albert and Tom laughed, then Tom cried out, "Norma, Norma, how about take out spaghetti for dinner?"

Norma stopped twirling, sucked in air, and shouted, "with balls."

Laura was still turning and spoke up, "Balls for me, too." Albert muttered, "Good Lord."

Doc stated, "I think the excitement has morosely captured the women folk. I would suggest Laura rest that arm!"

Albert said, "Laura . . . doctor's orders, this has been a wondrous day. How about a bite to eat on the *Magnetic*?"

She stopped with a "Whew. . .I'm getting dizzy."

Norma steadied her, and spoke up, "Don't forget our purchases--and our new husbands, then a little nap-py."

Laura said, a little giggly, "Okay, let's go . . ."

Norma said, "Hubby, have you squared up with Wade?

Tom smiled and replied, "Sure have."

Stan said, "Well, I better get on. I want to thank you folks for a memorable time that I'll not soon forget. With all the pleasantries, I neglected to collect license fees, $22.00. . ."

Tom said, "This is on me," and dug out his wallet.

Albert said, "I'll spring for the spaghetti."

Upon hearing the word, the girls chimed in harmony, "With meat balls," and giggled. Laura said in a low voice, "Lots of balls. . . ."

Albert shot a warning glance to Tom, he intercepted it, and they both helped their wives off the worn and tilted tuffets.

Doc said, "I'm going to get a bite over at Wong's, if you want, I can put in your order, and seeing as how this is your wedding night, Wong has four young'uns, and one of them would deliver, I'm sure."

Albert said, "Dr. Ward, I'll take you up on that generous offer. Thank you. Four spaghetti with meat balls."

Doc asked, "Soup or Salad."

Tom said, "Salads usually make the journey more successfully than soups," then added, "What about cheese?"

Doc stated, "He has jack and cheddar."

Tom said, "Have them both on the sides, Albert's buying. Thanks, Dr. Ward."

The newlyweds had their arms around each other's waists. Tom had his pockets stuffed suspenders, panty hose, and the folded yellow sailcloth, as they left Mattie's Emporium.

Norma stated, "God, I feel good, how about you blue eyes?"

Tom responded, "I'm practically down to the troposphere from the ionosphere."

Norma said, "That's nice, Tom, just five miles to touch down. I've got a humdinger of a question for the Cap. I'm going to save it for that special moment."

Tom said lightly, "Jesus, I hope it's not an embarrassing question."

Norma replied, "Maybe, just a teensy-weensy."

Albert said, "Look out, Laura." She appeared not to adjust her footing for the approach to the dock. She giggled and said, "I saw it, Teddy Bear, my legs are a little out of kilter."

Tom asked Laura, "How many cups of Wade's coffee did you have?"

She thought that over and said, "over four, . . . and I need to go to the head."

Tom grinned and said, "Albert, she'll be fine."

Albert said, "Well, here we are, careful through the transom door."

Tom spoke up, "Let me go first, then you pass Laura to me."

Laura said lightly, "You're talking like I'm - - uncompromising -- oh, oh."

Tom stated, "I got her."

Norma remarked, "That's my Tom. Thank you, you're a good husband. I want to lie down, after I take a trip to the head."

Tom asked, "You going to be all right?"

Norma replied, "Hell yes, I'm fine, I just got married to you, blue eyes."

Tom caught Albert's smile, and said nonchalantly, "Yes, I know, Mrs. Spoones. I'll be with you in a minute, I'm going to the galley."

She said as he turned away, "First it was the Wizard, now your galley, I'm not used to sharing. . . ."

Tom was out of ear range. He plucked a lemon out of the crisper, rolled it, then sliced it in half. He extracted the tarty juice and poured it into a glass. He then pawed through until he found a box of Arm and Hammer baking soda. He took a teaspoon of the white powder, leveled it off with his index finger. Both hands were occupied. He used his toe to tap on the door of the forward stateroom, and shouted, "Albert, Cap -- open up." Albert did just that, and looked questioningly at Tom, who said, "Here, take this for Laura. Make sure she has the glass in her hand, then put the soda in and stir like hell. When the foam nears the top, have her drink it, right down--all in one continuous swallow."

Albert respectfully said, "Thank you, Tom." Tom added, "Less than two minutes, she'll be fine."

Tom returned to his stateroom to find Norma sprawled across the bed, apparently in dream land. Tom had to cradle, and snuggled close to her to lie on the bed. He was tired. He would just close his eyes for a minute.

Tom heard a tapping, a long way off, and wondered what the hell it was, but was buffeted by Norma. He turned the opposite direction, and yelled as he fell out of bed. The tapping was louder. He got up, noticing that Norma was none the worse for wear, and made his way aft, toward the insistent noise. The rapid sounds were being dispensed on the glass, by a colorfully dressed, clean-cut Chinese lad. He picked up the four styrofoam containers, with whiffs of steamy Italian odors emanating from the closed barriers. The bill was clearly taped to the top carton.

The Chinese boy said, smiling broadly, "Are you one of the newlyweds?"

Tom smiled at the thought and said, "Yes, I am." Tom added an extra ten, and captured the warm savory containers. He hummed while making his way to the galley. He glanced at the watch on his left wrist, and his eyes focused on the bright gold band around his fourth finger. He fondled it, thinking with warm admiration of Norma. He smiled again, and returned to his cabin. Norma was stretching and yawning which infers a sensual stirring act in the male and perceptively brings forth the fantasy of a warm, cuddling female. Her catlike limbs, exuberantly feministic, and Tom's loins were becoming creative. He tried to play down the welling urge, and asked, "You hungry, sleepy head?"

She said, transposing her body, "Lovely food first, then you, Blue eyes. . .!"

Tom said, judiciously, "Oh, Jesus. . . " He was trying to get grips from his illusions and said, "I'll check on the other newlyweds, I'll see you in the lounge."

She said, with soft power, "Hold it, Tom!" He stopped, turned around and saw her come here finger. He moved to the little indicator. She arm necked and pressed, and received a mutual kiss. He sighed and broke his magnetic pull and moved to the forward cabin. He spoke through the door. "You newlyweds ready for dinner in 20 minutes in the lounge."

Albert reciprocated, "Thank's Tom, we'll be there.

Tom said lightly, "Okay, Teddy Bear" and Albert responded, "Okay, Blue Eyes."

Norma freshened up and went to the empty lounge. She sat down, looked at the Wizard's ring, and smiled with affection. She romanced the adornment with her fingers. She looked about, enjoying the luxury of the cruiser. She sighed deeply, rejoicing in the constant acceleration of life. She pulled out her pack, tapped one out, lit up, and cracked the adjustable window. She was having a benevolent smoke, got up and switched on the FM radio, until some peppy tune filled the lounge and lower helm. The syncopated jazz rhythms of "Ain't Misbehavin' . . ."

Tom saw the wandering swaying of "Blue Hue," as he approached the lower helm, and said excitedly, "You're surrounded with a blue aura."

She called to him, "Maybe, but I would rather look into your deep blues."

He was carrying utensils and napkins, and said smartly. "Hi, Gorgeous. Dinner less than ten minutes. Albert and Laura will be here shortly.

She asked, "Need any help?" Tom was moving and remarked, "No thanks, I've foiled it, and heating it in my Princess Oven." He made a quick trip with four glasses of water, and said, "Coffee wouldn't be appropriate. We're all coffeed up anyhow. Would you get some ice, and put it in the glasses--please."

She smiled and said, "You bet, Chef hors' d'oeuvre!"

Laura came in with Albert trailing, both smiling, and said "Hi."

Norma patted a spot next to her on the settee for Laura to sit, and spoke up, "I still can't believe we're married."

Albert said, "In judgment, I hope it'll always be that way."

Laura said, "Is Tom going to fix a Mattie's coffee?"

Norma grinned, and replied, "That's a no, Laura. Tom made it clear he thought we had enough coffee today," to which Laura responded, "Those cups were sure good."

Albert said, "Yes, they were that. I'll probably never know the secrets of that invigorating potion."

Tom had discarded the paper insulated cartons and placed the contents on the thick oval steel platters, covering them with foil in the oven. The girls picked at the salads until Tom appeared with the fragrant spaghetti, making two trips. Tom sat down and stated, "There's cheese in those packets, and I set our grated cheese to give us a choice."

Laura and Norma were saying, "Mmmm good;" Albert rolled his fork, drew it off, and inserted into his mouth declaring, "This sure is, Doc was right. How're your meat balls, Laura?"

She replied, "Oh, fine, how's yours?"

Norma let out a hoot, and couldn't stop laughing. Tom was grinning graciously. Albert was stunned. Tom had to laugh, then said, "Cap, you asked for that -- remember at Mattie's?"

Albert replied, "Yes, I sure do. I was quite surprised at their expressions."

Tom said, "Don't be, they had a little help!"

Norma questioned him saying, "What do you mean help, . . .did we embarrass you--is that what you're saying?"

Tom said, "These meatballs and sauce are good."

Norma said, in a low humming tone, "Tom, ... "

Tom chewed, then said, "You gals had a few more cups than Cap and I."

Laura quickly spoke up, "So?"

Tom put down his fork, slowly sipped his water, then stated, "I swore on my name about the goodies. But I will tell you this much. . .to clear the air, just one component, it has a considerable splash of alcohol in it."

Norma said, bobbing her head, "No wonder I was feeling so good."

Laura said timidly, "I had, I think, five cups of those delectable drinks."

Albert said, "Good Lord, no wonder! I should have guessed."

Tom spoke up, "Not necessarily so, Cap. Several additives and camouflage and conceal the real intensity of the ingredients. Generally, I would have selected a fine wine to complement this din-ner, but this is our wedding night, and, ... "

Albert interrupted with, "Honeymoon."

The girls grinned, and flashed a sexy smile to each other, but just a little embarrassing. For they knew no two women have the same fantasies. Norma, feeling her oats, cooked up a funny little smile, scanned each face, and rested confidently on Albert's features. She asked, "Cap, do you mind if I ask you a personal question?"

Tom hooted, "Oh, oh!" Laura glanced at Norma, but she wasn't looking her way, still wearing a contented smile.

Albert responded in a stately manner, "Norma, this is our wedding night."

Tom spoke up, "Norma, Cap's right, maybe tomorrow, huh?"

Norma crunched on a piece of lettuce, "Sure, it'll keep--you're right, you know how our curiosity sometimes gets the better of us."

Laura said, "Damn, Norma, you've pecked my interest--is it a good question?"

Norma said, "Mmmm, that all depends on who answers it!"

Albert said, shaking his head, "Norma, you and Tom, are like the sudden clash of two fronts, one cold, one warm, only the intensity changes the potentiality. I will yield to the dexterity of your skills. You win, Norma, ask away."

Norma was edging off her seat, her eyes were bright, and she had a keen cat-like look on her face. They all looked and saw her mask.

Tom said, "Norma, look at me now!"

She did, and the fallacy fell from her face. He patted her hand, and said, "Albert says ask away, but I want you to know . . I know your question, but not the answer. It might have better been directed at me, your husband."

Albert and Norma looked into each other's eyes, wondering what this question might be.

Especially since Tom's wizardry had plucked that unsaid question from her mind. Norma looked at Tom and asked, "God, you really know. . .Hmm, Okay, presuming you're right, would you ask that question?"

They all looked at Tom, and he quickly said, "No, I wouldn't ask, as I didn't ask you, but it isn't sacramental either."

Laura said, "Damn it, you two have me betwixt and between."

Albert said, "I concur, with me being, the between. . ."

Norma said, "Okay, I just thought that you, Cap, and this luxury yacht, and being an exemplary leader in your field, well the fact is, I'd be sexist, so I'll say -- neither of you considered pre-nuptial agreements!"

Albert let out a sigh, "Good Lord, is that it. . . ."

Tom said, "Yeah, that's it, ridiculous. . . ."

Laura looked funny at Norma, and Albert said, "It's of no consequence."

Laura said "Amen to that Albert," then asked, "Norma, let me direct that question to you?" Albert and Tom grinned and Norma said, "I didn't mean to come off as the wicked witch, really I meant, we're all salaried, and. . .Tom. . . get me the hell out of his situation, please!"

Tom said, "The concoction of those coffees really do have sadomasochism properties, each nervous system simulates this potion, with varying intensities. Norma was more adaptable for assimilation. Believe me, I'm not just making this up, because she is my wife."

Albert said, "I sure as hell know you haven't the capacity to lie or offend your friends, Norma. Everything is fine."

Tom said, "These coffees of Mattie's, will be out of your system in the morning."

Norma said, "I love you, Tom," and Tom said, "You were drug induced."

Laura was having trouble trying to wind her spaghetti, it was falling off, just before being accepted into her awaiting mouth. Albert had been watching, and became a bit nervous.

She giggled, gave up, and stabbed a meatball. She just caught the side and sent it flying off her plate. Laura said unnecessarily, for all were watching, "There goes one of my balls," and giggled.

Norma laughed and said, "Watch it, Laura, or you'll be ball-less."

Albert, using his fork, wrapped a generous bite of noodles around it, and moved it toward her open mouth. She graciously smiled with a "Thank you, hubby."

Tom asked, "Albert, I know that Wade is going to check our under hull tomorrow, but what next?"

Albert cut a meat ball in two, and was feeding Laura, to her delight. He stated, "We'll check our stores and then sail through Sheilkof Strait, south by southwest, latitude of 160° - longitude of 55°. He laughed, then added hopefully, "I know I speak for all of us, we began as four individuals bounded by salaries. The exciting adventure to possibly unravel, or detect, a fore-runner to one of nature's magnetic energies. This constant bombarding and storing in the troposphere seven to ten miles above the earth is a magnitude of electrical energy. This has horrendously built-up, massing like a violent cataclysm, releasing surges at times, but is at the verge of upsetting our fragile cohesion that relies, support, of each, in entity."

Norma was looking grim, "God, you send goose bumps through me, every time I come to the realization of what may happen."

Laura asked, "Albert, are you certain that this holistic act will materialize?"

Tom broke in, "It will happen as sure as you are sitting there eating spaghetti and meat balls."

Laura said uneasily, "J..e..s..us."

Albert said, "It's inevitable, the question only remains as to the intensity, rather than total annihilation." He napkined his lips and went on, "I just want all of us to realize how each day of our lives is not to be taken for granted, but a gift."

Tom stated, "Another reason our Captain has brought us together in this specific location in the world. . . ." He smiled and covered Norma's hand, then added, "This could be the birthplace of this strike force, but like an eye of a hurricane, we might get lucky if we can be in that solarium eye, when it chooses to erupt, and set up the chain reactions."

Laura stated, "I just want to say that this has been the happiest day in my life, with my husband and two good friends."

Albert said, "Let me add to that, day and night." Norma and Tom laughed, and Laura giggled. Albert grinned in anticipation.

Norma said to Laura, "We are certainly two fortunate gals to capture these two virile wonderful males."

Laura replied, "You're damn tootin' we are. . . . Mmmm, these meat balls are good, too!"

Tom said, "It's nice to have take out that's tasty, and I won't have much to do in the galley."

Albert said, "By the way, Tom, no set time for breakfast. . . ."

Tom smiled and remarked, "Thank's Cap, Wade won't be here till one in the afternoon."

The girls flashed smiles of apprehension to each other. Tom asked, "Laura, how is your arm?"

Laura looked up and replied, "Close to being on top of the world."

Norma asked, "Are we to have a tub schedule?"

Tom fondled her good ear and said, "Let's play it by this. . . ."

Norma said, "Okay, but you know how I love to soak."

Albert said, "Oh, yes, Norma, I believe you have brought that very subject up before," and continued grinning.

Tom said, "Her tub is like my galley. . . ." Norma interrupted, "Both places are refreshing."

Albert said, "Well, Tom, once again you have provided us with a fine dinner, although, not of your hand.   You carry the responsibility of catering.   There was one supplement missing -- Italian French bread."

Tom said, adolescently, "You said you'd buy dinner, remember -- you owe me thirty bucks."

Laura and Norma both laughed and Albert said, "I'm sorry, Tom, you're right again," and found his billfold. Laura said, still laughing, "You guys!"

Norma finished, patted her tummy, lit up and said, "The noodles have oodles of tootles." She then blew a cloud toward the vented window. She watched the layers of smoke being pulled, then swirling and escaping into the night.

She added, "I thought our wedding was magnificent, our March -- our song -- ceremony, walking on yellow -- silk flowers -- Tom who played that wonderful harmonica?"

Tom replied, "Doc Ward." She said, with bright eyes, "I wouldn't have changed it for the world."

Albert stated, "That's another remarkable feature of the human race. Our new friends gave us the time out of their lives, to enhance ours, in the solace of such an occasion."

Tom said, "The elusive power of destiny has brought us together. The four of us have been wandering this earth, for scores of years. In three days we have been brought together, not only as colleagues, but lovers, then matrimony." L a u r a stated, "I have a statement to make, not tonight, I will tell all of you tomorrow."

They all locked eyes with Laura, but conveyed a blank -- not even a hint of the mystery. Albert stated, "It has been one grand day, but a nice hot shower, then to bed, is the logical, and desired course."

Norma remarked quickly, "Albert, if you and Laura double up, there will be more hot water for Tom and me to soak."

Laura had a bite left and was busy chewing. She smiled and left the answer up to Albert. He said, with complete grace, "It would be a pleasure, right dear?" Laura looked up at Albert and spoke, "That's the first time you've called me dear. . . ."

Tom stated in a strange monotone, "Enlightenment can only be predicated by alleviating the known. . . ."

Norma broke the tension, "Jesus, Tom, those words excite the hell out of me."

Laura stated, "It's like a wondrous child, who just asked an oracle a question of great magnitude, and received an unqualified answer."

Albert said to Tom, "You never cease to astound me, is that a quote, something you remembered, or. . . .

Tom remarked, with his blue eyes flashing, "That -- that's just a comment, that enigma. Just for one previous moment, the epitaph began to open, the seminal, it's all there, merely for the asking, but as I reflect on what to ask -- it's gone . . ."

Albert stated, "Gracious, that must be an incredible fascination, to hold for even a second, the manifestation of life itself. . . "

Laura said dreamily, "Like a crystal that emits splinters of colors from each facet--with no ambient light."

Norma said, queerly, "Come on, you guys, you're spooky -- yeah, get on with your shower, so Tom and I can soak -- soak -- soak!"

Albert got up and said, "Shall we?"

Laura replied, "By all means, Teddy Bear!"

Norma giggled. Tom smiled and said, "Our 26th President would have been pleased."

Albert stated, as he helped Laura, "All I can say, with proximity, is that this was one hell of a fine day, good night all. . . ."

Tom asked, as Norma lit up, I won't be long. I'll clear off the table and wash a few spoons, Mrs. Spoones." She thanked him, for the chance of life. She was literally swept across the world, and thrown into Tom's aura. She had often thought, as a little girl--that invisible and specters determined your situations and destiny. The latter was all shattered when they told her that her first husband was killed. She remembered the radio playing, "Drinking the Lonely Wine," such a beautiful song, such a devastating shock. She had cried alone, until the Lex Baxter rendition had trailed off. She thought, "My God, was he the sacrificial lamb in the scheme of life with this intense love, she was now experiencing?" This was the second time in three years that she had thought of him. Destinies, she knew for certain, were going to be thrust in front of her. We have only to reach out and grasp them with courage. In this chaotic, unpredictable world, she truly realized she had plucked the gold ring from the carrousel of life. Life was never so precious as it was now. She would do anything that would predicate, or become an obstacle.

Tom came back, she got up and went to him, rushing with a wrap-a-round embrace, and held him very tight.

She said, "Thank you for the life raft to-night. That is one of my Achilles heels, talking without reasoning. I was way out of line."

Tom's eyes sparkled as he responded, "You're right, you know, we both are victims of those mortal words -- loose lips -- sink ships."

She said demurely, "I can hardly wait until we have my tub surround you and me. Remember, the conversation we had in the shower?" He smiled and said, his blue eyes sparkling, "You bet I do. You called me a Shine-a-Poo, and I was to be the judge as to love in the shower, compared to a tub."

She flashed a sensual smile that all but froze Tom, and said pleasantly, "I'm going to wax your ears!"

Tom said, "What do you mean?"

She replied, from now on, think, to when we will slip into the warm, cuddlesome, sensitizing liquid, our flesh upon flesh, two embryos, surrounded by a snug cocoon, romancing its mate."

Tom stared straight ahead and said, "Jesus . . . you said something about ears. . . ?

She said, "Yes, I'm going to wax them."

He thought of what the hell that meant, and knew, he hadn't the least idea. He stated, "You know the old proverb . . . seeing you have only one good ear, an ear in time, saves nine."

She laughed and said seductively, "We'll see Tom."

He was antsy as a beehive with the arrival of the Queen. They both heard Albert sweetly call, "It's all yours, newlyweds!" Tom swallowed, and uttered, "J-e-s-u-s!"

Chapter 18

Devlen Dawson received a fax from Port Graham. He was feeling his oats because of two unexpected buyers from Hong Kong, situated at the mouth of the Canton River. One of his captains, Dork Ninkeki of his freighter *Baracs*, had anchored with his catch in the hold. It was an extremely valuable cargo. Five gems, with the hardness next to a diamond, these fine stones had made their way at Instart, in Buimall. They had three temporary owners, until they arrived, in one specially marked bottle of Manchukuo soy sauce. It proved an excellent ploy as there were many fine Asian restaurants in Anchorage. The rubies were represented to him as cardinals, among gems, the finest is the color of a pigeon's blood. Carat for carat, compared to a flawless diamond, they were far more valuable and fetched enormous prices. Upon hearing about these five marvels, he decided that he would be one of the wanton owners of these precious stones. He had paid a handsome price, but he garnered one of the finest jewels in the world. He was merely handling, through interim buyers, and smuggling this small horde of five carats. Thirty thousand dollars out of his pocket, plus his twenty percent for safe transfer, had bought him one of the five jewels. He knew he had taken a calculated risk on having Captain Dork ram the *Magnetic*. He was humming as he took the fax to decode the message.

He jeered, staring into the screen modular. The printing appeared. "Laura McAgneti, had married Albert E. Dryfuss, another couple on craft, also was married, Tom Spoones, to Norma Moon. The Bayliner was dockside at Port Graham. Appears to be seaworthy. They are having the hull inspected for damage tomorrow. Signed Number 42."

Devlen swore, "God damn it!" He punched the key that represented 42, which in turn was reversed. The name popped up on the screen. Stan Marcus. He deleted the message, and punched in another code word, which was his list of ears. Each ditto he struck represented one hundred dollars that he would wire to their account. He struck the ditto six times, then scanned the screen for dates, and money sent, or to be sent. He discovered he was on top of his sentries. The exceptions were Stan Marcus, and Captain Dork Ninkeki. The latter was due at his home in two hours. He would stay here one more hour. He thought about what to do about Laura, the last string that could tie a knot around his throat. He would discuss this dilemma with Captain Dork.

He grinned and his jowls swashed up like two waves, then settled. Behind his flashy, silver plated buckle, with two sea nymphs embracing, he stowed his new prize--the five carat ruby of excellence. Captain Dork's eyes would pleasure to see such a fine gem. He grinned as he briefly fondled himself, knowing full well the ruby, was in the hidden pouch zipper behind the buckle. It felt good and gave him an excuse for doing the same.

Captain Dork leaned on the rust scalloped rail, and looked down, with mild interest, as the last of his cargo was being hoisted out of his ship. He turned his attention to the sea birds, then upon finding his momentary favorite, concentrated on one of the finest masters of sailing in the sky. He noted, as with himself, that all captains were not as skilled as others, or sea smart. Just as the gulls that navigated the skies some were just a bit more masterful than others. The shrills that emanated from their throats and out their beaks, were music to his ears. Most of all, they didn't mind that he possessed that permanent scowl on his face, they cared only for the bits of food he tossed to their screaming pleasure. He was always pleased when the cargo hatches had been opened to permit some of the fowl smell to escape. The stench, over scores of years, had permeated the old steel plates. He had the holds disinfected twice, but the rank scent, held its prominence. He was wearing his working black captain's coat. Three more were in his cabin closet. One was a new white, another worn just twice, black dress, and a third black working coat, and hats to match. The *Baracs* had been anchored but two days, however, he was itching to weigh anchor, and seek out a new port, or some old friendly ones. Many times he was sailing high, no cargo, just to take on goods. The holds of most freighters had to be full, or at the least, a minimum of 80 percent, or risk being labeled as a dealer in masked smuggling of materials, or humans. He reminded himself he was to meet D.D. shortly, a hot, salt water shower, was on the agenda. He grinned, or tried to, with his scowl, pulling on his facial muscles. His fulfillment was to get papers from D.D. and his bonus. He mused through his mind the location of the three blocks of Asian cultures, of sorts. He would take his pleasure, after the meeting with Dawson and a hearty meal.

The *Magnetic* did an overlay of gentle rocking, leeing and heaving, like a cradle of conformity, conveying a rhythm with its occupants, the newlyweds. The waves, and the *Magnetic* were having a romance of their own. Norma woke up after having a dazzling love affair of mutual sensations, exhaustion, then common sense peeked, just enough for realization, that rest and complacency was the manner to savor those lovely moments. She thought a lifetime with Tom was to be the most exciting, and rewarding experience any woman could desire--then the hammer fell, a stark flash of realization. The sudden reversal of polarities, resulting in a massive overload of the permanence, then the reciprocal of the magnetic reduction would trigger an electrical reaction that would annihilate millions of dreams in one sudden swath. The stone truth was, in her mind, what if she and Tom were in that net of destruction. The passion for living had never been as imminent as it was now--as crystal clear as Tom's blue eyes. She told herself she hadn't persevered in past miseries that every woman usually endures, to seek happiness, then to have it snatched away in one final termination. Her skin was marbling with goose bumps popping up, and a cold swell emanating inside extended to all her extremities. She feverishly grabbed Tom and shook him almost violently. His blue eyes popped open and he reacted nervously from his deep sleep. He stared up into the horror that was transfixed in her brown eyes, with her body trembling. He was about to shout. She touched his lips with her finger, then her wet lips. She moved to his shoulder, and he knew he had to choose his words wisely. She was irresistible, captivated by--God know's what, he didn't. He stroked her gently on the side of the face, and stated with ginger in his voice, "Honey, my ears are still ringing from that waxing you gave me." She was starting to calm down and replied, "You're right, you know and if you dare change, I'll kill you."

Albert woke up suddenly, ruffled, "What . . . "

Laura being a starboard girl, had her bum arm at his side. She had inadvertently flayed it, and hit his chin. She had cried out in pain. "Ohhh"--and apologetically said, "I'm sorry Teddy Bear." She stretched with her left arm and yawned widely, then spoke, "Ahhh, what a grand sleep!"

He flexed his arms and said in jest, "Your right wing is healing, that was quite a shot to my chin.

She said playfully, "Let me tenderize it," and she did.

He glanced at his watch and remarked, "My God, it's ten after ten."

She smiled, fluffed sheets and said perkily, "So the world's not going to end." She was sorry she stated that phrase and added, "Well, any how, you're the boss!"

He stated, "And you are the bossiness."

She laughed freshly and remarked, "We sound like cows."

He smiled and said, "I won't udder a word."

She said, "I feel so much better now that I have my facilities back. If I wasn't so happy, I would contend that you took advantage of me."

He stated, "As my wife, I would have done anything to gain your favor."

She said lovingly, "I worked up an appetite my love, how about you?"

He said graciously, "I'll slip on a robe and fix us some toast and orange juice."

She said, mockingly, "You are serious, you mean breakfast in bed?"

He said, "I don't believe Tom and Norma are up, but I'll fix us a snack. It'll hold us till Tom whips up his sumptuous breakfast." He waved, left his cabin and was surprised to see Tom beginning preparations for the first meal of the day. The coffee pot was on, but had not yet erupted, to send surges of pugnaciousness through the interior of the cruiser. The galley was a busy place and Albert had to move about to plead his case about toast and juice, all to no avail. Tom was assiduously defending his skills, but insisting that it would spoil their first breakfast together, seeing there were spouses. They parlayed, and Tom shifted to a higher gear. Tom told Albert, in no uncertain terms, that the only way he could help, was to find a tray.

Albert thought, and said impatiently, "Damn, Tom, I can't think of anything that would serve as a tray."

Tom told him with indifference, "Cap, take my bread board."

Albert smiled, and remarked, "Good Lord, you are a master under pressure."

Tom spoke like a prophet reading for a class, "I'm cracking your and your's eggs now. The spuds are a twinge away, the bacon is relaxing, ready to be plucked. Get two glasses of orange juice, Cap. Then take that, eggs, hash browns, and bacon on my board, with utensils and napkins. I'll follow shortly with toast and coffee. I spirited the spuds up a little with lemon juice for you and yours."

Albert loaded up his makeshift tray and made his way to his cabin, much to the surprise of Laura.

Tom, after a pesky, few minutes, completed his role as caterer. He rushed back, piled in, so his eggs could catch up with the bacon and spuds. He carried the main entree to his cabin, the steaming fragrance catching up as he plated her scrumptious dish, smartly before her.

She was smiling broadly, and stated, "Christ, you're a dream come true." Tom grinned and replied, "You're right you know and you are in my fantasies. I'll get our coffees, dive in." She said, "Thanks, Lover, hurry back."

He heard her faint voice thanking him for dicing her toast. He repeated his path, and re- trieved the steaming cups of java. The quarters still hung with the generous perfumes of bacon and hash browns. Tom placed her cup on the counter head- board, and asked, "Norma, do you sew?"

She crunched a long strip of crisp bacon and said, "No," then added, "Do you finesse the nee- dle?"

Tom was forking up brown gilded steamy spuds, and said while chewing, "A little!"

She asked, "Is there anything you can't do?"

He smiled and replied, "Yes, I couldn't possibly wax your ears."

She laughed, "Maybe if you have a second life."

He asked, "Do you believe in reincarnation?" She stabbed her egg with a point of toast, and put it to her awaiting mouth.

She nibbled the yellow graciously enjoying the adventure, and said, "No."

He said eagerly, "Then what?"

She finished chewing and stated, "Just checking. Neither of us has discussed our religion, passions in life, or our past-affairs, or. . . ."

Tom interrupted here, "There you go again, with that feminist most often used word--affair. I don't care if you seduced, or were seduced by half the male population, it only matters that we love each other now." He emphasized his last word and she reacted to it immediately. "What the hell does that mean?"

Tom sipped his coffee, giving him time for a credible answer, and stated while blinking his blues, "Now, is now, the synopsis, is present time, or moment." He bit a piece of bacon, and the snapping crunch sounded like a rifle shot. He looked down, right, and left, and then his images locked into the fury that her brown eyes were witnessing. He tried to chew quietly the already inserted bacon, his busy mouth.

He said gingerly, "Norma, do you know that moving the letters about, you know, mix in the word <u>now</u>, the word <u>own</u> pops up."

She replied smartly, with somewhat gritting her teeth, "It also spells <u>won</u>."

He said, "That's marvelous, you said it all, you've won my heart, my love!"

She said, "Albert stated it well when he called you a marvel, you are that, and more." She added with doubt, "Did you just set me up?"

He smiled and replied, "No, but a prone position is much more exciting."

She laughed and said, "You'd just like to have your ears waxed again, that's all."

He grinned, and stated, "Jesus, that's enough!" They both enjoyed the moment.

She kissed her coffee cup, then suctored it in her mouth noisily.

Tom looked up and remarked, "You're making me jealous of your cup rim."

She had the ability to flare her nostrils, and it sent his hormones flowing. She did just that and seductively remarked, "Come here, you galley slave."

He obligingly embraced her.

She held both his ears and asked, "Are they ringing?"

He replied, "Jesus, can't you hear them?"

She smiled and said, "We'll have something to look forward to tonight. You know, you guys seem to shy away from talking about your scarlet past."

He was a little annoyed at the subject matter and stated placably, "Would you like them in alphabetical order, or time zones, or talk about losing my virginity?"

She teasingly said, dunking a tip of her toast, "You're being a bit testy--although I would be interested in your first encounter with my sex-- was it her first time too?"

He remarked coyly, "If you pursue this line, I'll deny you one of the pleasures of life. . ."

She stated, "Look at me, Tom. . .what do you mean by that?"

He quantitatively stated, "The pleasure you have with the increments of the English language. You know I did rescue from an embarrassing situation -- you never really clarified. . . "

She suspiciously spoke, "I said I was sorry . . . Mmmm, this is a lovely breakfast, actually, it was intended to be a joke. You know, Cap would not have to pay her. . . ."

He was enjoying his limited prominence of this conversation and stated, "You know that Laura's and Albert's assets are now equitable."

She took a bite of toast, and replied, "Yes, I know, how are you fixed, blue eyes?"

He answered, smiling, "Just a dab in the bank, a little stock, no real estate--that's about it. What about you, my sweet?"

She replied, "A small bank account, I stored my car, other than that, just me -- blue eyes."

Captain Dork would have been impressive with the new and expensive white captain's coat and hat, except for the permanent frown, and only 5'5". He tipped the cab driver one dollar. He planted his tanned skin tight finger on the fancy door button. Chimes rang out on three separate levels in the impressive home. Devlen Dawson had excused his three housekeepers, and answered the door, wearing a plump smile. He stared at the little captain for a moment, then said, "Come in, Captain Dork. That's a handsome coat. Come, let's go in the study."

Dork placed his heavily embossed captain's hat carefully on a magnificent ornate teakwood, heavily spooled, entrance table.

Devlen headed straight for his elaborate bar, and asked, "Straight or mix?"

Dork also appreciated the good life and asked, "Can you make a quiet, colored Singapore sling?"

Devlen's jowls moved, and he replied with a smile, "Probably one of the best concoctions you've ever tasted. I'll join you."

Small patches of moisture laid claim to the underarm of Devlen's silk shirt. That was a chemical action that was always prominent when he was nervous. Both men had serious statements to make to each other, each biding their time, and waiting for a generous moment to strike and reveal their pleas. Captain Dork took command of his dazzling drink from Devlen's pudgy fingers. He tried to smile, but nothing changed the displeasured look.

He said, "I'm impressed. This looks like a clear rainbow." His turned down mouth clamped around the rim of the exalted elixir, and he stated, after a generous swash, "This is a butterfly mix. Congratulations, D.D."

Devlen took a hefty swig, and wiped his lips with his sleeve, and uttered, "Ahhh...that is a butterfly with a stinger--that packs quite a wallop. Ah, Captain Dork, you did indeed ram the suspected cruiser, however, unfortunately, the craft, and the occupants survived. They are anchored at Port Graham."

Dork was listening carefully, but watching the man's little black eyes, that were revealing intensity of excitement. He learned very young -- that was the real reflection of a man's soul. The fencing continued, Captain Dork stated. "I gave it my best shot, the *Baracs* is a hulk, and the steering is that of a tired dinosaur. But I did strike it twice, despite the high wake and sea surge of the cutting bow."

Devlen asked, "Would you like another?"

Dork stated, "When I finish this drink, then I'll be able to assess my capacity, thank you."

Devlen avoided Dork's uncompromising prune face. He caught himself losing his train of thought when he zeroed in on his guest's drawn frown. Devlen smiled, and picked up a thick envelope, and handed it out for Dork to take, making certain that Dork had to come to him to retrieve it. He stated, "This is your bonus for the attempted strike."

Dork moved closer but did not capture the white packet. He clearly declared, "I did strike -- twice!" Devlen abdicated his strategy and said, "Yes, a poor choice of words on my part."

Dork relieved him of the stuffed envelope, and slipped it into the inner pocket of his majestic coat. Devlen turned around and fondled the hidden pouch and extracted the gem in his closed fist. He smiled and faced his guest.

"I know you appreciate and entertain the finer culminations of life. Feast your eyes on my latest conquest, it is an extraordinary gem."

Captain Dork shouted, "Cause-celebre." He plucked the radiant red from his damp palm, and held it up, then slowly turned it a full 360°. He then pressed it to his forehead, and mumbled, "The praised one -- Mt. Hira -- Hegira!" Devlen was pleased with his focus and performance on the blood ruby. He asked hesitantly -- "What was that you said, or chanted?"

Dork shook his drawn skinned head, as in non-compliance, and spoke one work, "Religion!" Devlen was pleased that Dork may have let his guard down and decided to strike out, with his quandary.

He tried to relax, and stated, "As you know, Captain Dork, the loose thread in my tight spool is aboard the *Magnetic*." Dork listened keenly as he detected a hint of desperation in his voice.

Devlen continued, "Being you're a true seafarer, I thought you might lend your experience, for disposition of a certain person--for a handsome bonus, of course!"

Dork pondered, then stated, "There comes to mind two possibilities -- each depend on your take of human lives!"

Devlen, pretending not to follow, spoke up, "I don't know what you mean?"

Dork stated, his scowl more prominent. "Of course you do, Mr. Dawson, it must be quite collie-hangie -- when you accept the taking of four lives to assure the departure of one."

Devlen tried to smile, but it didn't come off, and remarked, "I would pay a higher prize, if just one memory, and tongue was no longer a threat to me -- but, if the taking of three or more is the only option, then -- Salaam!"

Dork stated soundly, "What would you pay for such a difficult task?"

Devlen had retrieved his ruby, and turned to the bar, while hiding his stone. He set about to mix two more of the delicious, powerful drinks. He said in a matter of fact voice, "It shouldn't be difficult. Just one person in a world of over a billion." He was immediately sorry he made that statement.

Dork sensed he held the reins and spoke up, "What might be your approach to solve your predicament?"

Devlen attempted to laugh it off and said, "That's not my pedant!"

Dork decided to squeeze a little more and stated, "Of course it is, that's how you and I have risen to enjoy the pleasures of life." Then he added, "These two men who are accompanying your victim, do you know anything about them?"

D.D. finished a fresh pair of drinks, the same mix, and passed the rainbow liquoring drink to Dork. He replied, "No, not really, except they seem to gather from the States, and their craft has no lien on it -- they stocked up stores to last weeks." He hinted, that cruiser, being new, must be quite a trophy. . . ."

Dork stated, "You know as well as I, temptations of snatching that craft would have as much chance of lasting success as missing the loud buzzing of bees around their nest." Dork squeezed his liquor through his staunch lips, with gratification, then released it to his throat, cascading into the stomach acids, and the friendly liquids romancing.

Devlen quickly remarked, "Ten thousand."

Dork sipped his rum laced drink and spoke

up, "That figure places a very low esteem on such a precious life. It is unacceptable," then added, "Like trying to retain water in a sieve, you can recruit many to do your deed, but you and I know, that would compromise your identity, just as the smallest of a slip." Dork decided to take a chance and stated, "I'll pass -- only one prize would I accept this challenge. . ..!"

Devlen was exposing his desperation on his immaculateness, of leaving nothing to chance. Tiny beads of sweat began popping up on his forehead, and jowls, and the back of his pudgy hands.

He spoke out demonstratively, "Name it!"

Dork directed his energy and responded with one word, "*Baracs*." That caught Devlen off guard. He looked like a sacrificial fat lamb.

He sputtered, "Now, Captain, let's be reasonable."

Dork cut him off and stated, "I could have asked for two ships, or one ship, and that wondrous bloodstone, and remember, Mr. Dawson, I am responsible for your obtaining those two sister ships. . .think about it . . .you will have no threads dangling from your spool." He knew he had him.

Devlen sloshed down his drink, some of it swilling down off his rounded chin. He asked, "What about my commitments. . .?"

Dork stated, "I will honor them -- but the ship registry is to be in my name, within ten days."

Devlen tried to smile, intimating that he had the better of the trade, but his mouth was dry at the thought of losing half of his ships in just one deal.

He drained his glass, then said in a bullying voice, "All right, you have nine days to rid me of Laura Dryfuss. She was just hitched yesterday."

Captain Dork stood as straight as his little frame permitted. His second dream in life was about to become a reality. He felt generous and stated, "I'll maintain your phantom names, for one year from the day I receive my papers. The authorities will never really know which ship is where. Just notify me when you want me to reverse the letters of my ship. Who knows, maybe we could strike a bargain, after the free first year, and then we'd be back in cahoots again!"

Dawson thought that over and remarked, "I knew I picked the right man, I'm sure we can be in cahoots once more. It has been a profitable relationship for both of us!" He added, as he went to this desk and retrieved a photograph. "This is the knot in the thread," and handed over a picture of a smiling Laura.

Dork briefly glanced at it, then stowed it away, in the inside of his neat coat. He asked, "When may I ship out?"

Devlen replied, "There is a cargo waiting to be picked up in fifteen days in Port Wenchow, China." He retraced his steps to his desk, and withdrew a large 9" x 12" packet, including all the papers of clearance and the cargoes to be stored. Dork finger saluted from his forehead.

Devlen responded, "Done, you need not call me when the thread is severed from the spool. I have eyes. . . "

Dork said, as majestically as possible with his face, "Pleasure to serve you, Mr. D.D." That brought a smile worthy of an old derelict, badly in need of major repairs. Dork could hardly hold the excitement to return to his ship, and pull the long, low, lonely ship's whistle.

Wade Freedman, owner of Mattie's, was dressed in his wet suit, and with his flippers tucked under his arm. He was leaning over, tapping on the glass and shouting, "Ahoy, ahoy there!"

Albert loudly remarked from within, "Coming Wade." He scrambled onto the deck, leaned over, and shook hands. Albert said, smiling, "Thanks for being so prompt."

Wade nodded and stated, "Please shut off your bilge pumps, and I'll slip down, and about." Wade donned his flippers, knelt down, and scooped up a cool handful of seawater and splashed his face. Then he put on and adjusted his face mask and checked his oxygen intake. He slipped in and disappeared in the blue green waters.

Laura came topside looking for Albert. He saw her and stated, "Wade in. . ." She interrupted him and determinedly stated, "I'm not going in the water."

Albert laughed, then said, "No, no, it's Wade from Mattie's. He's in the water inspecting our hull."

Laura laughed, "I love that word, our!" Tom and Norma made their way topside, talking about ears. Laura picked up their conversation and asked, "How is your ear this afternoon?"

Norma smiled, realizing the misunderstanding, and replied, "Great, how's your flipper?" Laura answered with buoyant vigor, "Oh, it's much better!"

Norma said, "Don't forget, exercise time tomorrow morning--maybe we can sneak in a soak. . . .!" The men laughed, and underwent a flash of jealousy, and fantasy.

Albert spoke up, "Thanks again, Tom, that was an exuberant breakfast our wives had in bed." Norma remarked, "The bed is the best place for pleasures."

Tom grinned and said, "She's right, you know. . . ."

Albert gestured and said, "Look, that's Stan coming towards us, the law of Port Graham.

Stan waved and shouted, "How're the old married couples?" Waves and smiles were returned.

Tom slipped through the transom door and they shook hands firmly.

Stan said, "I thought you might want yesterday's newspaper from Anchorage."

Tom sensed an uneasiness about Stan.

Stan asked, "How's the inspection coming?" Tom replied, "I don't know, I just came topside, thanks, for the paper."

He looked above and to his left and stated, "Looks like some nimbi stratus boiling up northeast."

Stan shifted his weight and stated, "Yeah, looks like a pretty good storm moving this way. You folks might want to watch yourselves." The gulls were screaming and wheeling, more and more were gathering, hoping for a feast. Three bobolinks were strutting their thin legs, their beaks pecking into the cracks of the aging timbers, and occasionally plucked out a wiggling insect from the piled driven logs. A settling calm was all about, another of nature's tricks, the calm before the storm, and the air pressures would collide. Splashing noises were prominent close to the trim tabs, mixing in with shrieking and gliding gulls.

Albert thrust his hand out to help him aboard onto the transom's long aft strips. Wade shoved his mask up, and shook his head, and firmly grasped Albert's awaiting hand. They locked and Wade was pulled from the foamy water. Two ospreys were observing the deliverance from the waters as they circled high above the screaming gulls and bobolinks. Wade shook himself, removed his foot forward, drew his hand over his healthy face, stating, "I detected traces of sea lace, the remains of a sea mouse. . . ."

Albert interrupted and excitedly stated, "Good Lord, man!"

Laura let out a refreshing laugh, then stated, "You're good, Mr. Freeman--Wade," she happily remarked. "Your sea lace is sea weed, from the Chorda Filum family, and your sea mouse is a large worm, Aphrodite." Norma asked, laughing, "Does that sea mouse have whiskers." Laura was contextually enjoying her field of marine life, and replied, "Actually, it's covered with hair-like setae. . . ." Norma said, "Well, I'll be damned, but I didn't bring any mouse traps."

Wade smiled, enjoying the conversation, then stated, "Nary a crack."

Stan said to Tom, "Well, I'd better get, are you folks leaving today?"

Tom replied, "I don't know."

He went back through the hinged gate and Wade stated, "I'll tell you this, thanks for paying me to dive, there are some interesting, jutting, irregular sea beds that don't conform with the contour. I'm going to investigate the floor, when I have more time."

Chapter 19

The brick walls are 20 feet thick, and 25 feet high. There are 213 gates and two water gates on the Pearl River, two outstanding pagodas rise 160 and 170 feet respectively, near the western gate. The city is nestled inside the protected walls and is very modern. Canton, China is a busy community port.

Dork was elated with the presumption of being a master of his own ship. He cabbed to a small shop located at the apex of Third Avenue and L Street. The proprietor was Dang Kota, an old acquaintance. He was a Mongol of sorts, a breed between Cantonese and Malayan. His back room was as large as his display room. Six tiny bells jingled when the door opened or closed, reverberating a pleasant high pitched ringing. The back room was closed off with thick Mandarin drapes that hung down to the floor. Embossed on those red drapes, were fearful yellow dragons poised. The mate was a hinged and seamless steel door. Dork entered the quaint shop to the musical bells and the smells that piped aromatic scents, like stepping into a refreshing rain forest, without the screeching of birds.

Dang heard the alerting high delicate tingles and pushed aside the ornate drapes. He recognized his customer, grinned, bowed and spoke, "Ahh, good to see you, the face that expresses."

Dork did a short bow and replied, "I have good news for you, Old Friend. May we negotiate, fairly, in our room of rooms?"

Dang gestured towards the door opening, walked to his window, turned the little sign around, and locked the main front door. Dork had already seated himself in the high-back, stiff teak chair. His face was close to the top of the plain but sturdy black table. They exchanged more personal greetings, and tuned in on each of their prosperousness. Dork accepted tea, a delicate drink laced with Ginseng, from the roots and leaves of the genus Panax, that flourishes so well in most of China. Dork sipped from the small, ornate cup, then smacked his lips, much to the enjoyment of Dang.

Dork said with grandeur, "I feel that all my ancestors have joined together and willed, my second greatest prize in my life, before the great journey to be with them. I also wanted to share my joy and good fortune, with an old and good friend." Dang's old eyes glistened and his grin was as permanent as Dork's scowl. An excellent portrayal of Ying and Yang. Dang noisily sipped his drink, and asked, "What is your pleasure?"

Dork held up his small tanned hand, and stated, "Business first, old friend, then I will have pleasure!"

Dang still smiled, and said, "Yes, yes."

Dork spoke up, "I want two plastic jelly devices, time set, to just penetrate, a poly hull. I wish no bodily harm for those aboard. My second request, is two plastic, throw away, hypodermic syringes, with three capsules of curare. Not to kill -- about 130 pound frame, to just facilitate motor paralysis. . . ." Dork sipped his tea, then continued, "Can you arrange this in one hour?"

Dang replied, "Anything can be arranged for a price, but I must tell you, old friend, it will be expensive."

Dork replied, "That is why I came to you and not the streets. I know you will be fair!"

Dang replied, "Is there anything else?"

Dork sated, "Yes, I want a pleasure, the finest that you have to offer."

Dang widened his grin to its extremity and spoke graciously. "I have a peach from Canton, not a virgin, but only plucked once, a blossom of 15 years, fragile and flowering"

Dork said, "I will take my pleasure, while you gather my goods."

Dang rose, bowed, and stated, "I will return shortly."

Dork got up and complimented him with a bow, then sat down in the hard chair to finish his tea. Dang retreated through the back door, locked it carefully, and hustled down the alley. Dork finished his tea, got up, and slowly removed his white coat, and hung it on the back of the chair. He could hear a faint clicking of metal clashing, then the sound of pleasant bells. Dang held her hand and brought her close to the table, then released her.

The young, thin Chinese girl stood straight, wearing a pretty black and white wrapper of silk, with a black fringed sash. Her face was flawless, and she continued to look down at the floor. Dork moved in front of her, and with his hand, he gently placed it on her smooth chin, and raised her head to peer into his eyes. She was fresh and delicate. She blinked once upon seeing his scowl, then retained her unconcerned stare. He gestured to her with his hand, indicating to open the wrapper. She undid the sash, and using both hands, grasped the hemmed borders and opened her robe. Her youthful, lean body was nude, she spread her arms wide, for him to see her nakedness. Dork took his time, noting her budding small breasts, and just a brush of black hair, above the neatly tucked envelope. Her eyes showed no revealing or apprehension. She remained in her stance.

Dang moved to Dork's side and reviewed her loveliness. He took out a small plastic compact, opened the holed lid, and extracted a wiggling, busy insect from the Gryllica family, and placed it in her navel, and had her release her right hand, and use her fingers to trap the insect in her round belly depression.

Dang clapped his hand gleefully, and said in a high voice, "Listen -- the cricket does not chirp, it is of the female gender, you need not worry, about planting your seed." Dang then gestured for her to pluck the cricket from her navel, and place it in the open little caged that he held in his palm. She did, and then grabbed the hem of her wrapper, and slowly resumed her spread arm position.

Dork took both her hands and lowered them to her side. Dang went through another door that was a small cubicle, containing a modest, neat bed and one chair. He turned on the wall lights, the two bulbs burned brightly, one yellow, one blue. He then withdrew, a small rectangular colored box with two green and orange fighting dragons across the top in exciting print--an oriental dragon, and assorted incense. Small letters at the bottom indicated that it was made in Japan. He removed the top, and little cardboard dividers separated three colored small pinnacles from the 24 in all. He selected two different colors, one red and one black. He placed them with spikes up, then lit both of them, after setting them on a hand carved wall mounted incense burner. Little swirls of dark smoke rose as the slow burning began and the puissance odors mixed with licorice and cinnamon. The glow gave off a potent, fragrant, and pleasing incense. He left the tiny room with door open and bowed. Receiving two in response, he left through the back door, locking it, testing the brass knob meticulously with a firm pull.

Dork indicated for her to remove her wrapper, and hang it over one of the chairs. He sat down, turning his chair from the small table, and raised one foot. She quickly dropped to her knees, not looking up, and removed his shoe, then his black sock. He kept his eyes on her, and she repeated the same gesture. He then indicated his shirt. She slowly, but skillfully, unbuttoned each disk. She undid the last button at his waist, and gestured for him to stand up. He did so, she unbuckled his wide belt, and unzipped his fly, carefully, and slowly, still not looking up. She then unbuttoned the last two buttons on his shirt, and help him remove his trousers. Then she knelt back on the floor and slid his short cut-off underwear, down to his ankles. He stepped out of them, took her hand, led her into the small, colorfully lighted, perfumed room, and gestured for her to close the door.

Wade scanned the sky, then pointed, north by northeast, and stated, "Folks, that looks like a heller coming. You'd be wise to settle here until it blows over. It's moving pretty fast, and those dark clouds means it'll be show time this evening. Our inlet will pat and soothe the fury. But if you leave, and continue, south by southwest, you'll play directly into the force at Kennedy entrance, looping up with Cook Inlet." They all stared at the distant black rolling clouds, as they crept nearer.

Norma broke the silence, except for the crying, baying and shrieking gulls, and said, "Cap, if it's all the same to you, it'll be nice to be married for two days, and nights."

Albert said, "Thank you, Wade for your concern, we'll certainly take that into consideration. In case we decide to weigh, I want to pay you for the inspection."

Wade said grinning, "Make it easy on yourself." Albert withdrew his wallet and selected a Franklin.

Wade perked up as he accepted the bill and tucked it up in his wrist, under the rubber suit and said, "It's been nice doing business with you folks, I should really pay you. If those cluttered, raised sea beds under your boat prove to be a catch."

Tom shouted, above the diving birds screams, "Don't sell the Wizard, I would like to give him another chance--some day!"

Laura piped up, "I love your coffee, Wade, I think it has medicinal characteristics!" She tried to raise her right arm.

Albert shouted, "Be careful, dear!"

Norma remarked, "Laura, tomorrow is our exercise day, if you're a good girl, we'll have our last soak!" The clamoring of the gulls grew louder as if sensing the changing weather patterns. Wade gave them a last wave. He looked funny with his bushy red hair in that black rubber husky suit, getting smaller and smaller.

Laura said joyfully, "What about coffees -- I'll make 'em."

Tom grinned and said, "Thanks, but I can take a hint, and what the hell, it's a good idea," then added, "Maybe we better have a pow wow."

Norma said, "I've never seen a blue-eyed Indian!

Laura asked, "What about Hiawatha?"

Norma said, "That's a girl Indian."

Tom stated, "Not so, my love, it was a he, Chief of the Onondaga Tribe. He was also believed to possess magical powers."

Albert said, "If I had to choose between a computer and my choice of knowledgeable men, you would be my choice, Tom. Your mind has an extensive capacity of storage, and recalling of great magnitude. Then there's that magical, spooky sequence that inadvertently slips in and out, like bolts out of the blue."

Norma said, "Out of his blue eyes, into those deep pools, Mrs. Spoone is so cool."

Laura remarked, "What! Are you going mystic, Norma?"

Norma laughed, and said, "No, it's just jogs my memory about the poet, Longfellow, who wrote poems about immortalizing Hiawatha. They joshed one another, as they made their way to the lounge, except Tom who headed for his galley to set up the pot. Norma sat down with Laura next to her and Albert adjacent to Laura. Norma spoke so both could hear, "Laura have you seen those two pesky double-headed cobras your husband sports?"

Albert said, "Good Lord!" Laura tried to grin in case it turned into a joke, and replied, "Are you serious?"

Norma countered back, "You can bet your bra on it."

Laura looked at Albert, "What does Norma mean, do you have snakes....?"

Albert laughed discretely, "Laura, honey, it's just a tattoo!" Laura asked, "Where is it Albert?"

Norma said, "You know, honey, that's the same line I used!"

Albert said, "For heaven's sake, you're making something out of. . . ."

Norma stated, "Do you want me to relate about you stepping out of a shower, with the towel around your middle, and the *Magnetic* lurched, and the towel and me. . .. "

Albert said, "Thank God, here's Tom. He'll straighten out this misunderstanding." Norma laughed and laughed.

Tom set the steaming coffees down, and asked, "What did I miss?"

Laura said, "The time Albert came out of the shower with a towel around him, and the boat surged, and the towel . . . ."

Tom laughed, and said, "Yeah, that was some story Norma made up. The Cap had to tell us where his nest was that contained his two headed cobras!" They all had to laugh.

Laura nudged Norma and stated, "I should have known how you guys mess around, but I'll tell you one thing, my Albert is going to introduce me to those two fellas tonight."

Norma laughed and said, "By golly, Laura, that's one thing I forgot to ask Albert. May I?" and Laura replied, "Be my guest."

Albert said, "Thanks a lot, dear."

Norma remarked, "Cap, what sex are they -- are they in a compromising position?"

Tom was enjoying this bombardment of spicey conversation. Laura said, "Okay, I'll find out, and tell you both tomorrow."

Albert sipped his coffee and tried to avert their testy eyes.

Tom said, "Seriously, what's your decision, Cap?"

Albert spoke up, "Let me say this--this approaching weather front appears to be a dandy, electrical spectacle. According to the Coast Guard, warnings are up, and if the winds sustain their course, we'll be moving with it."

Tom said, "That would give us an excellent opportunity."

Albert remarked, "Yes, indeed, it would."

Laura stated, "I'll not interfere with your decision. Whatever you choose, but let me also say this. As a marine biologist, I've studied the cold blooded chorditis and vertebrates in our seas. Six months or so into my studies, it became relevant that their sanctum depended on food, sea bed life, and the atmosphere above, not including the pollution of waters."

Norma asked, "What are you driving at?"

Laura added, "Just this, I also began to study all complements including how weather patterns related to the pertinence, and found some of the most violent storms in the world are located here in the Aleutian chain." They searched each other's faces in search of enlightenment, and found none.

Norma stated, "We need conductive electrical charges of high intensities--a lightning ball would be a prize."

Albert stated, "If we could see 40,000 feet or more, and have the opportunity to scan the top of the storms, and measure the upper currents, we could predict hurricanes."

Laura firmly stated, the major shift of weather patterns is raising havoc with the overflowing of rivers and seas. Various areas have been rising with torrential rains, heavy seas, and winds are slowly rotting and skewering lands, where sea sand land joins the mutual complacency. The mighty seas are certainly winning the battle of digesting the land!" She looked about and continued, "Major changes in the earth's strata and faults have increased our underground temperatures!"

Tom winked at Albert and said, "You have one smart gal. Her input is invigorating . . .you'd better take good care of her."

Albert beamed, and with a smile said, "You're right, you know!" then quickly added, "We have great potential for witnessing and measuring the conductance of a corposant--or so called St. Elmo's fire."

Laura said, "I've never seen one, what does it look like?"

Albert replied, "The electrical massive forces of energy are concentrated in a ball like image, thousands of times more powerful than a stroke of lightning. It can be witnessed during violent storms, at the tip of a mast, or mountain top. It's like a hissing tongue of white and blue extraordinary light."

Tom stated, "The master of electricity, Nicholas Telsla, died before he perfected the scheme to extract electricity from those balls of lightning, to present to the world, a cheap supply of current."

Albert stated, "Electrical discharges, traversing the atmosphere, cause the sudden expansion of particles due to heat, and then with a sharp compression of those beyond, causing the thunder, by a wave of motion, thus initiated." Albert took a long swallow of coffee,, licked his lips, then added.   "These massive amounts of electrified particles are gaining in intensity and join the magnetic disturbances, which is increase and overload the monumental. Every flash of lightning, releases a burst of radio energy at kilocycle frequencies which travel to the north and south poles. The surfaces of the ionosphere act as the charged plates, a condenser storage for an electrostatic field of thousands of volts per foot. The magnetic field, above us, is a strong electromagnetic field that is swelling every day and saturating in a violent build-up, of electromagnetic properties. Remember, our galaxy is held together, in principle, that all principles remain stable."

Norma said, concerned, "Jesus, and we know that amplified energy, interacts with the heavy resonating in the bloating magnetosphere, that in turn, added to the magnetized particles in the Van Allen Belts, produce heat, and the fallout of charged particles."

Laura said, questioningly, "Then what you all are saying, is that our atmosphere is becoming critically saturated with electro and magnetically charged particles higher than ever recorded."

Tom said, "That's right, you know, and therefore, storms, lightning, earthquakes, and winds will be more severe than ever recorded." He tapped the newspaper and added, "Stan was good enough to pass me this from Anchorage, a day old, but let me read these items." He sipped a bit of his coffee, then continued, "These excerpts, Antarctica, more violent light bombards, over 300 million tons a day are being dumped in our world, that cannot be absorbed. . . .here's another, ultraviolet radiation rose by 30 percent during the last few days, along the central coast of Chile. Researchers at Santa Maria University in Santiago stated that the rise in U.V. readings is due to a record depletion of the stratosphere ozone this season over the higher latitudes of the southern hemisphere."

Norma remarked, "Jesus Christ."

Albert shook his head and spoke up, "I knew it was getting worse, but not at this rate!"

Tom continued, "Yeah, I'll say -- and look here's another item. Laura was telling us what was happening, referring to the rising water levels. In the Caspian Sea area, waters claimed 80,00 acres of farmland in Northern Iran. A Mazadarn official said that the heavy rain falls, overflowing rivers, and encroaching sea, had inflicted heavy damages to roads, bridges, homes and farmlands."

Laura said, "You bet, the seas are swelling and claiming more sovereignty, and what with the transgressing, in our heavens, is clandestine."

Norma asked, "Damn it, isn't there any good news?"

Tom looked at Albert and shrugged his shoulders. Albert looked at Laura, then back to Norma and stated, "I don't know, I haven't read the paper, but the news is negative because we have upset the delicate balance too long. Mutual cooperation is almost unattainable, due to third and fourth countries, unable to afford the exorbitant costs. It's difficult enough to squeeze any moneys from the leading countries of the world."   They all looked haggardly into each other's eyes, realizing the consequences of what was happening. Seemingly very few humans in the world, were ill informed or presently uncaring, without understanding the magnitude of the awaiting disasters.

Albert said, "I'll add another diversity, approximately every eight days, the sun's rotation brings a new region of interplanetary, solar magnetic field opposite us, and the earth's field is slightly changed in response to the erratic flip-flop in polarity.   The sector boundary's passage, also induces a day or two of violent turbulence in our earth's field."

Norma began trembling from the absorption of the conclusions, said as cheerfully as possible, "Cap, are we going into the storm front . . .or?"

Tom inserted, "Yeah, Cap, what's your decision?"

Laura looked into Albert's eyes and said, "I'm with you, dear."

Albert looked into their faces and stated, "If it was my personal choice, I would stay in a safe harbor during this storm. . . but I have a commitment to S.A.W. and as a scientist. Knowledge is power and power may save our lives when the melt down of overcharged electromagnetic fields collide and began commending the chain reaction of electrical and magnetic forces. I say, weigh anchor, and attract!"

They clapped lightly, and grinned, each attempting to give support to one another. They now were caught up in the exciting drama.

Tom said, "I'm going to topside and un-shackle the *Magnetic*."

Albert moved down to the helmsman's chair and started the twin engines. He then moved topside to the flying bridge. Norma grasped Laura's good hand and gave it a squeeze, and received one back.

Norma said, "I"m going to batten everything down and make certain where I stashed some of our emergency equipment." She laughed, then added, "Laura, why don't you monitor the radio storm warning data?" Laura replied, "Sure, Okay, be careful."

Albert engaged the engines, and the *Magnetic* settled aft, as the shrieking gulls displayed their aerodynamics and screaming expressions of grandeur.

Powerful magnified lens enlightened the determined eyes behind them. He watched the departure of the cruiser. Stan noted the time, and went to his desk, and unlocked the bottom drawer. He secured a fresh sheet of paper, and transposed what he had just observed, into the relationship of the code. Having completed the deceitful message, he typed on his fax, to the home of Devlen

Dawson.   He leaned back in an old oak swivel chair, and drew his hand over his wrinkled, sun-tanned face, pulling his cheeks, with his tense fingers, then moving down to his chin. Still looking up at the dull ceiling, he wondered if it was worth it, selling his data.   He tried to console himself that no harm would come from providing the information.   After all, he was just passing on events of the day, but what if. . . ."shit, forget it," he said aloud.

Laura wrote down the areas likely to be affected and the probable intensity of this large electrical storm. She bumped her bum arm in the excitement, and said softly, "Damn it." She then went to her stateroom, and secured Albert's cap, and made her way to the fly bridge. Albert beamed at seeing her, and his hat. She smiled and said, "Here, Teddy Bear, cock it to the left."

He smiled and did as he was told, and said, "Come here, a little closer." They kissed warmly, and held their lips a trifle longer than usual. The black storm clouds were getting a little closer, but the aggregation was amassing in size. Less than two hours of day light left. Laura jumped and turned to her left as a strong roll of crabbed thunder could be heard. Its menacing echo was equal to its power and approachability.   Tom   found Norma on her hands and knees in their cabin checking the locker stowage aft. Tom grinned and said lightly, "Praying already?"

She looked up in apprehension and said, "Yeah, blue eyes, I'm praying. I remember where I stashed some of our emergency equipment."

Tom said, smiling, "Come on--off the floor, you're rattling my fantasies, Mrs. Spoones."

She smiled, and asked with complacence, "Tom, will you promise if you inherit a signal from the cosmos--or wherever, no matter how devastating it is, promise you'll tell me--I'm just a little hesitant about this storm."

Tom replied, "Sure, I promise."

Devlen's fax machine ignited with a message. He ripped it off and went to this desk, retrieved his decoder, and transcribed the information. He was aware of the storm in that area, through his Coast Guard channel, that was in his study, and tuned in 24 hours a day. He smiled when he discovered who had sent it, and that it complied with his information. He relayed the message to Captain Dork on the *Baracs*. He finished with a smile, and thought what a great life, and began whistling, while fixing a strong Cuba-Libre.

Dork was summoned to the small radio room. A thin, brown Filipino named Dassat, from the island of Palawan, handed him the message. Dork took it and asked, "Did this just arrive?" Dassat replied, "Yes, Captain." Dork had acknowledged, he had signed on for a little over two years, and had kept his nose clean. Dork also knew that this man had more talent than he perceived in his behavior. Dassat, as a seaman, had known for over two years that the messages were coded. They made absolutely no sense, to anyone with nautical association, yet he had said nothing.

Dork scowled and tramped all the way back to his lonely cabin, and translated the communication. He swore out loud, "Damn -- damn." He tried to sit back on his much to large worn leather chair that squeaked when swiveled. His mind was reacting on what to do, what course to take. They were coming to his anchor. He had planned to attack while they were moored. His dinosaur, *Baracs*, was no match for their speed or maneuverability. He withdrew an old pocket watch he had bought from a waylaid Portuguese mariner, who had no money for rum. He pressed the knob, and the hinged case snapped up. He hastily noted the time and went to the overhead chart containers. He found the one he wanted, and realized there was less than one and one-half hours of honest daylight left. Dork pulled his code converter file and wrote down the message he would write to Devlen Dawson. He took it down to Dassat, and asked him to send it immediately. He then informed him he would be in his cabin, awaiting his response.

The *Baracs* had sailed out of the outer fringe of Cook's Inlet, and was anchored just off English Bay in deep water. English Bay is a small settlement that lies on the edge of Kennedy entrance, and port to Kachemak Bay, inlet of the English Bay River.   Dork was anxiously awaiting the return message from D.D.

His intercom activated, and Dork responded with, "Bring the message to my cabin." Dork met him at the open bulkhead door and thanked him. He quickly deciphered the message. The translation informed Dork that Laura had a brother, first name Linnel, living in Eugene, Oregon. Dork's radio was issuing constant inserts, every fifteen minutes, from the Coast Guard updating the progress of the storm. The wind had changed slightly, and Dork would have to rudder, then re-anchor, to maintain his bow with the swell of the deepening troughs of waves. The waves were just beginning to cap, and commencing to heave. He was sitting in his big chair, his feet resting on an old oaken box, six inches high, tucked next to the large swivel captain's chair. He realized the storm front had expanded. He complemented the situation. The *Magnetic* was heading for his anchored freighter. He just had a little over eight days to dispose of Laura, and fourteen and one-half days to reach his new port, Wenchow, China, located on the east China Sea. Dork thought through the limited possibilities. Only one seemed tangible. The storm could possibly be turned into a friendly catalyst in his need for dramatics. He was forced to act. His prey had made the first move that was not contemplated. He intercommed to his radioman,

Dassat, and instructed him to recruit seaman Adda. Both reported to his cabin promptly. He pulled his large pad before him and began writing a message that would be put on the radio to the unsuspecting crew of the *Magnetic*. Dork had no reason to believe that the captain of the cruiser would have his broadcast turned off the Coast Guard band due to the impending developing storm. He finished, and was rereading it for accuracy when he heard the tapping at the steel door. He got out of his chair awkwardly and shagged it to his door. Then opening it, and saying as he turned, "Come in, come in." He recaptured his chair and adjusted his feet on the little worn box. The two thin, lean Filipinos walked up to his desk, and stood silently with a noncommittal look in their eyes. Dork had selected these two specimens, not just because they would act, but their race were good seamen, and excellent swimmers.

Dork had laid two wide waist belts, that had many evenly divided pockets on his desk. Dork stated, "These belts both contain the same goods. A plastic explosive, with timer, and a hypodermic syringe, and one capsule containing a remedy for motor paralysis," then added, "Do you know what I just said?"

Dassat said, "To boggle the mind?"

Dork replied, "Yes, she will appear drunk of sorts, just so she can still walk, but will have to be guided and held." He went on to explain that the cruiser, *Magnetic*, would be docking at English Bay. It was critical that they be waiting in hiding, and accomplish the placing of charges, port and starboard, before the passengers disembarked the vessel. There will be two men and two women aboard." He handed the photograph of Laura. The two men took turns looking at her smiling face. Dork stated, "If possible, attempt to dismantle their battery system."

He left the timing of the charges up to them, and indicated they would only blow small holes in the hull, not powerful enough to injure anyone on board, including themselves. Adda laid the picture on his desk. Dork stated, "I know a storm is brewing, but that will be our friend, take the dinghy with the motor. The actual snatching of your prey will be your scheme. You will be acting as a team. Set your explosives so that you can disturb their concentration. Then stick her, whoever is in the position to do so. The sinking will be slow, because the charges are light. You will have the storm on your side, surprise and confusion." He waited, then added, "Bring her to my cabin--anything else?"

Adda looked at Dassat and Dassat said, "If we choose to do this act, what is our reward?"

Dork hesitated to gain their attention, then strongly spoke up, "A year's pay for each, when you bring the prey to me. You must act quickly, and try not to be followed." He added one word, "Acceptable?"

They grinned at each other, and Dork gestured for them to inspect, then put on their belts. Dork stated as they checked out their goods, "Dassat will be in charge of the team. Hurry, wait in hiding for the craft to tie up." Good luck, will see you soon. I'll have your reward waiting."

They tried to salute, it was sloppy, but they grinned, with the anticipation of their hoard of money in one lump sum. Dork knew these two men were very resourceful, creative, and were determined to collect their prize. It was a credible risk, but well within the grasp of success. The rolls of thunder became more prominent as the two lean Filipinos were hand wrenching the heavy rusty side gangway walk stairs. Lowering it toward the tethered rocking dinghy. The gears of the old cog and teeth meshed noisily. The sea fowls were diving, catching the wind, and soaring to new heights. The black boiling clouds, and the melodrama of the deep powerful rolls of bass thunder, echoing and reverberating, in a soulful manner, only to be replaced by a more resounding, sonorous depletion. Crackling could be heard from a distance as the thousands of volts of electricity began their random targets of discharging from one cloud to another and then to land or objects on the sea. The winds were blowing the preceding differences of pressures. The old hulk of the massive freighter barely bobbed, as the waves had turned to white caps. The wind was intensifying the frivolous surface. The two slim Filipinos made their way down the aged steel ladder, against the shrieking gulls, and the unsheathing spasmodic authoritative roars and sequences of intense power.

Dork engaged his running lights, and returned to the large chair. He estimated how long it would take for his two crewmen to dock. He summoned his cook via intercom and told him he would eat in his cabin. The pattern was repeated at each mealtime. The cook scratched his head, as Dork ended his conversation. He had always asked what was the main ingredient at evening mess.

Chapter 20

Albert throttled forward, the heaving waves were cutting about 30° against the bow. The soaring birds were displacing in wake of the coming storm. Thunder and majestic flashes of lightning were acting as overtures. Tom was in his galley, busily preparing his up and coming dinner of salon Coulibiac, with lemon parsley sauce. He had thawed out the frozen patty shells, had one egg hard boiled, and was beating a fresh one. The sauce pan was on, melting the butter, as he dumped his whipped egg in and followed with a can of mushrooms to saute, until tender. He added lemon juice, salt and pepper, then added thinly diced onions and celery into his busy pot. The juice from the can of salmon also entered the perfuming sauce. He rolled out his patty shells, then waited for this textures to saute. He then poured a generous filling in each shell, closed them, and sealed the edges. He placed them on a greased baking sheet, ready to pop in the oven. He would serve lemon parsley sauce to romance his shells. His blue eyes were flashing and he was whistling their song, "Shine on Harvest Moon."

Norma slipped in and the fragrant net spread over her. She remarked, "Hubby, that smells damn good. Need any help?"

Tom ceased his musical rendition, smiled and replied, "No thanks, it's all ready for my oven to do her job."

She said, "I like when you summon objects you love, as she and her. That makes me feel regal, Blue Eyes."

He grinned and stated, "Yeah, two days ago, you took me to the carpet, about the feminizing of just such words." She laughed and remarked, "Oh, heavens, that was long ago when we were courting -- remember?"

He replied, "Yeah, I know, you just don't want to talk about old times."

She responded, "Yes, you're right on track, Tom boy. I noticed the friendly seas are getting restless."

Tom slipped his loaded tray in the oven and set it at 425°, then he answered, "Yes, but remember, I told you we would have the finest location of cabins, mid-stateroom, has its advantages," then added, "Christ, a ship rammed us, and the *Magnetic* took it twice and never flinched --well, maybe a little. He remembered he was upside down with Laura on top of him. He cringed silently at the thought.

He said, "Come here, Mrs. Spoones." They met, and kissed passionately. She grasped his buns, and he accommodated hers--both of them were muttering, faint whining, and cunning caresses. Roars of menacing thunder pierced the ears of the compromising couple, as they held each other tightly.

Laura stated, "Whew, that wind is just a little chilly."

Albert smiled and stated, "The more voltage will be generated, my dear," then added, "You better put a wrap on, look!" He gestured, a high voltage of electricity struck in the surrounding mountain. He said, "That was a dandy discharge. Hurry Laura, and come back and enjoy nature's mighty display." The thundering existing time became shorter. The pernicious disturbance would soon completely surround them, becoming more violent, as the flash that preceded, its deep pestilence.

Laura had to enter the galley to get to her forward stateroom. She stopped as her eyes captured them hugging so tenderly. She piped up, "What smells so good?"

They untangled from their embrace, and both smiled at Laura. Tom said, "I thought I would appease the Sea God, Neptune, and have our dinner the center piece one of inhabitancy of his or her huge garden.

Norma grinned and said, "God, listen to that attractive narrative."

Laura smiled, and said, "Both our men have the gift of English."

Tom said, "Hey, you gals aren't pikers with your diction."

Laura stated, "I have something to tell all of you at dinner. It has to do with the context that Norma brought up about having a prenuptial agreement."

Norma, startled said, "Oh, oh, I'm sorry Laura, I was just intending some innocent fun."

Tom stated, "I remember--you said you had something to say on the subject."

Laura said, "I'm going to get a light sweater. It's a little cool on the bridge."

Tom burst out, "Damn. . . ."

Norma stated, "I told you I was sorry. . . ." He said quickly, "No, no, not that, I forgot to take my two pills this morning with all the. . . ."

She cut him off, and remarked, "Yes, Tom, spit it out, with all the. . . what?"

Tom's eyes shone, he smiled and stated, "Romancing." He added, hold down the fort, while I get my little darlings."

He heard her say faintly, "Jesus."

He came back, but missed Laura's trip to topside. He smiled and said, "Here honey, I want you to start taking these babies--go ahead. They are, what you would call, garter snappers--yeah!" His blue eyes shone as he took her right hand, and placed two large soft jells, one yellow and one tan inside them. He had his two, and drew a glass of water. He tossed the two jells down his throat and said, "Mmmm, now it's your turn, honey," and handed her his unfinished glass.

She asked, "What did I just ingest?" He smiled and replied, "Vitamin E and Bee pollen--- actually the busy one is Royal Jelly -- Bzzzz." He laughed.

She said, "Okay, but if I get fat, not one damned word." She added, "Is that stuff the that makes you spooky sometimes?"

Tom looked dejected and replied, "I told you before, it just happens." He continued, "You'll see, a few weeks of those little honeys and you'll BEEEE, like a new woman."

She placed one hand to her hip and re-marked, "What the hell's the matter? You tired of me already?"

He shook his head sideways and said, "Come on, honey, you know. . . ."

She interrupted him and said, "I'm full of BEE regurgitation....!"

He smiled and held his arms open and said, "Come on, be my Honey Bee!"

She grinned and said seductively, "You want to buzz my nest?"

He said, "Jesus, don't talk like that, unless you want my antlers." She laughingly retorted that bees didn't have antlers.

He said, "I know, they have stingers. She smirked and said playfully, "Girl Bees Bzzzz and have stingers, too."

Tom said, "Let's talk about something else." She said, placing her finger on his chin, "You're the one feeding me bee food and calling me honey."

He said, nonchalantly, "You know, the bee community functions show an even higher order of intelligence than ants. One solitary bee species is called cuckoo bees."

She smiled, shook her head and said, "My, my, my Tom!"

Tom said jokingly, "God, just think if Cap was taking these royal jells. He has two nests, one of the double headed cobras!"

She smiled, then said seriously, "Tom, what do you make of the statement that Laura just made?" He replied that he didn't have the slightest idea.

"Beee-quest, I gotta check my oven."

The Coast Guard channel had and still was announcing areas that affected small craft warnings to avoid. The *Magnetic* revolutions increased to maintain speed and maneuverability. Laura had joined her husband topside amid the elaborate electrical storm. The Coast Guard announcer had just finished his official declaratory and then Albert and Laura were stunned as they listened.

"Attention--anyone aboard the Magnetic, or knowing the whereabouts of the craft notify Laura Dryfuss that an urgent message awaits from her brother Linnell. Contact officials at English Bay. Repeating, this is a forwarded message."

Albert profoundly remarked, "Good Lord." Laura thrust both hands to her face, as lightning cracked no compassion, the arrhythmic thunder bellowed widely, and pressed the echoing mammoth orchestrations.

She cried out in defiance, "God, Albert, Linnell--my brother!"

Tom and Norma also heard the startling plea as speakers were wired to every area. They rushed up to the flying bridge. Norma won, grabbed Laura and patted her neck gently in consolation. A huge bolt struck but a few miles from them that initiated a brilliant bash of millions of volts of energy, and they all reacted. Lightning bolts have been clocked at speeds up to 100,000 miles per second, and as hot as 55,000° F. This bedazzled strike could be seen for miles.

Captain Dork was aware as well as the crew of the *Baracs*. The two vessels were less than four miles from each other, unbeknown to either captain. The crack that tore into the white night, and almost simultaneously slammed the worldly bass drums, belched dramatic thunder, causing Albert to oversteer on his new target, and the *Magnetic* slammed and lurched, having been hammered by a full high wave. The craft fell into the violent dip of the passing swell and the *Magnetic* shook.

Laura screamed out, "My God. . . it must be. . . ." She began crying. The sky was bright with bolts of cloud to cloud lightning, cloud to sea, and terra firma strikes. Three gulls had just been electrocuted, their scorched feathers sizzling, as their lifeless carcasses crashed into the leaping, fretting waters, sacrificial flyers, now grounded, and waiting to be injected by the scaled gilled swimmers. No longer could you time the brilliant strike of light to the repercussions of the bashing thunder.

Norma and Laura screamed as a healthy lace of sea spray had swept up and was thrown over 20 feet, taking them all by surprise. Their hair and faces were drenched with saltwater, and they were flagrantly slapping their wet hair, except for Albert. Just his eyes were streaked with a malicious stark terror for just a moment. Tom's blue eyes were filled with seawater. He clinched Norma's shoulders and shouted to Albert, "Are you all right?"

Albert yelled a reply but only part of it came across in contest with deep horrendous expelling of apologetic thunder. Although they were huddled together, Albert yelled, "I'm s...", then added, when the raging noise subsided for a moment, "That was my error." He gestured to Tom to come closer, and shouted for him to hear his instructions. Each looked into each other's watery eyes. Albert's hat had saved his hair from the wrath of the sea.

Tom reconveyed Albert's command. He put his arms around the two drenched women and cried out, "Come on," and gestured down. They held hands and Tom went down the stainless steel ladder first. Because of Laura's sore arm, Norma indicated for her to go next. The craft was pitching and hawing, as design and power clashed, in the highways of displacing waters. The mighty sea was displaying its rights. Tom clasped Laura's waist and eased her down, then repeated the same for Norma. Once inside they flayed and stomped some of the tears of the raging sea. Tom waited until they sat down in the lunge, then stated, "Stay here, I'll get a couple of towels."

Laura stated, "He and I are just about all that's left, you know!"

Norma said, patting her hand, "I know, I've been devastated. . .maybe he got married, and is waiting to surprise you!"

Laura said, "Thanks, I'm sorry, I'm acting like a child. Norma, you'll never know how much all of you have meant to me. . . all you guys!"

Norma smiled, and said, "Now come on, don't get feminine on me. We've got some tall soaking to do."

Laura managed a small smile. Tom came back with two plush bath towels, and Norma quickly set about drying Laura's face and hair. Norma said wistfully,

"Tom -- blue eyes -- how's your oven doing?" Tom shouted, "Oh, Christ,...be right back."

Albert came in, shaking his wet hat and took his place in the helmsman's chair. He checked the running lights, they were on, glanced at the dual controls, and everything seemed normal.

Norma spoke up, "Cap, catch" and tossed her towel. Albert caught it with a thanks, as the *Magnetic* rolled, dipped, sidewashed, and fell. The clamoring had not let up, and the seas were attempting to match the intensity of the storm. Rains began falling, and Albert put the wipers on, vigorously increasing the magnitude of the operation. The waves lashed out predominately on the port side of the bow. His trim tabs were being subdued by the languishing of oncoming stress on his port stabilizer. Discharges of elaborate flashes
were in full view and the enclosed 360° lounge area. The glass protected them from the onslaught of the collision of rain drops that came crashing and were being recaptured by the authoritative sea.

In reference to a synonym, "Ashes to ashes, dust to dust," vapor - vapor - rain - rain, commanding over three-fourths of our earth's surface. Albert said stately, "Good Lord, there's enough juice out there to light New York City for weeks on end. Just imagine, if someday we could harness that tremendous power. It even increases when induced with moisture."

Norma said, "Yeah, and we could eliminate that noisy part, too."

Albert asked, "Are you all right, honey?"

Laura replied, "Yes, I'm better now, I'm sorry."

Norma butted in, "Stop it, Laura, you didn't do anything that I wouldn't have done." She moved over and massaged Laura's neck and shoulders, and Tom watched somewhat jealously, sort of transfixed. Norma said, "Tom, honey, coffee would be nice."

Albert added to her request, "Yes, Tom, we could all use one of your flavorable cups." It won't be long now until we can see the lights and reference buoys."

Norma asked, "How large is English Bay?"

Albert put his cap back on, then replied, "I'm not sure, except it's more likely smaller than Port Graham."

Tom came back and sat down on the settee.

Norma asked him, "Is your oven Okay?"

Tom smiled, "Yeah, thanks to you, I shut the Princess down. Dinner will be late. We'd better dock and take care of business, then maybe we will set another plate."

Laura looked up and smiled at the thought of such an occasion. She perked up and stated, "Albert." He replied, "Yes." She answered, I want all of you to hear this, you know I was going to tell you at dinner, but I got to thinking that there is a logical explanation for my brother to be at English Bay."

Norma quit massaging, and sat down next to Laura. She interrupted, "Jesus, Cap, may I light up?" He waved Okay, Laura and Tom nodded. Laura continued, "It just might be, . . .remember, Norma mentioned, kiddingly, of course, about prenuptial agreements. Well, with the excitement of the wedding and all I completely forgot that my brother and I are the only heirs to our Aunt Margaret's estate. Maybe something happened to her and he couldn't reach me."

Albert remarked, "He couldn't possibly have known where you were unless you told him."

Laura smiled widely, and said, "No, no, I didn't tell him, you told me this was a very confidential cruise. I'm sure this is the only way he had to contact me."

Norma asked, assumingly, "Was she rich?"

Tom flashed a look at Norma, but she didn't receive it.

Laura looked down in her lap and said, "Yes, she had told us our inheritance would be well over three million."

Tom said, "Jesus. . ."

Norma said, "Wow," then added, "Cap, I bet ya that your wife has you outgunned in shekels!"

Albert stated, "She has in loveliness already." Smiles were all around their eager faces. Tom said, "I can smell my fragrance, gotta go, be right back."

Norma got up saying, "I'll lend you a hand, sailor."

Tom grinned and remarked foolishly, "I could take that two ways--remember."

She kiddingly responded, "Yeah, but we weren't married then, blue eyes," and laughed. A bash of lights in powerful, but brief moments, traversed the skies with the sounds of roaring thunder of a troubled romance in the heavens, laced together for a brief moment. Rain began filling in the ocean, and soaking the land.

Tom and Norma played hands, laughing, as they captured each other's body, and romanced their waiting lips. The galley was lush, with scents of celery, lemon and salmon. Tom fumbled with the switch that influenced the heat in his busy potato fragrances. They cradled their heads on each other's shoulder and held each other firmly. Tom signaled with a warm squeeze and she responded. They said nothing, their intense flow of sensations and body language did all the talking.

Albert said to Laura, "Those mighty rivers of electricity are probably being recorded for their frequencies from a cloud to ground through predict-able areas in our world. Our storms are becoming more violent, but by monitoring these nonstagnates, we may better understand their secrets and one day, they may serve us masterfully."

Dork complemented his frown, alone in his large creaking, slightly moving chair. His small stature and light frame were unable to supply the ballast, to render the swivel stable in the heaving of the moving ocean. The dinghy holes in his massive old derelict freighter were empty, adding to the buoyancy, thus giving those turbulent waters more influential power. Dork had to hold on to the desk, or be escorted, at the whim of his swelling watery highway. To add insult to the battered old sea traveler, rain began punishing its flakey, reddish, brittle black oxidized paint. Dork's fixed frown laminated grotesquely with every electrical flash of a magnitude that pierced through the rain spattered port holes in his lonely cabin. He pondered the thought of how he wished he could witness the snatching of his prey. He knew the sea was violent and cold, and his crewmen would have to strike fast or risk failure. He had purposely failed to finalize the face of his victim to be, but rather to see and sense instinctively, then draw from that person, to decide the finale. He had to render her harmless from ever revealing the information that she apparently possessed. His next step was the temperance Port of Wenchow, China, where scrupulous races of the world lay in solace, trading, buying and selling goods of unimaginative concepts, one of his favorite ports. Many men would delight in his pleasure of dilemma of having such an attractive female to fondle while deciding what course to chart. Bondage was his preference, rather than just killing and dumping the body in the briny graveyards of the deep. But, by God, he would not sacrifice his colossal prize, his ship, at any price.

His cabin door opened. His cook spoke up, "I tapped, but the thunder, . . .I have your dinner, Captain."

Dork tried to straighten up in his chair, but slid back, and responded. "Thank you, bring it to my desk." There was a clean, but dingy, faded white towel over the plate. The Portuguese cook placed the heavy tray on his desk backwards. He shuffled out quickly and closed the steel door. Dork whipped the towel off, and scowled at his reversed dinner. He turned the tray and tucked the white fraying cloth, in his cleft, grey shirt. He grabbed his fork and contentiously gratified his frowning face. He quickly laid his fork down, and heavily salted everything. He resumed his iodized dinner, holding onto the desk with one hand and forking with the other, as the storm howled, and peppered, the familiar old rider on its turbulent highway.

Dassat and Adda were close to the menacing fingers of their churning supporting liquid. Dassat was maneuvering, at times, with the rolling whites, crisscrossing, as the next swell carried them on. Just minutes from leaving their steel haven, they were drenched with the fretting froth, spraying them at every capping wave. The black boiling clouds were still being held captive. Tons of fresh water waiting for that moment when the trap would be released, then the bombarding would present a maze of targets. Below, lightning and bellows of thunder rocked the profane skies. The sea was attempting to match her lover's rival above with the agitation of madness. The authoritative surface was spewing and beginning to roll violently. They jitterbugged their dare devil craft over the pitching altitude of the insolent waters. Neither man spoke. They both had deep respect for the mocking, deceiving, roller coaster, that seemed to envelop them, then suddenly tossing them up and over. The sea spray had their eyes stinging, but the defiling waters kept them aware of the immediate danger. The constant spinning of the old brass propeller chirped loudly, as the blades furiously spun, and only cut the misty frothy air. When the dinghy rode an under crest, it would dive, and the chopping blades would be swallowed, again churning and propelling them toward their destination. The next pitching hull of water would present the same contour of exposure. Their small craft was rapidly becoming heavier, taking on water from the turbulent spews of sprays of the incoming waves. The sea inside was already over their shoe tops. They were nearing the exciting, cascading resilient shores. The horrendous breaches suddenly lit up with a stark light on their capricious shoreline. A patch of sand, gravel, and rocks loomed up unexpectedly, as their struggling boat rode the crest of a long swell. Dassat shut off the tired hissing engine. Adda was soaked thoroughly, he grabbed the tailing rope that was attached to the bow ring, and bailed overboard. His sodden soles slipped on the unstable slick rocks, and the tricky conforms, where the sea and the land challenge for mutual, territorial rights. His fist clutched the nylon tether as the swells expanded from his chest and then receded to his ankles. His cinched preserver was a detriment, rather than an ally, at this juncture in time. Adda spit seawater from his mouth as he trudged up the slick, rocky shore. His feet then dug in the terra firma. He aspired to a forceful pull on the rope. An incoming wave had run under the heavy water soaked belly of the surging dinghy and literally jettisoned toward the shore. Adda screamed mercifully, as his horrid white eyes captured the dripping high bow and undercarriage that was hurdling toward him. His following wailing cry died in his throat as a bash of crackling lightning was followed by pressing rolls of thunder that masked all sounds for miles around. Adda was the only recipient of his death scream. The hysteria froze on his face as death was instantaneous and his chest was crushed as the craft continued its thrust. The thin keel was tearing through his chest to his head, peeling his face, and inducing massive pulsation of blood, that was diluting quickly with the salty sea. His ripped muscles and sinews were flaying from the once symmetrical, aesthetic face. The thunder roared, then pounded, with echoes of bellowing. A new electrical bolt, then another and another.

The clamoring became excessively unreal. A chaos of unleased energy--Dassat scurried out of the water laden boat. He desperately groped for the rope, the end of which was hidden beneath the bronco hull. He gave it a thrust, tightened the drooping rope, waited for the next wave, pulled, and moved quickly. The dinghy surged further on shore. He lashed the vibrating rope to a jutting pinnacle. He clinched his teeth as he tied a stevedore knot. The excited vessel was tugging and lurching with each swell and attempting to discourage him. He cringed, as each crashing bolt streaked and struck, searing and exploding trees nearby. He nervously looked about, screaming Adda's name again and again. The flashes emitted complete daylight, as his hawkish eyes searched for his mate. Dassat had never witnessed a storm that sustained a cataclysm, with such voluminous energy. Bolts of electricity such as this were an awesome display. It was as if the heavens were going to reclaim all that lay below. The profoundness and relentlessness of the electrical authority, and the sudden disappearance of Adda had left him temporarily comatose. He crouched, pulling his wet legs close to his thin frame as the rains pounded and spread a ritual purification about him. He dared not look at the menacing spectacular. He had convinced himself that something in the dark boiling waters had snatched Adda, and his body was being dragged, scraping along the bottom of the sea bed into deeper waters. His thoughts were confused; anarchy was driving his usual logistics out of proportion. What else could it possibly be. He did not find his body, he did see him leap out of the boat, his fist clinging to the tethered rope--then nothing. Reality crept back. He got up slowly, totally soaked, and the water sloshed in his shoes as he made for the navigational lights ahead. The bold heavy, but sturdy old dock with slips, stood fast, as each bashing wave tried a knockout punch. Nine crafts of varying sizes were tossing in aggravation, their harnesses snapping, but grudgingly resilient to the rupturing waves. He crouched as he made his way onto the dock amid the stunning torment. He suddenly sat down on the water dock with rain slashing his water soaked frame. His eyes were filled with water, running off his thin pinched nose. He couldn't see, but fumbled with his shoe laces until they succeeded in their task, then in a rage, he hurled his shoes into the churning sea. He felt the bulges in the packs around his waist. There were still two bulky protuberants. He had no idea how long he was to wait for the arrival of the marked cruiser. He chose a slip, with accessibility to observe incoming boats. He watched the predominating rhythm of the heaving craft, then leapt into the struggling boat's deck. He screamed as he felt a searing pain in his toe. He sat down, and pulled the naked foot up close to his eyes, and made out an ugly tear with blood spurting out. A metallic glint caught his eye and with his index finger, he confirmed it was a large double barbed fishing hook. The rain was flowing off his forehead and into his blurry eyes. He briefly saw and felt the warm blood, but had to feel and gauge at what angle the hook had gaffed and lodged in his ravaged toe. He gritted his teeth and screamed as he tore the flesh open, but possessed the sharp double barbed hook between his bloody fingers. He tossed it overboard and pulled a damp handkerchief from his pocket. He hastily stretched it out, spinning it and laced it around the torn and bleeding slash. He tied a knot, and winced when he squared it. He rose and tried the cabin door. It wasn't locked. He limped over to the worn cushioned chair, sat down carefully, and moved his injured foot to the metal foot support. He wrung his eyes out with his wet knuckles, and tried to peer through the rain dashed windows. He waited, his eyes bulging in anticipation. He pulled his wet shirt up and unsnapped the flab of his webbed catch that was around his waist. He withdrew the water resilient packet, with the syringe and the capsule and facilitated the exchange. He then returned his catch. He saw some dim lights and leaned forward and strained to catch a better sight through the rivulets of water making bastinades against the glass windshield. He began to become excited, this must be the craft that held his victim. A clear bolt of lightning crackled, and lit up the bow splitting craft as it approached.

Chapter 21

Albert stated briskly, "Tom, you had better slicker up and make certain the bumpers are down on both sides. This could be a little tricky."
Tom got up and gave a thumbs up to the gals, and stated, "On my way, Cap, that's one hell of a display out there."

Albert throttled down to feel the valence between the swells and the *Magnetic*. Laura and Norma looked around nervously, their eyes unable to focus on any one object, except the relegation of the turbulence. The fury of the unleashed, catastrophic, electrical bash had Albert frantically clutching the chrome throttles, revving and reversing the powerful diesels, in one horrendous moment. The women clutched the glass table to maintain their balance. Albert shouted, "Good Lord, that reciprocating vertical wave just about launched us in, or over, the dock. Everything's all right. I just misjudged the intensity."

The lightning charge released its momentous electrical energy, exploding at the top of a 60 foot tamarack tree near the ravaging shore line. Its tremendous strike seared and scorched as it split the tapering trunk. Relentlessly, the sound was ostentatious, as was the scene. Laura, using one hand, pointed as daylight was relevantly so bright.

Norma stared toward's Laura's gesture and cried out, "Jesus Christ," and had to shut her eyes against the brilliance. The sudden change of direction and the churning down at aft, had sent Tom sprawling on the aft deck.

He cried, "God damn it" and his flaying hands were searching for anything to grasp. It was brighter than day for a hanging moment. His excited hand hit, and grasped around one of two extensions of the ladder jack used for access to the flying bridge. He hung on, as the rain peppered him.

Albert, attempting to remain calm, said in a steady voice, "Norma, why don't you don your slicker, and see how your hubby is doing?" There still was terror in the eyes of both women. Laura sent a frantic message to get with it, and just had time to touch Norma's fingers.

Norma sprang up and said, "Sure, Cap, be right back with my Tom." She high-tailed to get her gear, and see what the hell happened to him.

Laura stated apologetically, "If anything happens to any of us, I'll never forgive myself. Albert, you're risking everything because of me. This cataclysm is too violent and overwhelming to attempt docking--get us away from here--now!"

He had heard her plea, and dared not interrupt her. He rallied all his navigational skills, and the *Magnetic's* schematically tested deliverance. Its powerful engines responded, joining Albert, and the *Magnetic* was one convincing adversary. Tom and Norma came in, dripping, and nervously shaking the water from their attire. The horror in their eyes was unequivocally present. Laura captured their naked fear and called out, "Albert--get away from here!

Tom waited for a response from Albert but none came. He asked, "What the hell happened, Albert?"

Albert replied quickly, "Sorry as hell, Tom, that incoming wave was more dramatic than antici- pated, it was virtually commencing to catapult our craft either into, or over the dock."

Norma tried a bit of levity, and spoke, much to fast, "Yeah, my Tom was laying down on the job!" No one said a word.

Albert finally asked, "Tom, are the bumpers out, port and starboard?" Tom glanced at both women, the replied, "Yes, Cap, they're in place -- is there anything I can do?"

Laura grimaced and stated, "Albert, I've asked you twice to forget this madness!"

Norma took off her slicker, gave it a shake, and chanced a look into Tom's eyes. Tom saw her plea and acted on it. He said, "Albert, it's not worth the risk, why not lay back till this heller dies down, then we can safely dock, and search for Laura's brother?"

Laura let out a pent-up sigh. For just a brief moment, Albert was determinedly wanting to challenge the angry elements, then the culmination of reality set in, his fingers relaxed as he turned the

Magnetic bow dead center on the incoming white frothed waves.

Tom removed his rain gear and asked, "Anyone hungry?" Norma went over and wrapped her arms around Laura, their cheeks pressed together in solemnity. The intensity of negative and positive electrical colliding particles was still demonstrating their powerful longevity.

Albert looked up from his lower helm cockpit, and said, "The sea still retains its many wonders. Nearly nine tons riding on her surface, and we were treated with her fury like a fallen leaf on a windy day. Thank you all for snatching me out of complacency in my private vendetta with the mighty sea. . . ." He let out a long sigh, and added, "A personal thanks to you, Tom, I have one request."

Tom looked down at Albert and replied, "You've got it, Cap, What is it?" Albert replied, "A hearty meal!"

Norma got up and spoke, "I threesome it."

Laura smiled and happily stated, "That makes it a solid quadrant." She got up, stepped down to the lower helm, and kissed her husband on the neck.

Norma said, "Come on, blue eyes, I'll be your K.P. slave."

Tom smiled upon hearing her tantalizing lingo. He lead, and was just in the galley, when she caught up, physically spun him around, and sharply stated, "You wonderful son of a bitch." She threw her arms around him. Tom said, nuzzling, "I was just doing what your lovely radiance projected." He laughed softly, then added, "Just like your old two playmates, Wheeler and Woolsley, huh?"

Norma smiled and said, "You're damn right, blue eyes, just keep on doing what you're doing, cause, Honey, I love what you're doing to me."

Tom said lovingly, "You sure make me feel good." They exchanged a sensuous kiss, and Norma said, "God, you're good. You got Cap back on track, Christ, you're better than 911!"

He slipped their salmon couebiac in the Princess oven and said, elegantly, "Let's get our little spoons and set the table," then added, "Please, no candles."

She pinched his butt, smiling, and stated, "We don't need them, you're my torch."

He began to get nervous and said, "Jesus, you sure know how to ignite a guy."

She replied, "No, Tom, not just any guy, my hubby."

His wizardry was evident as he said softly, "You're certainly the silver Spoone in this marriage."

That did it, she fell apart and said, "Christ, what a cardinal you are, Mr. Spoones." Tom said excitedly, "I'm getting, you know, what . . ."

She pulled him closer and said seductively, "Damn you, Tom, you've already lit my fuse."

He said in bewilderment, "Jesus what are we going to do?"

She said calmly, "What any gal would do under these circumstances. . . ." He uttered, "Oh Jesus, Oh Jesus!"

She exclaimed tantalizingly, "God, Tom, you are ready to soar!"

Tom cried out in frustration, "Not in my galley!" She smiled and undid his belt, and said,

"Yes, Tom, right in front of your Princess oven. . .She'll probably be as jealous as hell."

Dassat was shaking for many reasons. He was cold, wet, his toe was throbbing, and the wrath of anticipation of having to be enveloped in the confines of these Hades waters, to plant the charge, then the sticking of the woman. He thought the latter would be the choice, this angry war between the heavens and the sea, had his confidence rocked. He thought if he could get her alone, she would be putty in his hands. Women are vulnerable when a loved one is the target. He recoiled, in his slouched position, as the lightning seemed more violent, and closer. He wished one of those strikes would spear their craft, then he quickly thought about what if he was aboard when it happened. He cringed as it rolled in his spiritually distressed mind. He could barely make out the sharp bow, cutting cruiser, but it was getting closer. His bulging and burning eyes could not believe what he was witnessing.

The vessel suddenly made an aberrant move. The sky lit up and roared as he made out the first three letters on the transom, M - A - G, as it headed away. He felt drained, his eyes stinging as he peered through the muddles of careening, splattering rain drops, until the craft disappeared. He raised his hankerchiefed toe, and fondled it, incurring pain. He knew then that they would come back in the morning, or after the storm had passed. He had two choices--stay here, then leave just before daylight, or give them a chase. The belligerent storm had cautioned him. The wise inhabitant of a foreign confine, steps lightly, or flays in the bog of his enemies. He would wait one more hour and then decide. The Godhood above was remorselessly relenting its power and targets. He knuckled under, his tormented sockets. Hunger pangs invaded his thin intestines.

Norma disturbingly remarked, "What the hell is that?" Tom said, with a light smile, "That's my Princess' womb, advocating a signal that our dinner is inseminated with the desired heat."

She disgustedly swore, "Christ, you've just incarcerated me, and my bell didn't have time to ring!"

He sadly stated, "I'm sorry, Hon, the Princess didn't know that."

She pulled a long smile and replied, "No, but you did!"

He countered, "I was a little busy, you know!"

She sighed and remarked, "Well, at least you, and your Princess, initiated your galley."

He got up nervously, and said sheepishly, "I have to shut her down and crack the oven slightly."

She smiled and shook her head, "Jesus, blue eyes, you're something else. Do all you guys take bee pollen pills?"

He replied concerned, "No, not that I know of, but it's catching on more and more."

She remarked, "I still enjoyed our passionate floor play." He laughed, then said, "Now that it's over, that hard teak parquet floor, has taken a toll on my knees."

She got up as Tom extended his hand and she spoke, "I'm a little tender in spots too. Your Princess stove has little or no chance of witnessing this floral display again. I'm sore!"

Tom said, jokingly, "We are officially an old married couple. I have housemaid's knee, and you have bunions."

They both laughed and she said, "I had planned for you to slip on your elastic panty hose fashion show, for a 16 hour stretch, but that would aggravate your pink knees."

He said, "Thank you, Mrs. Spoones. Let's get the table set, my salmon is struggling upstream." They hustled, and Tom elected to serve a 1985 cabernet sauvignon. It was a fine year for grapes worldwide.

Laura was at the wheel, Albert was coming in from aft, after releasing the two anchors. He asked, "Please shut the engines down." She smiled and did just that.

Norma stated, "Hurry up, you two, dinner's on the way, and Tom's antsy about the timing." They both grinned, freshened up, and returned. Norma filled the ice bucket and placed it on the glass topped table. She hiked back to the galley to help Tom deliver the seafood dinner. Tom had used his galley mitts to bring in the hot tray, and dished up his steaming brown filled patties of salmon coulibiac with lemon parsley sauce. He also prepared asparagus tips in melted butter and had lemon-lime jello for dessert. Smiles were all around, and rightly so, declaring how good it was.

Tom stated, "I'll modestly decline to pour the wine, after the chastising I received at Charlie's Saloon."

Laura stated that she would consider it an honor to do the pouring.

Tom announced "Our environment has provided our main course." The red wine was auspicious with spicy, intense focused currant berry, and cherry, crisp acidity, carrying the flavor's firm, but not overpowering. Tannings were gracious with the flavors. The storm was moving on. The winds were subduing, and the rain had diminished to a steady drizzle.

Dassat regretted throwing his shoes into the sea. He had intended to swim underwater, but his shoes would have been a cross to bear. He stared at his cheap, leather strapped watch, waiting for a flash, to verify the time. He was accommodated. The hour was close. He slipped out of his dry asylum. The intensity of the rain was gasping, with the contracting torrential display. Again, he crouched, and played blind man's bluff. He desperately tried to avoid the aquacade of the heaving and undercutting waves that were attempting to pull any unsuspecting creature into its coffers from the rocky shores. He was disgusted that it took so long to retrace his path, having no relevance, because of his horror. A lingering flash confirmed his dinghy was still tethered. He was soaked, as he peered into his waterlogged boat. He pathetically saw in a flash that the water was nearly running out of his tilted transom. There wasn't even a can to bail it out. It was now impossible to launch. Nearly a half ton of water had congregated in the aft of his boat, forming a small aquarium. He dropped to his hands and knees, and let out a scream. His ravaged toe had struck a sharp rock. Blood flowed, and then seeped out of his makeshift bandage. He reacted like any animal, using his uninjured naked foot to balance, and leverage, as he groped, dead reckoning, as practicability dictated, under the draft of night. Over the rocks, he would step like an ant and use his flaying hands as his eyes, on the congested surface. Then he would move, and repeat the same, through the steady drizzle. His fingers suddenly responded as they romanced a half encompassed, slivered, debarked pole. A less intent flash indicated his find was speared in a congested snag of timbers. It's nest of extremities was disarranged in a porcupine fashion. Excitedly, he crawled under and over the uprooted trees. He lashed out, "Son of a bitch," as his back hand and wrist smashed into a sharp penetrating pinnacle rock. The crystal of his watch splintered, and his little time indicator, flew into the darkness. He instinctively drew his bleeding knuckles to his mouth, and found himself sucking his warm sweet blood. He proceeded at a cautionary pace. Groveling and every so often, reached out tenderly, following the peeled pole, until he asserted its length. Then he set about finding rocks, rolling and lifting four stones, placing them to form a crude pyramid.

While touching and underneath, the long spar, Dassat then made his way back to the upstart spar, and touching the end, stood up. He climbed up on the pole and balanced himself, as he wavered, he jumped straight up. If he missed, he would fall about three feet in a maze of rocks, and brush. He kept his legs rigid as he descended. He screamed as his sanguineous swelling toe smashed into the solid wet wood. It cracked, then snapped as they both fell to the rugged, erratic surface. He collapsed in a clutter of rocks and trash. The heavy drizzle relented to a fog-like misty lace. He grasped the slippery pole on one end, hoisted it on his shoulder, slowly and patiently limped toward the noisy, assiduous seashore. The pain was obviously directing his favorite leg, supporting his thin frame and spar, but the footing was precarious. Now the dark night was his blindfold. The trailing end of his pole was striking and bounding behind him, making it difficult to control on his shoulder. The skin underneath was reddening with vicious blows as vibrating, resonance, continuing to vacillate, then tearing his skin. The intrusive shoreline was splashing and sloshing, looming resoundingly. He lowered his pry, and positioned it with care, from the leeching, and suctorial action of the pulling sea. The discommissioned shoreline was littered with washed up trash. The backwashing waves attempted to pull and digest its unbroken covenant. He placed each foot tenderly in the cold moving water, and hand followed his dinghy. He squatted, feeling for an imbedded rock to set his pry on, closed to the aft or far aft transom, to turn the boat sideways and drain its unwanted cargo. The moon cycle was a new moon, meaning there was no reflection of light, or presence. Serious sounds of the waves was the crowning mandate in the pitch black night.

His body wavered as each claiming surge challenged his light frame. The cold water irritated his slashed and swollen toe. He hung onto the transom with one hand, crouching, and flaying his free hand until he found a sound slippery polished boulder that was sedimented with the sea, exposing then covering with sedimentation. He had found his fulcrum. He was working like a blind man, to all the present laws--he was just that. He was listening, touching, then feeling. He stepped gingerly, each step, paying his dues. He recaptured his pole, and edged down the gunnel of his waterlogged vessel. His pole was on his shoulder and he attempted to place it soundly under and against the small keel. He had to grasp the pole with both hands to overcome its buoyancy. He screamed, inhumanely, his extended mouth emitting a prolonged horrid wailing. He released the submerged end of the pole, it popped up and struck his lower lip and chin. His lungs, unable to respond to another hideous cry--he tasted his own blood for the second time. He had felt the devil's pods moving on his hands. He had felt the cold flesh of a human hand and had cursed the blackness of the night. He did not know it gave him a reprieve from witnessing the wrath of hell. Adda's hand, with his onyx fingernails was moving with the rhythmical movements of each wave, and it was summoning Dassat to join him.

They talked about the revelation of the display, and the sanctimonious releasing of millions of volts of energy. Laura had a second glass of wine, enjoying the complement, with her tasty dinner.

Albert spoke up, "Tom, you certainly reveal romance in your culinary."

Norma laughed and started coughing. Tom looked nervously about and passed her his napkin. Albert continued, "You have seduced my palate, and my digestive system."

Norma regained her composure, smiled and stated, "The women of the world, envy my premise, with my Tom--in more ways than one."

Tom grinned, and Albert sent him a stout fellow, good job look. Laura expressed, "Whew, that wine was delicious, but bold. I'm going on deck to catch a breath of that fresh air--it's always so fresh and lovely after a storm."

Norma got up saying, "I'm going to my cabin, I'll pack a few dishes on my way, and Tom, I insist when you get ready to wash, call me. I volunteered to be your galley salve."

Albert said, "Well, I'm going to take a quick hot shower. Laura, don't be too long." She replied, "I won't, Teddy Bear." Tom grinned and stacked some more dishes of spent salmon bones, and packed them to the galley. Norma snatched a pack of cigarettes from her catch, and celephaned her pack. She tapped one out, got her ashtray, and lay on the bed. She double pillowed her head, then lit up, and enjoyed her smoke.

Dassat must get away from this internal womb of hell. He frantically floundered, then found his pry and savagely thrust it under his dinghy. Groaning and cussing, he leveraged the heavy boat, and it began spilling water back into the sea. It became easier as the waterfall widened. He grinned. He was becoming the master rather than the puppet, with Satan controlling the marionetting, spectacular, demonstration of his power. He lowered his pry slowly, less than a foot of water remained trapped. He shouldered his pry, and drug it out of reach of the calling waves. He released his tie, and hunched under the bow, and strained to dislodge his captured craft. His lip had stopped bleeding, but his toe was obviously disdaining his limited stealth. The next wave loosened the keel and was suddenly reacting to the liquid surface. He stepped in carefully, and set his oars. He pulled tenaciously from the merciless shores.

Dassat was again saved by darkness from the devil's cloak of tricks, or he would have seen Adda waving goodby, goodby. . . .His oars cupped into the waves as he headed towards the direction in which he thought that the *Magnetic* had absconded. Each steady pull in the black night was bringing him out of his debacle of diversion to his sanctuary of purification. His wits were approaching his complacency. He listened attentively, reducing the sound of his autistic oars. Silence was essential as he quietly cut the black surrendering waters. He repositioned his cutting blades as he quieted his ride. He licked his tongue, feeling the rough scab. He made out distant lights that were bobbing, apparently an anchored craft. His oar locks, were squeaking. He scooped up, trapped water, and laced them as he closed in closer. He saw the sea foundling, its sleek hull. He nudged up to the transom of the *Magnetic*. Call it a sixth sense. Laura turned and saw an offensive man with a swollen cut lower lip. She started to cry out.

He put his finger to his scabby mouth, and in a hushed voice said, "Your brother sent me, don't be frightened. He's waiting for you."

She asked cautiously, "What is my brother's name?"

He replied, "He told me it is Linnell, Laura!"

She gestured, and caught her mouth with her hand, and uttered, "My God."

He stated, "You must come with me alone --now."

She nervously remarked, "I must tell my husband!" He spoke calmly, "I'm leaving, he told me, no one was to know. He is in trouble. I'll tell him you refused to come."

He began retrieving his tie to the transom ladder. She saw his intention and made a decision.

She outstretched her hand to the Asian. He extended his, and she entered the sloshing dinghy.

He said as he pushed off, "It was at great risk that I ventured out into such a night of nights.

The putty said, "How is Linnell?" He stroked and dipped his busy oars, and replied, "You will see for yourself, very soon, Laura."

He rowed well beyond the ability for anyone to hear. He roped, and fired up the outboard motor and headed for a faint glow of lights. Laura was trying to keep her feet out of the bustling, captured water, and found herself unconsciously biting her lip in anticipation of seeing her brother.

Chapter 22

Dork received an urgent message from Devlen Dawson to switch his letters on the bow of his ship to *Scarab*. Presuming the identity of her sister ship, once they cleared the Gulf of Alaska, then resume to his designated Port of Wenchow, China, in the East China Sea. Dork was nervously waiting for his two crewmen to bring him his prey, thus assuring his ownership of the *Baracs*. His dream as a boy was to own Captain's Papers, achieving that, the dream of dreams, his own ship. There was only one human being preventing such a feat to occur. He would have the world by its balls and could hardly contain his thought.

Dassat tied up to the rusty landing dock that was hinged to the scabby old iron, staircase. Laura was very excited. He gestured for her to go first. He would delight watching the ellipticity of her butt as they slowly made the long climb. She was cold, and unnerved, and just slightly suspicious that her brother was not waving to her from the deck. She thought that maybe he was hurt in his cabin. She was unable to identify this old freighter, but she knew it was very old and smelly and not well maintained.

Dassat rapped loudly on the door and shouted, "I have Mrs. Laura Dryfuss."

Dork fought to get out of that clumsy chair and hustled to set his eyes on his prey. He opened the door and saw his victim.

She noticeably flinched, seeing his ugly frown. She excitedly asked, "Where is my brother?"

Dork asked, "Where is Adda?"

Dassat replied, "Down below, I'll talk with you later, Captain, I'm hungry."

Dork spoke, "Yes, and get checked by our healer, then added, "What happened to your shoes?" He replied, "I'll tell you after I eat," and closed the door.

Dork indicated for her to be seated in front of his desk, and he recaptured his husky chair. He stated, "You are as beautiful as I was informed about you."

"She placed her hands on her lap and stated impatiently, "All right, Captain. . . ."

He interrupted her saying, "Dork, and I will call you Laura."

She continued, "Captain Dork, where is my brother?"

He said, "Your brother must have some powerful friends. He evidently got himself in some serious trouble in a little city in Oregon. I was told he fled in fear of incarceration. He is waiting for you in a foreign port."

She nervously stated, "That's impossible. He's not the sort to get in that kind of trouble."

He answered slowly, "All I know for certain is that they say he killed a young woman."

Tears began to squeeze out of the corners of her green eyes. She stammered, "How is it you, as the captain of an old freighter, have such ridiculous information, even if it is true?"

He spoke out with confidence, "The owner of this ship, also has seven others, we were all informed to be on the lookout for you, indiscreetly so as not to alert the authorities."

She countered, "But the Coast Guard released the message of my brother being in English Bay."

He said in a logical tone, "The message given to the Coast Guard gave your name as Laura Dryfuss, neither his name, nor your former last name, was intentionally deleted, to protect your brother."

She was fretting, "Is Linnell all right?

He replied, "He was, and I am instructed to take you to him."

She started perking up, "I know you have a radio, please notify my husband that I am all right and am going to my brother."

He shook his frown, and said, "I cannot send such a message for all to hear. My instructions forbid it. . ." He knew he had set the hook, and she was gilled. He added, "If you like, I will have you returned to the *Magnetic* and report that you were contacted and refused to help your brother."

She asked, dejectedly, "Where is he?"

He slid back by accident in his deep chair and replied, "I can't reveal that information. When we dock you will be taken directly to your brother by those on shore, where he is hiding in our next stop." Then he added, "I was told that if any word leaks out as to his whereabouts, the people holding him will turn him over to authorities and have him deported, under guard. Murder is a very, very serious crime in any country." He was enlightened at how this unique deception had developed, no guards, and knowing she wouldn't jeopardize her brother's life. He asked kindly, "Are you hungry?"

She replied, "No, I have just eaten, when will we be at this port where my brother is being held?"

He pushed his oversized and creaky chair from his desk, and stated, "Just a few days. I'll give orders to weigh anchor, and I'll show you to your cabin." He hurried to open the door for her, and gestured for her to precede him. This is the one time in his life when he wished he could grin broadly.

He stated, as he opened the door, "I must ask you not to mix with the crew, you are beautiful, and. . . ."

She nodded and stated, "I understand."

He quickly remarked, "You have the free-dom of the ship, and if you need, please call me. This is a bit small, but it is clean, and as you know, this is a freighter," then added, to set the hook deeper, "Your brother must have influential friends to spread such a wide net to bring you and he together again."

She asked, "Who is this man that has eight ship captains helping my brother?"

"I do not know, please excuse me, while I attend to getting us on the way to your brother's port." He closed the door on his bewildered prey. He gave orders to hoist anchors, and summoned his second, and informed him as to heading, destination and speed.

Tom called and Norma came out smelling of smoke in her hair, taking her voluntary spot next to Tom. They kidded one another, and finger kicked drops of water aiming them at each other, playful-ly, Albert had finished his shower, and slipped on his short robe. He popped in the galley and asked, "Have either of you seen Laura?"

He heard two, "No Cap's." He said, "I'm going topside."

Tom spoke up, "Okay, Cap, we're going to turn in as soon as my galley slave finishes up."

She laughed and Albert heard her say, "I'm way ahead of you, pink knees."

Tom looked about as he rinsed out the stainless steel double sinks and remarked, "Thank you, honey, everything sparkles and ready for our next preparation."

She grinned, "Come on lover boy, let's hit the sack, this has been one hell of a day."

He said, "My Princess Oven had a good bake."

She laughed, "Are you intentionally trying to make me jealous of your Princess?"

He said, "You're my Queen....bzzz." They began undressing in their cabin when Albert burst in and uncontrollably shouted, "Laura's gone, she's nowhere on deck!"

Tom said awkwardly, "Did you check the heads?"

Albert devastatingly replied, "I've looked everywhere, she's vanished!"

Norma, not even bothering to cover herself said, "Jesus Christ!"

Tom said, slipping his pants back on, "Let me check it out."

Norma realized her breasts were uncovered, found her robe, then guided Albert to their bed and made him sit down. She asked, "Do you want a drink?"

He leaned over and clinched his face, and just stared.

She stated with amenity, "We'll find our Laura. She just didn't disappear!"

He spoke on unconscionably, "Was she real, this is a nightmare. Why didn't I go with her on deck?"

She said with authority, "Stop it, Laura is a smart, big girl, We'll find her. . .and remember, the sea is one of her lovers."

He said, pleadingly, "What if she fell?"

She patted his hand, quickly replied, "My God, she can swim, and we're anchored!"

Tom returned and Albert was holding his head. Tom gestured with outspread hands. Norma stiffened. Tom moved, and sat next to Albert on the bed. Norma quickly left their cabin. Tom didn't know what the hell to say. He was stunned, and was relieved, when Norma returned.

She lifted his head, and said, "Come on, Cap, drink, then we'll set our strategy and get our Laura back." He downed the hefty drink to the ice cubes, that tried to invade his throat. Norma grabbed his glass, and forced his hand down.

Seeing that he was not in control, Tom stated, "Look Captain, she didn't just fall. . .she had to be taken. . . ."

Norma was watching Albert's eyes.

He continued, "I didn't see any sign of a struggle. . .not that there was one. Either way, she had to have been abducted."

Albert and Norma were hanging on his every word, hoping he had the ability to capture the unknown. He went on, "Our next question, which direction did the kidnappers come from, English Bay, or from another craft. . . " He paused, then added, "The logical reasoning would seem that they came from shore. . .English Bay. That message just may have been a trap to draw us to this precise area."

Norma asked, "Why?"

Albert excitedly exclaimed, "Maybe they knew of her inheritance!"

Norma remarked, "Jesus Christ, god damn that money!"

Tom then said, "If the Cap is right the best of this mess is that she is alive. They will probably be sending us a ransom note!"

Albert and Norma traded nightmares.
Tom conditionally stated, "Albert, Norma and I think the world of Laura. We must be level headed, and not dash recklessly about just to be on the move. There is no moon, it's dark as hell and it's nearly midnight. We wouldn't be greeted at this hour by the residents of English Bay for questioning."

Albert said pathetically, "They can have it all, everything. I just want her back, "Oh, Jesus!"

Norma stated to Albert, "Tom's right, we need our wits about us, if we're going to get our Laura back. By God, we will."

Albert started to get up and Tom gestured for him to stay seated on the bed, and stated, "Look, Albert, I have some sleeping pills. I'm going to get them." Tom flashed a sign, indicating that Norma was to keep him contained. He hustled out and retrieved his by-by's. He returned and put them in Albert's hand.

Albert popped them in his mouth, and washed them down with the melted water in his mixed drink.

Norma stood up, faced Albert, offered her hand and spoke, "Come on, Cap, I'll tuck you in."

He didn't object, and they left with her squeezing his hand for comfort. She helped him remove his robe and pulled down his covers. He slipped in, still disoriented.

She kissed him on the forehead and said softly, "Tom and I will do everything possible to get Laura back. The three of us make one hell of a team." She bent down and whispered in his ear, "Remember what Laura said, that makes it a solid quadrant!" She touched his forehead with her index finger, it tailed, as she turned, and left with wet saline on her finger tips. She cried after leaving his stateroom. She was dabbing her cheeks as she reentered her cabin. Tom saw her washed out eyes.

He said, "If you don't mind, honey, I'll shave tomorrow, let's go to bed." He then added, "Did you get Cap tucked in?"

She sniffed and replied, "Yes, on both questions. . . God, I left him crying."

Tom said, "Jesus, I don't know what the hell I would do if I lost you. . . ." She stopped undressing, and they both hugged each other tightly. She talked over his shoulder, still clasping each other.

"Don't lecture me Tom, but if you could possibly summon the devil himself, make a deal, I'll abide by your decision."

He didn't say anything, but patted her back affectionately. Tom said, "I'll tell you one thing, we'll see her again."

She gasped quietly in horror and was too frightened to ask the question, dead or alive. They finished undressing and hurried to bed to hold one another, thankful that they were spared and able to give solace to Albert. They finally went to sleep.

The screaming birds and the bright sun shone on the *Magnetic*. Although it was cool outside, the ravaging storm had passed its displaying anger. Albert had shaved, dressed, and was apparently going over charts in the helmsman's chair. Tom didn't wake Norma. He got up, shaved, dressed and decided to go to the flying bridge. He glanced at his watch, 6:35. He had to pass Albert to go aft topside. He walked over to him, looked, and captured the stark reality of a naked mind entrenched in doodling. He thought, "Jesus, he was plotting courses in four different directions." He said, "Hi, Cap, can I get you anything?"

"Christ," he thought, "what a stupid statement he had just made."

Albert, bleary eyed, looked up and said, "What did you say?"

Tom said quickly, "I'm going topside. where are your binoculars?"

His voice subdued, Albert replied, "In my cabin, hanging up."

Tom did a turn-around and found his quarry. He returned and neither man spoke, as Tom hustled into the morning sounds, and coolness. He went up the ladder to the flybridge, and carefully scanned the full perimeter. The shrieking sea fowls were gliding and gathering in the hope that this might be a fishing boat and their search for food for the day would end here. He was hoping for some sign, any indication, to explain her disappearance. He felt flat, as he searched the aft deck, not a trace. The perception seemed to indicate that she just vanished. He squatted down looking for anything, a button, or. . . He flinched, swore, and feeling white bird shit was splattered on the back of his hand. He reluctantly retreated. Albert didn't even look up as we went by to the galley. He cleaned off his environmental swash of white streaks. He cooked with indifference, it was a solemn affair. He sat the table, just three place settings. The two men remained silent, no response by Albert. Tom went to his cabin and kissed Norma till she awoke.

She slipped her arms around his neck, then

reality flowed back to her menacingly. She asked softly, "Did I dream about. . ."

He stopped her by touching her lips, and said, "She's vanished. Breakfast is ready, but here, take your E and Bee pill." She popped them in her mouth and he handed her a glass of water.

He added, "Get dressed, our Cap is spellbound, and he needs you to retain what little he has left of his sanity. Hurry please, Hon."

The three ate in silence, only the screaming cries of the gulls and the chopping of waves romancing, the hull was prevalent. Albert played with his eggs, bacon and spuds. He ended up using a half piece of toast, as a little bulldozer, just moving everything, displacing it, but not ingesting. He did drink one cup of coffee. Tom cleared off the table. Norma was talking to a stranger. Tom returned and clearly observed Albert's drained and opaque eyes, and the loss of his intensity. He sat next to Albert, and viciously grabbed his wrist, sinking his thumb fingernails painfully deep.

She said, "Jesus Christ." Albert's eyes changed from their fixed iridescence.

Tom stated firmly, but softly, "I'm pulling our anchors, do you want me to engage the engines, and make for English Bay, or would you like to do the honors? It may give us a lead to where Laura is being detained."

He responded, "Yes, weigh, I'll take us in!"

She kissed them both on the lips, startling Albert, and lightly remarked, "Now you're talking, Cap!"

Albert stepped down and sat in the helmsman's seat. Tom turned, heading topside, silently, and frantically kept pointing his finger at Albert. She waved, and Tom did his chore.

Tom came back quickly and said encouragingly, "Hit it, Cap, just maybe this will give us a notch in this mystery."

Norma gave Tom a "come here finger," and he followed her to the galley.

She said softly, "I have to go to the head, I think Cap is going to make it."

Tom replied, in a doubtful mode, "I sure as hell hope so, he has the wheel."

That statement swept doubt on her observation. She started to leave, then turned and stated, "You are one hell of a guy, blue eyes."

He grinned and slipped back to observe Albert's rationality.

They nestled up to the dock and slips at English Bay. They all wore light jackets, it was still early in the morning and the dampness and cool air hovered complaisantly. Norma and Tom took note of the small boats, laced in their slips, like racing horses, before the release of the gate. They walked to the off ramp, and Tom stated, "You two go ahead and inquire. "I'm going to take a look a little ways down this beach."

Norma stoutly declared, "That's not a beach, it's a maze of rocks and wood garbage. Be careful, Hon, I mean it."

Tom responded, "Yeah, I'll be careful, you know me."

She tried to grin, but stated, "Yeah, that's why I'm asking you to take it easy."

"Thanks, see you, Cap!"

Albert spoke, "Take care, Tom, that lagoon is treacherous."

Tom waved and descended a slight slope, and then began picking his way over and under, moving toward the incoming waves, then further away. Not by choice, he found himself moving further away from the shore. He stopped and decided to go directly to the edge, and work from there. The surges of sea water were louder as he surveyed the congested sea shore. His hairs frantically rose on the back of his neck as he stared down, and across, at the body of a man, laying face down. The corpse was being manipulated by each surge and retreating wave.

He uttered, "Jesus Christ." He quickened his pace and snagged his foot, hooking it under a gnarled fork of a broken limb. He fell in an undignified manner in a nest of trash and rocks. He felt a slashing sensation of searing pain, just above his right sock. A jagged, broken branch had ripped and torn a gash. Blood was spurting out. He cussed, "God damn it!" He untangled his foot and examined the bloody split and quickly pulled his sock up higher, then pressured the wound, until the bleeding subsided. He waited until it was beginning to coagulate. He was furious and excited, then goaded himself, to get to the sprawled victim. He moved at a slower, safer pace. Reaching the water he squatted over the partially rigid man, aimlessly flaying figure moving in lock with each wave. He gritted his teeth, and grabbed a handful of wet hair, and lifted the head up, and screamed. He dropped the ravaged face. His scream was swallowed up with the sound of the surf and surges. Half of his savage face was literally ripped off, the other half was split, bruised and grating. It was being torn with each wave, activating his face against the sharp edges. He had kept his vomit in check, and began the task of searching the pockets of the moving corpse. The waves had claimed and captured him up to his shins with seawater. He noticed a wide web belt under the life preserver of the victim. He couldn't get to it, unless he physically turned him over. He thought, "Jesus Christ," and knew that he would have to look at that inhuman expression, or rather what was left of it. He forced himself to flip the wet body. The color of the skin of the deceased was tanner and had a light frame, indicating he might have been of Asian descent. He undid the preserver, and removed the webbed belt, then he removed the goods from the pockets of the helpless corpse. He also recovered an item, from his own shirt flap, and shoved them hastily in this pocket, without inspection. He got out of the cold water, unzipped his fly, pulled his shirt and blazer up, and strapped the captured belt on. His shoes sloshed, spewing water out, as he nervously traversed the trashy shoreline. He climbed on to the approach to the main dock and stood up clumsily. His excited blue eyes scanned in both directions to see if he was being watched. He walked peculiarly, his dexterity was expressing the lingering horror of his find. Unsettled, he desperately tried to control his mannerisms and emotions. It seemed to take him so long, to eventually board the *Magnetic*. He was the only occupant, and went straight to his cabin, took off his wet shoes and socks. He felt an awe, of insanity, like he had committed a bastardly crime. "Jesus," he thought, he had! He had frisked a dead man, stolen all his property, and told no one. He then remembered not what he took, but that he did, and relieved the dead man of his goods. "Christ," he thought, he wished Norma was here to hold him, and then fix one of those Tom. . .something drinks that straightened him out when Albert had vanished. He pulled up his pant legs, and retrieved a Vitamin E, Soft Jel and pricked the end, and squeezed it generously, spreading it on the crusting gash. He attempted to gather his wits back as he stretched out on the bed. His face was drained and his blue eyes were still darting, and his heart was still working overtime. He waited in hell.

Norma rushed in, out of breath, and a horrendous look on her face. She cried out, "You bastard, where have you been. . . Cap and I thought you had been snatched too!"

Albert came in, and exclaimed, "Good God, I thought they had gotten to you, Tom!"

She saw his wet shoes and socks, and the gash on his shin, and remarked, "Jesus, Tom, are you all right?"

He stammered, "I'm okay, now come here!" He clamped his arms around her, and she whimpered, as he unintentionally hurt her. She was aware of his excessive strength, and his antsy blue eyes. She pulled away, and nervously asked, "What is it Tom...."

He tried to smile first, it slipped then he asked, "Please fix me one of those things Tom drinks. . . ." She spoke, "Coming up, Hon. She galloped back, not bothering to add ice.

Tom greedily swallowed the contents, and cried out, "What the hell was in that scorcher. Christ, that tasted awful!" She smidgined up a little smile and explained, "I know, but we haven't any gin on board, that's your department."

Tom let out a "whew," then asked, "Did you have any success with the folks at English Bay?" "Every word was a negative, not even a maybe. That message had to be a set up, and damn it, we played right into their rotten hands." She was concerned and touched lightly around Tom's sticky wound.

Tom stated, "I put the good stuff on it--I slipped!" Tom got up slowly, and sat on the side of the bed. She watched his coordination. Tom was still unnerved and slightly trembling. Albert moved and placed his hand on Tom's shoulder, and Tom looked up in appreciation.

Tom remarked excitedly, "I found proof that Laura was kidnapped. There was a dead body, an Asian, I think!"

Norma and Albert were intent on every word and drilled each other's eyes in anticipation.

They both spat out, "Jesus Christ!" Tom nervously detailed the horrible find, and described the remains of the desecrated face. Tom's relation had them captivated. He exclaimed, "I frisked the body!" Norma grimaced at the very thought. He added, "I didn't report finding the dead man, I don't know if anyone saw me, but Cap, we'd better get the hell out of here, fast!"

She gestured for Tom to stay and stated on the run, "I'll untie --- Cap you get this bugger moving!" She beat him out, and they set about hastily doing their chores. The troubled threesome and the *Magnetic* sliced through the sea at high speed. The sea fowls screamed in defiance.

Chapter 23

The *Baracs* was steaming at 57° longitude and 152° latitude, just leaving the Gulf of Alaska, and entering the mythical line separating the so-called Pacific Ocean.

Twenty-two years ago, this tramp of the seas, was converted from crewmen manning heavy wide shovels with scoops of coal and transferring the lumps into the fiery furnaces, that heated the water in the old boilers, to its present conversion of oil. The manipulation of this fuel was 79 percent faster in building up steam pressure to drive the old screws. This change was also doctoral practical, as it relieved more space for cargo. However, the speed of the ship was nominal, but by all standards, remained the a dinosaur of the oceans. Like trains, there still is a niche to haul the massive bulky products on the highway of water.

Captain Dork had just given orders to his second, to alter the ship's name, that clearly stood out in the aft and both sides of the bow planes. Captain Dork had suggested, years ago to Devlen Dawson upon his successful bid of both old sister ships, to have the names of both ships on plates, with each letter attached separately, so that implementation of a name change would be relatively simple. The possibilities had intrigued Dawson, and he immediately drew from the ramifications, of the ability to change the ship's name, almost instantaneously. Two identical ships, with the same name, or their own, or switch names without arousing curiosity, was a perfect blend for the masking of cargos, and papers for transporting of such goods. The second in command assigned two crewmen to the task. A bowman's chair was lowered, with a crewman in it, and an additional safety line, to attach to the heavy plate. Only two very large steel nuts, with internal screw threads, held each weighty plate. Each of the individual letters was bolted with much smaller holes, the round being exposed. The changing, from start to finish, normally takes about four and one-half to five hours, requiring two crew men. The sequence of letters would change the name *Baracs* to *Scarab*. The crew men were simply told the registry had been changed, at the new owner's request. All old hard life preservers and lifeboats were always unmarked.

Dassat rapped on Captain Dork's steel door. The original had long since gone, and been rendered to a mere shell of a determent. The scores of years of use, and very little maintenance, had left the once paneled teak door like a door of hanging beads. Replacement of the original sham was a high priority on Captain Dork's list, when he became captain of the *Baracs*, for Devlen Dawson. Dassat heard a faint "come in." He hated this knuckle buster.

He entered and Dork stated, "Ahh, come in, sit down, where's Adda?"

He sat in a plain wooden chair, directly in front of the old worn teak desk. Dassat looked about quickly, assuring himself they were alone.

Dassat stated, "He's dead!"

Dork's scowl fell even further, and he asked, expressing displeasure, "Did you kill him?"

Dassat nervously replied, "No, I wouldn't, I needed him."

Dork countered, and asked, "You have knowledge of the test of the tongues that wagged, never tell, never tell?" Dassat's eyes clicked erratically and shifted in his hard chair, and hurried his words out of his contemptuous mouth.
"Captain, I swear, I didn't harm him. That devil storm had its mark on him. He didn't see that devil wave. The dinghy tore his face off!"

Dork pondered, then stated, "Where is his body?"

Dassat began to sweat lightly, and tenuously stated, "I left him in the water at the shore line. . . about a quarter of a mile from the slips at English Bay . . .I didn't find him until I baled out the dinghy to race after your prize, and snatch her, for you."

Dork digested the brief account, and asked, "What of his wallet and web belt?"

Dassat sat on the edge of his chair and gestured with his hands, and remarked, "Look, Captain, they had left at high speed. I had no time. I had to catch my prize, or try to bring the dead body of Adda back."

Dork stared at Dassat, and he shifted his eyes away from that defective mask. Dork finally spoke up, "If you have spoken the truth, I have your reward." He reached down, and pulled open the bottom drawer of his desk.

Dassat tenaciously blurted out, "Fair is fair, I risked my life, and Adda lost his to comply with your wishes. It is by right, that I alone brought you your prize. By right, I should also have Adda's share."

Dork, taken by surprise, straightened up too quickly in the big chair, and slid back embarrassingly, as did his frown. He wiggled and regained some simplism on the edge of the big swivel seat. Dork declared, "I did not order you to do that deed. Greed is what has infected your body. That is why Adda was so brutally defaced, before dying. Greed, is that what you are demanding?"

Dassat stubbornly, but lightly stated, "I did the task of two -- what's fair is fair."

Dork spoke rapidly, "Are you implying that I am unfair?"

Dassat spoke apprehensively, "No, No, Captain, I thought . . ."

Dork interrupted him and stated, "Your plea has some merit, I will reward you one and one-half years wages!"

Dassat grinned broadly and exclaimed, "Done, done, Captain. I will tell all who should ask, you are fair."

Dork shuffled the bills behind his desk, and stuffed them in a large dark envelope, then tossed it onto his desk.

Dassat jumped out of his chair, and quickly grasped one end of the stuffed packet.

Dork slammed his little hand on the other end, and stated, "If any word of this slips out, I promise you, I will test you with the ritual of tongues that wag... Do you understand?"

Dassat had to look into his eyes and frown, then replied, "Yes, Captain, thank you."

Dork released his end of the envelope and stated, "Your actions are questionable, but you did succeed. That will be all." He added, "What happened to your shoes?"

Dassat replied complacently, "The devil storm took them. Adda and I wear the same size shoes." He left with his prize inside his shirt.

The cruiser had just passed Shuyak Island, on the port side. Albert made the decision to skirt the beginning of the Aleutians through the Sheukof Strait, 59° longitude and 153° latitude. Albert had regained much of his symbiosis, and began functioning, due to the assurgent of thought of recapturing his Laura. Tom and Norma moved to the lounge for two reasons, one was to keep an eye on Albert, and the large glass table top, to study the goods he had taken off the corpse. Norma had fixed old fashioneds, all around, sugar bitters, lemon and whiskey, and a subtle slice of orange on the rim of each glass.

The clouds were a mix of high cirrus, and cirro-cumulus. The bright sun had no trouble explaining its presence. The tinted glass, surrounding the lounge, filtered the bright rays with enriched, soothing and demur light. Albert remained in the helmsman's chair at the wheel, just down one step, and opposite the settee's lounge. They all sampled the tasty drink.

Albert asked Tom, "Any clues?"

Tom replied, "I don't know yet." He had started to empty his pockets. A worn thin wallet, a small seemingly old, but sharp knife, one large brown crypto-crystalline structured with a swarm of reds and colors. It was a beautiful agate. She picked it up and admired the beauty of the stone, stepped down and passed it to Albert. Tom was carefully removing all the papers from the wet wallet, Norma was separating them, and arranging them for logical convenience. Tom was examining a thin bronze medal--two viciously wide knives, the tips of which were interesting. It was dated 1946 and the words were Commonwealth of the Philippines. There, apparently at one time had been a ribbon attached, but only the round disk had weathered the years. Tom got up, stepped down, and gave it to Albert to inspect. She excitedly shouted, "Look, Tom, his Maritime card!"

Tom looked, and swore, "Christ, it's been outdated--over two years."

She said, dejectedly, "At least we know his name!"

Albert spoke up, "What was his name?"

Tom turned the card and read it off, "Adda Sananna."

She unfolded a wadded up $20 bill, there were also two fives and three one dollar bills in his fold. The silence was compelling, except the slicing through the water. She unfolded another, square, damp paper. Her brown eyes sparkled, and she screamed, "Jesus, Look, the name of the ship. . .and even the captain's name."

Albert cried out, "What ship?"

Tom looked and shouted, "the *Baracs*," then spelled it out."

She exclaimed, the captain's name is Dork Ninkeki," then added, "It's a pay voucher in rupees -- 4000 of them, payable on demand, at the East

Indian Bank of Bombay, through the account of Captain Dork Ninkeki."

Tom placed the clandestine paper on a napkin, and took it over for Albert to scrutinize. Albert looked up into Tom's flashing blue eyes, and exclaimed, "By God, we'll get that son of a bitch."

Tom said, judiciously, "Now, all we have to do is find that ship, the *Baracs*. Albert reminded him that they would all have to keep a sharp look-out for that craft.

Norma remarked, "We'll find it Cap, and our Laura, too!" Then added, "Jesus Christ, our first big break! Tom, you sure pulled one out of the hat."

Tom sedately said, "That man, Adda, was brutally slaughtered, it was a hellish sight."

She took his arm and said softly, "Tom, we have to think of Laura.

He replied, "I know, I've always had her intentions foremost, but I just don't know if I'll ever be able to censor out of my mind, the sight of that watery ripped face."

Albert stated, "Tom, I wish this information could have come to focus in a less evasive light, but you didn't cause Adda's death."

Albert and Norma both jumped when Tom screamed, "Jesus, I forgot about the belt." She was hyper after that unexpected outbreak, and sternly stated, "Damn you, Tom, you just about made me, . . . you know." Albert asked, "What belt?" Tom undid his belt, and zipped his fly, and was so excited he grabbed at his falling pants, and missed, but Norma caught them, just below his knees.

She said wistfully, "Your knees are still pink." Tom unclasped the wide webbed belt, and reached down as she was pulling up his pants. She took the belt and stepped up to the settee, and laid the old damp web out flat.

Albert said, "Be careful!"

Tom sat next to Norma, and they each started on one end, opening the flaps. She loudly sounded, "Holy Christ, look at that!"

Albert asked encouragingly, "What is it?"

Tom stopped his search, and stated, "It's a hypodermic syringe!"

Albert barked out loudly, "Good Lord." She jumped when Albert shouted, and yelled, "Can't you guys hold it down . . .that did it, I'll be back."

Tom asked after she left, "Are you all right?"

Albert said, "I'm sorry I yelled, but I just realized, I hadn't turned on the radio, and set it on scan. There are hundreds of eyes out there, maybe some one will mention passing that vessel. You know how the seas talk, they keep track of what's going on!"

Tom said excitedly, "Yes, you're right, why didn't I think of that? It's possible the *Baracs* will use its radio?"

They were both radiating an euphoria when Norma came back and asked, "Okay, guys, what's the latest poop -- what happened?"

Tom explained the scanner bit.

Norma clapped her hands and stated, "Laura, here we come!" They recaptured their seats in the lounge and resumed unwrapping the parcels in the pouches.

Tom said, "Well, I'll be damned," as he unfolded the plastic wrap to reveal a clay like putty.

She looked down at it, and asked, "What is it?"

Tom piped up so Albert could hear, "This could be an explosive!"

She unopened another item and looked into Tom's eyes.

Tom remarked, "Jesus Christ, that's a timer!" She stated, "You opened the last one."

Tom nodded and did what he was told, and stated, "God, that guy was loaded for bear. This capsule has to be the juice for the needle. I haven't the faintest idea what it is," then added, "What do you make of this, Cap?"

Albert replied, off-hand, "If that Adda was one of those bastards that took Laura, I'm glad he's dead!"

She seriously stated, "You guys realize, if Adda was part of a scheme to kidnap Laura, it all makes sense, except for one thing."

Albert interrupted her, "Two things." Tom asked, "What's that?"

Albert exclaimed, "There had to be at least two bastards, one, Adda got killed in the storm, the other stuck Laura and. . ."

Tom interrupted with a burst, "Jesus, if that's true, then we may have a bomb aboard!" Norma said stonily, "I'm going to fix another round of drinks!" Tom spoke up, "Yeah, that last one was a smasher."

They heard her say, as she left, "Yeah, so was what you said!"

She returned with the drinks and stated, "Tom, your conscience should take heed, that bastard was going to stick Laura, then kill all of us."

Tom nodded, and spoke up, "That would account for no screaming or struggle." She said, heartened, "Cap, we really know she's alive, they wanted her alive."

Tom agreed and replied, "You bet, and we know where she is. . .on the *Baracs*. She spoke soundly, "Yes, and they don't know that we know she's on that damn ship or boat . . .whatever."

Tom asked, "Norma, why don't you and Albert monitor the radio and sea lanes, and I'll fix us some sandwiches?"

She said, "That will be nice. Tom, I don't want to keep harping on your gift, but if you do get an ominous lead, remember you promised!"
Tom nodded and asked, "Cap, do you have a favorite sandwich?"

Albert replied, "No, no."

Tom, heard as he left, "Cap do you mind if I light up? I need one."

Tom made five generously stuffed sandwiches, and he cut them twice. He set his tray of warm snacks under layers of blue smoke hues that were snaking above them. Her eyes lit up, and she refreshingly laughed, "God, Tom, you never cease to amaze me," then laughed again.

Albert spoke up, "Good Lord, that's my line, what has he done now?"

She said happily, "I'll show you," and brought Albert a Spanish sandwich cut in four pieces.

He looked down and said, "It looks familiar and smells great." He picked up a quarter, chowed down, and sounded off with a "Mmm." He picked up another piece. She sunk her teeth, and a little steam escaped with delectable fragrance.

She peeked inside her little quarter and delightfully said, "Thank you, blue eyes, Mmm, you even put in olives." She added, "Are we just about to pretend that there is no bomb around?

Tom replied, "No, no, but if he did place the gob, it could be anywhere on the deck, or attached to the hull."

She said for all to hear, "If there's even a chance of a bomb, wouldn't it be sensible to slow down?"

Tom spoke up, "She's right, you know!"

Albert cut their speed, then stated, "We should be safely away from English Bay."

Tom remarked, "We are, if no one saw me."

Albert stated, "I believe if there were a charge, it would have been activated by now!"

She said, with a crooked smile, "You guys are reassuring. . ."

Tom added, "This timer is really a shorty and they are not triggered by remote control. I think the Cap's right, probably. He didn't have time to scuttle us deep in the aqueous blues," then smiled, and said, "These Spanish jobs are good."

Tom asked Albert, while chewing, "How's our fuel holding out, Cap?"

He replied, after a quick glance, "A little over half left, close to 200 gallons is all. That island is Kodiak, that's port of us now, we'll clear it in less than 75 miles. We'll also be out of the Strait of Sheolkof." She had butted her cigarette at half mast, when Tom brought his tasty sandwiches.

She spoke up, so Albert was certain to hear, "Cap, just where are we headed?"

He replied, rather sullenly, "We had to hightail it out of English Bay. I thought we could cover more territory using the Strait, then we'll swing east at Chirik Island, and, if our fuel holds out, we'll make a wide sweep and loop back at Kodiak."

Tom volunteered his thoughts. "It just makes common sense. That storm was so violent that we couldn't dock at English Bay, but a smaller craft could have been used, and docked on the beach, and also used to approach us. If it was a large craft, we would have seen or heard it."

Norma got excited and stated, "That puts the odds in favor of us, looking for a larger craft, and that the smaller craft's mother would be a ship."

Albert bluntly said, "The British Admiralty has defined a ship as being any craft used in water transportation, propelled by any other agency, than oars."

She said flatly, "Damn."

Tom said, still munching, " I think we can guess its centerline. . . of her." Norma smiled and he continued, "is between 100 to 700 or 900 feet.

Albert chimed in, "That does make our odds better, the bigger they are. . .I've a good feeling about finding Laura."

She smartly stated, "I have too, Cap, she'll be back, won't she Tom?"

Tom added, "Logistics are on our side."

Albert thought to himself, "If that pirate ship headed east by southeast, he would have, little chance, or none, of seeing Laura again. But had he not taken this route, there were just too many miles to cover on this ocean, let alone one boat. Even at best, it was one hell of a chance to find her quickly. He knew he would eventually find the captain and the ship. Time was, and is the substantive enemy. Also, he knew, distances would be limited by fuel and weather. He thought of taking on barrels of diesel on his next refueling at Kodiak and also remember to buy a drum pump for the extra barrels.

The tramp ship displayed its new name on both planes, and one of the aft. It was riding high in the water and was spinning its old screws tenaciously south by southwest. Dork heard someone kicking his door. He yelled, "Come in." Sedman Partridge had all he could do just to open the heavy door. He scampered in and moved to the front of the old desk. He was nervously wringing his hands, and stated, in a shrill voice, "Captain, we're sinking!"

Captain Dork struggled out of his chair and shouted, "Where-- what section?"

Sedman replied, "Section F!" Dork demanded, "Where's my second?"

Seaman Sedman replied while shaking, "He sent me to tell you he's at Section F."

Dork shouted, who's at the wheel?"

Sedman, very frightened, was shifting thin weight on each leg, and answered, "I don't know, Captain, are we going down?" Dork shoved him aside and hurriedly left his cabin, heading down to Section F.

Laura peeked out of her cabin and saw Captain Dork bent over and scurrying. His black captain's coat was open, and his scowl, ever depressing. She crossed to her hard bunk, and sat down and began to face the staunchness of her decision. She was beginning to breach the reality of putting her loves in prospective. She suddenly experienced a stab of horror. What if she had to choose between the two men in her life. What if she had the power to save only one? She desperately tried to shake the heart rendering dilemma. This must be a sorcerer's trick, impossible to equate. She was so lonely for Albert. In her excitement about her brother she had transposed all circumstances and feelings of how her husband must be experiencing her alone. She passionately wished she had not chosen to come as a volunteer. She knew now she had acted selfishly, like a child. She had acted on the premise of instinct, rather than maturity. She thought of what Albert must be going through. She then thought that Captain Dork had only said a foreign port--that could mean China, Japan, Singapore, or who knows? Only Captain Dork knew. She finally decided that she had acted foolishly, and wanted out, now. . .!

Chapter 24

She peered out, and seeing no one on deck, went out and looked about in an assuming manner of taking on the noon sun and air. She strolled leisurely to the starboard rail, and hand touched the blistered paint, moving along the specific section, that had a small safety gate leading down to the long steep gang-stairway. She looked down, and saw the dinghy with the motor that was attached to the tether of the platform of meshed and rusted steel. She slowly looked about, and seeing no one, took a deep breath as she attempted to open the gate.

Dork was scrambling, moving quickly, hellishly scampering in a bent over position, the tails of his black captain's coat, kissing the oil stained, metal stairs, as he descended to the lower level. He saw the ocean had filled the whole hull with about eight inches of cold, greenish-blue sea waters. He stepped into the agitated trapped sea and yelled out to his Second, who was busy on the starboard hull of the bay. The Second responded and they sloshed toward each other. Captain Dork ordered, "Number Two, get the pumps in here."

He answered, excitedly, "The pumps are working, but not fast enough, the seals are leaking, and I'd say we're pumping at 55 percent or less."

Captain Dork stated, "Stuff the inference point!"

Number Two replied, "Won't work, two long seams and plates have fractured." Dork shouted, "Show me!"

They waded shin deep, and Dork saw the nightmare. Not only were there two horizontal splits, but two vertical plates were spewing in water. Number Two stated, "If we had a cargo, her belly would have been filled by now."

Dork commanded, "Get some men down here, and build a pressure bridge against that fractured section, so it won't all give way. I'm going to the wheel, and head for shallow waters." Dork kicked the waters in disgust and headed for his stairway. His coat tails were sucking up the sea water. He raced up the steel steps, and went to the wheel house, which bridged above the main deck. He turned the old hull west, by northwest, and headed for the shoals of Trinity Islands, which rose slightly above sea level, and not inhabited. Dork could not call out an S.O.S. or Coast Guard with Laura aboard. His best chance was to save his ship in shallow waters.

Increasing the speed would induce more pressure on the cracked plates. This damage to the hull was approximately 9 to 11 feet below sea level. The *Scarab's* nautical speed had slowed two knots, than the previously set ten for two very good reasons. One, obviously, was the trapped tons of sea water, the other he was having to cut against two to two and one-half foot swells in his quickest course to the Trinity Islands.

The Second had ordered Dassat to get every crew member down to section F with tools to build a pressure bridge. He was very explicit with his order and he also instructed Dassat to have the cook report to Captain Dork in the wheelhouse. He stated, "Hurry," and Dassat went running to obey his orders. The crew showed up in the cargo department of Section F, and Number Two, was barking out orders as to how to acquire wood and steel to build the pressure bridges against the leaking plates. He snapped out that there was really no emergency, already they were headed toward friendly shores. He ordered them to get cracking.

Captain Dork's mind was filled with horrors of this, or that happening, just when he was endowed with his second only wish in his life--his own ship. The thought of losing his world as a stark absolution, and it was the permanence driven, like a stake in his heart. By some chance if they sunk, the insurance on his ship, this ship, really the *Baracs* would be nullified. The name was now the *Scarab*, and Devlen Dawson had the title to the insurance. He would collect on his ship, then change the lettering on his ship *Scarab* to *Baracs*, and he would come out smelling like a rose. He swore in three languages. Dork knew it would take three to five hours to change the three name plates and that there was little time.

Laura closed the metal gate with a small, high clinking sound. She winced, and moved quickly down the scaly old grilled steel stairs to the small platform that was tethered to the dinghy. She splashed with the trapped water as she undid the lines that held it, and moved to center seat. She used the oars, just to steady the boat's distancing from the moving giant of the ocean. She looked up, nearly 60 feet above, the sky scraping hull, and guided the little craft away from the turbulent aft waters that were churning and boiling, as the huge bronze screws spun, pushing the hundreds of tons through its liquid mass. She rested her oars, and moved to the tiller or transom seat to check on the correct sequences, to start the old sixteen h.p. outboard motor on the first try. She thought all was set, and gave it one hell of a pull. The old engine popped, but turned over and she put it in gear. Engaging the old propeller she turned the dinghy starboard, the chimes on the hull, peeled and flung the splashes of the sea into two parts. She had a determined look on her face as she opened the throttle. What Laura didn't realize, was that she had just escaped from a ship that could, in all essence, be ready to at last to retire on the bottom of the sea. She kept glancing back at the smaller giant and noticed it had altered its course drastically. In her deliberate haste to elude the old derelict, her only course, was further out to sea, due east from the Trinity Islands. The sun shone brightly, reflecting off the millions of minute reflectors of the conditioned waters. She looked all four directions, there was nothing but water, and she felt so relieved and so lonely. The old motor putted on.

Captain Dork had to increase his speed two knots to maintain his original speed. He estimated less than ten nautical miles for deliverance. The deep haunts of his ship were filling with tons of sea water, reducing the efficiency of the antiquated driving, twin screws. Also, his old marvel was settling deeper into the coffers of the dark green waters as it labored toward Trinity Islands.

She wished she had brought a hat, and a covering of some sort. The sun's rays refracted the water, intensifying the brutal ultra violet waves, that are shorter than light waves, but longer than x-rays. Laura would eventually realize, that she was heading east to the Pacific by the sun's position, which was part of her education in Marine Biology. Her immediate escape was the first priority. She guessed that she was moving at about four to five knots, and the ship was probably doing close to her speed an hour, and in one half, to two hours, it should be out of sight. A slight breeze was aiding her harborage, even though it meant the Pacific Ocean.

They all listened, and watched intently to the radio chatter, and the eyes behind the sweeping binoculars were blue. He lowered them and passed the heavy glasses to Norma. His arms were getting sore.

She said, "Guys, I'm going to the Flying Bridge to get a better see!" She climbed up the stainless steel ladder, and was met by a slight pick-up of the cool breeze. She captured a cigarette and took a deep drag, then positioned the powerful glasses, scanning, and traversing slowly. She saw a large ship, then returned to the turret sequences. She had one puff left, the light breeze had smoked a third of her white smoking stick. She drew it deep, and then finger kicked it into the sea. The next round of surveillance, and the freighter was more prominent. It had settled deeper in the liquid ocean. The bow was almost directly in their direction, making identification an impossibility. She climbed down, and shouted to Tom, as she entered the lounge and lower helm. Albert reacted nervously and exclaimed, "Did you see anything?"

She said, "I don't know -- Tom!" She gestured for him to follow her. Tom let her climb up the ladder first. Her movements delighted him in every sense. He touched her ankle. She pointed and handed him the binoculars, and slipped behind him, and locked her arms about his waist.

Tom noticed how low in the water it seemed for such a big ship. His mind popped up with the probability of an overload, to compete with the faster and more efficient ships, built within the last 20 years. He was just about to ask Norma to tell Albert to alter his course, when the heavily laden old derelict changed course. The *Magnetic* had passed through the Straits, and was cruising between Trinity Islands, and the southern tip of Kodiak Island. He spoke aloud, "The name is *Scarab*, then spelled it out "S - c - a - r - a - b. Tell Cap, will you, hon?"

She smiled and playfully said, "Aye, Aye, blue eyes." Tom was idly wondering what cargo that old ship had in its bays and its destination. A bundle of billowing isolated clouds, blotted out the bright rays of the sun. He thought it was likely his imagination, maybe the filtering of the sun's rays off the water, probably was perceived as an optical illusion, but it seemed the old freighter was settling deeper into the grasps of the ocean.

Tom hurried down the ladder, and rushed into lower helm, and excitedly said, "Cap, my eyes could be playing tricks, but I think the freighter, the *Scarab* is sinking. . . . Yet, if I'm not nuts, why hasn't she sent a S.O.S. I sure as hell didn't hear one."

Albert said sternly, "I didn't either. Give me the glasses, I'm going to the Flying Bridge, Norma take over and Tom come with me. . ."

Albert pointed to the compass, and stated, as he turned, "Hold her at that course, unless you feel me taking over with the duals." They quickly hurried up to the flying bridge. Albert was concentrating on the line, and proportion center line length. He barked out to Tom, "Half speed," and Tom responded, as did the *Magnetic's* powerful engines. Tom saw Norma's head, then all of her as she climbed up to be with the men, and sound out the judgmental call by Tom.

Albert was still fixed on the hulk and cried out, "Good Lord, Tom, you're right. I don't see any action on the main deck either.

Tom spoke up, "The name *Scarab*, is an ancient sacred name that represents a sacred beetle found on Egyptian coins, mummies and works of art. It's regarded as a symbol of creation and creative power."

Albert replied loudly, "Unless there is some divine miracle about to happen, that ship is going down . . . . The captain must be trying to make it to shallower waters."

Norma stated stunningly, "Jesus, the ship isn't gong to make it. I thought a scarab was a long sword."

Tom replied, "A saber is what you probably mean." Tom screamed, "Jesus Christ." They both responded with one word getters. He continued, "Christ, Laura's on that ship. . . a play on words, that's the *Baracs* spelled backwards, it has to be. . ."

Albert cried out, "Good God."

Norma said, "Sweet Mother of Jesus."

Tom excitedly stated, "Cap, let's circle it to see where we can board her."

She said, "I don't see any life boats being lowered."

Albert stated nervously, "Not from this side."

She put her arm on Albert's shoulder and he said in a determined manner, "You're right, Tom, we had better circle, and do it fast."

Tom spoke slowly, "Laura may be already off on the other side in a lifeboat."

Norma blurted out, "The two lifeboats on the port side are still secured."

Tom exclaimed with a "Jesus."

She asked hesitantly, "Do you think the ship will make it?" Neither man spoke.

Dassat was nervously working with the crew, the water was up to his waist. Number Two was still barking out orders, and the men were cursing, and the candor was increasing as the cold water crept up their bodies. The crew was at the brink of getting the hell out of there, orders or not. Dassat moved to the outside of the calamity, and moved slowly through the water, a fat rat was frantically trying to make it to the steps also, only Dassat made it, and ran up the metal steps. He knew the ship was going down. He raced to the crew's quarters, and retried his hidden, dark envelope, stuffed it into his shirt, and ran back up to the main deck. He looked about, then skirted the open deck, and ran to the starboard side to the safety gate. He unhooked it and looked down in horror, crying out, "Shit." The dinghy was gone. He was trembling, as he heard the growling voice of Captain Dork, from the bridge. "Get back in the hold, Dassat, or by God, I'll put you through the tongue that never wags, you'll never pleasure yourself, or any woman the rest of your scurvy life."

Dassat was attempting to respond, "But, Captain, the . . . "

Dork cried out, "That's an order. . . now!" He nervously hustled back to the stairs leading down to insemination of the ill fated dinosaur. He stayed on the steps and looked down and watched the frantic crewmen, despondent and lead by crazed Number Two, hopelessly trying to prevent the invariable sea from capturing its prey. He crept down a few steps lower, the water was up to most of the men's arm pits. They were screaming at each other and the heightened fear grew to a nightmare. He contemplated which of the four lifeboats was seaworthy, probably half would only serve as an escape. He would have chanced the dinghy, although there was little gas left. That devil storm couldn't wreck it, and it killed Adda, but he got half of his share. That dinghy was his good luck charm, and it was gone. Either the tether snapped, or more likely, one of the crew stole it, to escape certain death. He would choose death, rather than let Captain Dork perform the wagging tongue ritual. He swore, if he lived, he would never again serve under such a man, that had the power that was worse than death. The thought of that devil witchery hanging over him was unbearable.

Captain Dork had the cook stand by him in the wheel house to use as a messenger. His communication to the hold had broken down. He ordered the cook to get the large wrenches, and boatman's chair, and rig, and place it at the base of lifeboat (A), not in the lifeboat, but on the deck. The cook's eyes were blazing as he ran down the clanking old metal steps. He just about fell over Dassat, who was still crouching, watching the now undisciplined mayhem. The cook tried to scream to Number Two below. It was impossible for him to hear.

Dassat asked, "What's the message?" The blinking cook told him his orders. Dassat quickly realized the situation. If the Captain was afraid his ship was sinking, and wanted this equipment placed on the deck, there must be a mystery that could be worth a prize.

He said to the nervous cook, "I know where they stow that gear, come with me." The two ran up to the next deck, to the ship carpenter's stores. Dassat loaded the cook town, he took the large and small wrenches. They hurried up the steps to the main deck. Dassat waved to the bridge, making sure Dork saw him in useful work.

Captain Dork could see the rock abutments, and the churning waters of the busy shoals of Trinity Islands. Number Two cried out, "Everyone out -- now! Go to the main deck, now! They waded, waving their arms, hoping their flaying would speed them faster to the stairwell to escape this hell. Each man's fearful eyes had their own horror of being trapped in the holes of this monstrous coffin. The intense fear had an ugly odor that was already impregnated in the steel plates. Out of breath, they welcomed the bright sun, as their soaked bodies and terrorized eyes searched their fellow crewmen for solace. Some were looking up to the bridge for an answer. They looked like drenched marionettes, in the bright sun--soaked, kicking, and flaying water off their exhausted frames. Captain Dork was at the very edge. If he shut off the engines, they would be engulfed by the intrusive sea. They could lower the life boats, with greater safety, but he felt the water was still too deep. He had another choice. All stay aboard, and open the engines to hasten ship to shallower waters, then upon reaching the shoals, reverse the engines, and attempt to stabilize, preventing the hull from being ripped out. The men on deck were cursing, and shaking their fists. Number Two was doing his best to maintain discipline, but he too was watching the bridge, in the hopes that an order to abandon ship. Captain Dork made a decision.

Laura watched the slow, latent freighter fade, expanding the distance between them. She had the throttle wide open. She didn't know that the old ship was in trouble and sinking. She only wanted to get back to her husband and new friends. The dinghy was bouncing and slapping the waves. She reduced the speed just a little. Her tongue licked her upper lip, as the tramp ship was barely visible. All she could see were swells of water, water, water. Her feet were soaked, but they felt good, the sun had finally heated the trapped water, but the reflection from the rays was horrendous. She started to look up at the sun, and had to hold her hand to shade the glare bouncing off the busy surfaces. She then came to the conclusion that she must be moving in an easterly direction. She forced her mind to go back to that terrifying experience off English Bay, after the remnants of the ravaging electrical storm. "Damn," she thought, "she would have to guess what direction they went during the night, they tied up to the old rusting platform, then too, the ship's direction." She couldn't be certain of a damn thing. She dared not turn back, they might already have found her missing and sent out scouts to find her. She looked toward that little dot, it was directly west of her heading. She cussed herself again, for her selfish decision, and would apologize to Albert, willingly, on her knees. She had put herself in his place and knew the tearing hurt of not knowing whether she was dead or alive. She wished she had taken a drink of water before her escape. . . or was it. . .betraying, her own flesh and blood. She again looked in all directions, and saw waves and swells. She sucked in her lower lip and froze as the noisy motor began missing, it popped, then vibrated and stopped. She picked up the gas tank with its umbilical cord. It was light and empty. Her dinghy began to turn sideways, manipulating its sovereign right, that all upon its surface act with its accordance, or sink to its floor, and be decimated by the billions of its inhibitors, or the rot of time. She quickly set the oars in the water, and realized that this old boat was not designed for one person to row, because of the inaccurate placement of the seats, in relation to the oar locks. She straightened out her ride. The swells were still moving her in an easterly direction.

Captain Dork suddenly caught sight of a vessel, laying port to his ship, coming at high speed. He grabbed his glasses and saw that strange looking antenna and swore. If they found Laura, his prey, there would be hell to pay, and the loss of his ship. His whole life, seemingly, was being tested in one stunning moment of wrath. The frothy white splashing waters that were determinedly trying to slip by the rugged shoals could easily be seen by the naked eye. He shouted loudly to the men below, "Brace yourselves, men, I'm going to set her on the shoals." Number Two couldn't believe what he had heard. The captain had turned into a madman, not allowing the crew to abandon ship. The churning, laboring derelict, had swallowed hundreds of tons of sea water, and the engines were still onerously turning the vibrating screws. Their speed was less than two knots, and the inner trapped sea was swelling higher and higher. Number Two yelled to his crew, to run, grab and hold on. They were going to ram the shoals. Two of the crew just stood transfixed, and didn't make it. Captain Dork had judged the depth by the whites of the swirling waters. He reversed the engines. The momentum of his huge steel ship, with the moving mass of waters inside her, was uncompromising with the drive of the screws. The thousands of tons of inertia had totally overcome the pitiful thrust of the opposing blades. The tearing, and ripping of metal was clandestine, with the grinding, searing, and slashing of the undercarriage of the massive hull. The sea swollen, gored giant, still rising upward, but continued moving shoreward. The dying hulk continued to grovel on the sharp pinnacles and rocks. Its body was still rising and moved awkwardly with all the unyielding mass with a final lunge of gasping exertion. She then fell back slightly. The moving wave inside the ship reversed, and the tearing, shrill of metal, began again as the clumsy old wreck, pulled back from the sharp penetrating rocks. The steel dinosaur had become the prey of the treacherous rocks and the shallow sea as her guts were being ripped out.

The crew of the *Magnetic* had maneuvered to inspect the lifeboats on the starboard side and just caught sight, that they were also intact. Albert cut his engines.

Tom spoke, "Jesus Christ, they're going to ram the shoals."

She cried out, "Oh, my God," and thrust both hands to her horrified face.

Albert muttered, "Jesus -- Laura." As they watched, the once proud, hailed old king of the seas being torn and shredded, her underplates yielding to recapture her trapped waters. The high pierced wrenching of ripped plates filled the air, overpowering and frightening, the hundreds of shrilling gulls, nervously rising out of the collisions of the intense strike and the radiating noise. The volume of clashes reached its extremities. The high rise of the bow was like a final salute of its once mastery over the torment of the sea. The last flaying hand of a crowning victim, giving up in the subversive waters. No one on the *Magnetic* could hear the screaming and wailing of the crew, as it lammed up against the massive rocks. They witnessed the calamity of an immense proportion and the magnitude of grinding, peeling of the man-made giant, where the eternal sea and the land met unharmoniously.

Norma put her arms around Albert, and turned her back to the ugly spectacle. Albert had throttled to neutral, and his craft was slowly being carried away from the settling of the hulk on the rocks.

Tom dramatically stated, "If Laura is on board, let's get to her. We have to render aid to the crew and the captain."

Albert was still being held by Norma, and looked straight ahead, and stated, "If Laura's dead, I'm going to kill that bastard."

She squeezed him all the tighter. Albert asked determinedly, "Have either of you a gun?" Tom and Norma answered negatively. Tom said, "Cap, we better get. . . I would suggest we approach her port side!"

Albert said, "Yes, of course," and engaged the powerful diesels. Tom whispered in her ear and Albert seemed to take no offense to the secrecy. The grounded hulk slowly began slipping off into deeper water. Then suddenly emitted a metallic groan, became snagged, and the bashing began with the graphic waves, still slightly moving the torn wreck. The stern bow plates were still riding high, while the aft was just about overrun and under water. A new enmity was formed. The tangled tons of twisted steel became a breach to the terra firma and to the shifting lines of the relentless sea.

Two of the crew who failed to act on the screaming orders of Number Two were slammed backwards when the initial strike impacted under the bow plates. They wailed cries of horror as their arms and legs flayed uncontrollably. One man struck his head on the steel deck, his brown hair was changing color to bold red under the bashing rays of the noon sun. He blinked, his terrorized black eyes, and relentlessly tried to get up, only to be brutally slammed down again. The other crewman had managed to position his arm to cushion his fall, his humerus bone snapped under the forceful descent.

Dassat ran for a cargo bay cover and grasped the lift railings just before the first collision. He gritted his teeth, and bit his tongue, as his feet went airborne, and stretched out his thin frame. He gripped with all his might. Blood ran out of his mouth. Three others, including Number Two, ran desperately and reached the hangers that held the lifeboat above them. The lifeboat suddenly swung violently at the strike, and the sudden momentum snapped its weathered lines, and the 16 foot, lopsided lifeboat, came crashing downward. Death screams, and wailing were drowned out by the metal clashes, and the grinding on the volcanic laid rocks, and the crashing of the freed bulky, heavy old lifeboat, as it struck the three huddled grasping frightened men. It abruptly struck and crushed their bodies on its way to splintering on the steel deck.

Dork was thrown backwards against the aft bulkhead, in the wheel house, then was dramatically pitched forward, glancing off the wheel and over the instrument panel. His hatless head struck and crashed through the glass. His body hung half out of the teak frame, with just splinters of glass in its molding. The sun and broken edges and pieces erupted into a flashing spectacular of light. But no one witnessed the hundreds, if not thousands, of fractures of flashing with the acute slivers of glass. And for one brief instant, he was still alive, his arms were flaying to recapture himself in the wheel house, and the flesh on both hands was being sliced, with each struggling move. He hunched his wiggling frame back into the safety of the bridge. He acted on reflex, and pulled his hanky from his pocket and wrapped his left hand, tying a crude square knot. His black, handsome captain's coat was ripped, tattered, and covered with blotches of fresh blood. He looked like a weathered small scarecrow. He stomped down the unpositioned stairs, not realizing they were tilted upwards a good 20 - 30 degrees, as was the main deck. He floundered out on it, with blood running down his face and hands, and boldly began barking out ridiculous commands to his despicable crew.

"Change the name plates at once, hoist the bowman's chair!" Only two crewmen responded to him, they cautiously walked towards him, one with a busted arm dangling, the other with a slash to his forehead. Captain Dork's mind, for all purposes, was not functioning. His only thought, at the impact, was to change the nameplates, at all costs, so that he could collect the insurance and buy another ship of his own. He pointed upward a bit at lifeboat A, and shouted, "That's an order, start at the bow plates, then aft!" His eyes were mar-bled, as he turned and took a step, then yelled, and shouted, "Get to it, I'll be in my cabin when you've finished." He looked at his watch, his eyes did not register the time, he was just going through the motions of reality.

He turned, hesitated, then turned back and yelled, "Have Number Two take the wheel while you change the name plates and the name." He doggedly tramped on down the slanted deck to his open cabin door, and entered without closing it. His swivel chairs, on rollers, was up against the wall. He pulled it back to the desk, released the chair, began to sit down, and the old chair rolled and struck the adjacent bulkhead, then teetered, slightly moving back and forth. Two times he missed the busy chair. The third time he won the battle. He reached between his legs and grabbed the seat before sitting down. He plunked down quickly, and simultaneously thrust his arms crossed on the desk, and laid his frown on it.

Chapter 25

The *Magnetic* made a loop to deeper waters, thus positioning herself on the port side. of the wrecked and gutted giant. Her bow was still poised proudly. The long metal rusty grilled stairway was locked on the starboard side. The deep rubbing, clashing, and grating of metal sent vibrations throughout, resonating from the surface. It strangely sounded with its deep rumbling, like a wounded dying giant.

Tom asked, "Any ideas on how we're going to board that monstrosity?"

Albert stated, "Her aft is almost under water-- maybe if I reverse the *Magnetic* and settle close to her, I can jump...and . . ." Norma stormed in protest, "Yeah, and lose your ass doing it. Then when Tom and I get Laura off, you won't be around, you guys!"

She added, "Cap, you know the sea, and what sails on its surfaces. We all witnessed that dreadful wrecking, what's her chances, or how long will she maintain the positioning, before breaking up. Christ, you guys got to me, calling her a she!"

Albert replied, "I believe she hung up pretty good, but a storm of any consequence would tear the rest of the hull to a skeleton." Tom stated, "What Norma is saying, is that now that we have found the ship that holds Laura captive, we should proceed with caution and not be rambunctious."

She said coyly, "That last word was a knockout!"

Albert strained and stated, "It's the not knowing that's tearing the hell out of me. What do you both have to offer that's damned quick?"

Tom spoke up, "Cap, everything is on our side and with teamwork, we'll soon have Laura back. What if this ship had gone the other direction, what if this wreck took place at night, or a storm."

She cried out and pointed upward, "Look!" They waved back and two survivors and the three, watched a miracle happen. One of the crew was being lowered in a boatman's chair. The sea gulls by the hundreds were circling closer to the dying ship -- their high terse screeching and shrilling in search for food.

Albert smiled for the first time since his loss of Laura. His mind was charged with relentless fury. He felt so close, he barked, "I'll squeeze in close, we'll put that man in, and then I'm going up and send the other man down. The crewman in the descending chair was kicking the drab, quivering hull. He missed and soon spun until he regained his adjacent orderly control, and continued his downward prominence.

Tom asked quickly, "What if there are more men on board, what if they are injured, what if the captain is waiting with a gun."

Albert snapped back and stated, "Be damned with the what ifs!" She stated coldly, "What if they get aboard and overcome us, or kill us?"

Albert muttered, "Good Lord," and Tom said calmly, "Anyone of those what ifs, or combinations is displeasurable."

Norma give Albert one of those flare guns and charges. "How many loads do you want, Cap?"

Albert replied, one in the chamber, and I'll pocket three more, making a quadruple!"

She kiddingly gave Tom a light elbow and stated, "Jesus, I wish Laura could have heard that! Let's get our Laura back--now!" She looked up and shouted, "Jesus, that man can only use one arm."

Albert was cautiously maneuvering the *Magnetic*. He was using his transom ladder for the butting up against the steel plates of the vibrating ship. Although he was feathering the controls, the incoming waves made it impossible to romance the old hulk.

The crewman in the chair was but a few feet above Tom. Tom had scooted down the ladder, and placed himself partially on the aft trimtabs.

She watched Albert's tactfulness, the *Magnetic* was an extension of his fingers. It was a masterful coordination, with the many shiftings of the two crafts, and the different reaction, that each had upon the surface.

She suddenly broke her daze, realizing what she should have been doing and cried out, "Jesus, Cap, gotta go and get our first aid kit!" Albert heard her say as she went toward the latter, "Should I fix that man a drink?"

Albert replied loudly, "Ask him, after you guys take care of his needs."

She stopped going down the ladder, her head just above the deck of the flying bridge, and stated, "Cap, I have to ask, shouldn't we have radioed the Coast Guard?"

He grimaced, and said, "No, not yet. . . think, why didn't they call the Coast Guard?"

He heard her say, as she went down to aft deck, "Yeah, you got one hell of a point!"

Tom stretched out, and up, and grabbed the man in the chair and lowered him to the aft deck. The man groaned as Tom accidentally brushed his injured head and shoulder.

Albert waved, then gestured to the man at the rail as the low prolonged, repeated, grinding continued on the wrecked hulk. He indicated he was going up in the chair. He waved back in final communication. Albert called Tom to take over at the bridge.

Tom raced up the ladder, and stated, "Please be careful for all of us, and especially, your Laura!"

Albert exclaimed, as he went toward the ladder, "You two watch yourselves!" Norma helped to hold the chair for Albert to get in and tie himself in. Tom waved at the crewman that Albert was going up. Albert looked up to nearly 60 feet of vertical Steel.

Norma gave Albert a thumbs up, but he didn't see it. Albert kicked lightly, to keep from hitting the steel plates, as he slowly was being manipulated up the hull. He patted his pockets, making certain he had the flare gun, and loads. Norma poured a double straight swill of bourbon for the injured crewman. He downed it, grimacing, licking his lips with deep gratitude and satisfaction.

She waited, then asked, "Was there a woman on board?"

The seaman thought to himself, they had treated him with care, and the Captain had gone mad. He replied, "Yes!"

She screamed, and he jumped as she cried out, to Tom, above, "Tom, Tom, this man says there's a woman on board." Tom shouted, "Christ, what luck!"

Albert had made it up the scabby old rail, and fell down on deck lightly. He unharnessed the chair. The crewman pointed to the split gash on his head. Albert told him to lower his head as he carefully fingered the matted, clotting injury to inspect the wound. Albert asked, "Can you make it? I'll lower you down." He grinned from ear to ear and began getting into the boatman's chair. Albert helped him up and off the high rail, waved and shouted to Tom, who waved back. And his second crewman was abandoning ship slowly. The grounded hulk was making loud strange metallic bowel sounds, harsh and grating. Tom hoped that Cap had asked the question, she did. They could not hear each other with the noisy sea and gnawing of metal with each wave. Albert had not asked about a woman on board. Norma was excited now and asked the man's name. He looked into her eyes and replied, "Burlen, Miss."

Dassat was flicking his tongue about the inside of his mouth and swallowing a little of his own blood. He had only been on one wreck before, and they had plenty of time to get away in life boats from the disabled ship. But this captain was a madman. He cringed, still holding on the hand rail, to the large cargo bin covers. He heard glass shattering, actually thought he saw an explosion of flashing light, and looked up to see Captain Dork using his head as a battering ram, and had halfway fallen forward, hung up on the frame of the shattered glass. He watched him squirm and wiggle, and disappear back in the wheel house. He stayed, clutching the small hand rail, and then saw Captain Dork clumsily stride on deck. He was bleeding and his clothes ripped and torn in disarray. He tramped up the tilted main deck, passing close enough to him, but not looking around. Dassat heard those ridiculous orders, and his pausing only to bark out more of the absurdities. He waited until Captain Dork returned to his cabin, then he walked over to where Number Two and two shipmates lay lifeless, a mutual grave. He peered into their frozen and exploited expressions. He heard brisk talking, and slid behind the rubble out of sight. The two crewmen came closer to inspect the corpses. Both men mumbled, and quickly crossed themselves.

Billy looked down, and saw the boatman's chair, and rig, partly covered with debris. Burlen dropped to his knees to help, using his one good arm. Both men were enthusiastic about a second means of escape. The long scaled, hinged stair wall on the outside of the hull was a mangled entanglement of pretzeled steel. The moaning and clashing of metal still persisted with the slight moving of hundreds of tons of twisted, iron carbonated, members. He saw Billy and Burlen waving and yelling on the rail at port side. He stayed out of sight, reached the rail and looked down at the trim lines of a cruiser. The sun's rays were striking and refracturing light from a strange, large antenna. He painfully swore. The metal grinding was in harmony with the constant waves, grating of stubborn steel and rocks. Dassat kept in a crouched position and made his way aft, to have a look at a possible means of escape. A portion of the huge aft was partially under water and when the swells of the crest swept in, crashing, it filled with frothy white sea. He gauged it to be but 60 feet from the aft to dry rocks, that had boiled up in a volcanic surge at some time in the long, long ago.

He could swim that easily. The soberness of greed again gripped Dassat. That lower, left hand drawer in the captain's desk, held prize money. Why let the sea claim it? He further built his greed with the thought that Adda's half share was still there. By right that belonged to him. What's fair is fair. He patted the brown envelope in his shirt, and fondled the corner of it. He carefully worked around to the entrance of Captain Dork's cabin. The steel door was open, striking the bulkhead each time the giant ship shifted with a powerful wave and swell. He stopped suddenly, and realized he could move faster and quieter, without his shoes, if he didn't bump his swollen and tender toe. He sat down and took them off. This time he would tie the shoe strings together and sling them around his neck. The rocks on shore were certain to be sharp. He grinned as his mind was clearly in detailed command. This would also speed the race, if one occurred, and he would be ready for the swim. He walked lightly into Captain Dork's messy cabin. He saw Dork's uncombed hair, his arms crossed on the big teak desk, with his scowl hidden in this arms. His chair barely moved, but it squeaked with the timing of the steel door. Captain Dork looked to be asleep, or dead. He moved toward the desk on Captain Dork's left side. Knowing the steel door would strike the metal bulkhead, he nevertheless jumped, when it did clash. His judgment began to rise, as to the chance he would be caught. Another opportunity would never present itself--go for it!

Tom was nervously trying to be auspicious with the controls of the *Magnetic*. He simply didn't have the experience. He was constantly over or under compensating.

Norma was shouting to Tom, "Hold it, no, no, hold it. . . " The second seaman was just a few feet above Norma's reach. She was stubborn, and apprehensive as she reached to capture his feet. She lunged and grabbed, by one foot, however, she hastily pulled him down on the aft deck of the *Magnetic*. She shouted, but Tom couldn't hear her victory. She threw him a kiss, and he responded with one of his own. He throttled forward and positioned the cruiser approximately 40 feet from the scaly wreck. She went to the lounge, and fixed a similar drink for the second crewman. Burlen saw the drink and licked his lips, although knowing it wasn't for him. She handed the second seaman the stiff drink and asked his name.

He gratefully took command of the glass and stated, "They call me Billy." He gulped it down. Norma was fascinated, watching the predominant Adam's apple busily rising and falling, till the glass was empty. Burlen looked on, like a beggar, and was watching the contentment on his crew mate's face. Billy sleeved his lips.

He uttered, "Bless you, Ma'am."

She smiled and stated, "Call me Norma." She told them where the main head was located, not the one in Albert and Laura's stateroom. She inquisitively asked the two smiling men, "How many more aboard?"

Their eyes suspiciously searched each other's and there was a pause.

Billy finally said, "I think the captain is the last of the lot."

She cried out, "But what about the woman, Laura?"

Billy said, "The truth is, we heard there was a woman brought aboard, but we didn't see," he added, most positively, "where they put her!"

She tried to pry out more information and asked Billy, "Billy, what do you mean, by they…?"

Billy weighed his answer and stated, "Ma.., Norma, it was just a logical guess, I'm sure the Captain wouldn't have left the ship, unless in port, I'm sorry."

Albert saw bodies on the starboard side of the deck, and hurried over. His eyes captured the blatancy of any captain who thought more of his ship, than his crew. The three dead crewmen were jammed together, their bodies crushed. There were the agonizing contortions as each man met his sudden and personal death. Each face froze in a fixed horror. He felt sorry for the predestined men. He thought what a useless waste of livelihood. He looked around, and up, and saw the glass shattered in the wheel house. He rushed up the steps, and entered the cluttered, old command bridge. It was empty. He moved out to check the few cabins. He entered a small cabin, his heart skipped.   He thought he could, or somehow sense, that Laura had been here. He couldn't be certain, there was nothing to confirm his first impression. He heard a muffled clang of metal, above the regular rasping of grating steel. He moved out and saw a steel door swing from the bulkhead, but not close. He touched his flare gun and waited. The steel door jerked then banged against the welded wall. He released his feel on the gun and entered the large cabin. He saw an impressive old teak desk, and a man's head lying on his crossed arms. As he approached slowly, he saw the man was wearing a shredded, ripped, black captain's coat. The sleeping, or dead captain, had dried blood on his fingers. No doubt about it, this man had to be the captain. His thin short frame was in a large swivel chair that was moving, oh so slightly, in rhythm with the irritating banging of that damn door. He moved closer, and had to stretch across the desk to grasp his shoulders, then shook him.

Dassat was building up confidence before he attempted to sack the bottom drawer. He responded as a shadow intervened in the open steel door, masking the bright sun rays. He moved quickly to the private head, and watched as he saw a tall man, in his late 50's walk hesitantly toward the desk. Dassat was almost certain that was one of the men on the *Magnetic*. Captain Dork had told him and Adda that there were two men on the *Magnetic* that night of the devil storm. He watched as he bent over and reached across and shook Captain Dork's shoulder.

Captain Dork lifted his head groggily, and decreased his arm pressure on the desk. Suddenly he began rolling--faster, and slammed into the adjacent bulkhead. Captain Dork was ricocheting in his busy chair, and was scrambling to get out. Albert reached down to stabilize him and Captain Dork raised his head to see who it was. Albert gasped at the blood speckled scowl. He finally pulled him out and the chair scooted the corner and floundered, squeaking with each slight pitch of the ship.

Captain Dork spoke out, "Have you completed changing the name plates?"

Albert was mesmerized by that scowl. His grimace stayed set in an awful frown. Albert surveyed his face and saw to his horror Captain Dork's marbled eyes.

He blurted out, "Good Lord."

One brave sea gull gracefully glided in the bright sun, made a majestic landing, and awkwardly started strutting on deck. The extensive kaleidoscoping terns were circling and shrieking, wave after wave at different glide paths, searching, wide eyed for food below. Two more dropped down and glided to the deck. Three others perched together on the huge rim of the stack and watched below.

Tom was needlessly worried about the safety of Norma with the two liberated crewmen. But he was nervous as you never know about the actions of some of the human race. Albert had yet to signal; he had to act. He was ineffectual and. . . He hightailed it down the ladder and entered the lower helm, hearing subdued laughter in the lounge.

She saw him and excitedly asked, "Any sign of Albert or Laura?"

He took over the empty helmsman chair and replied, "No, not a damn thing, and I'm getting worried."

Norma said loudly, "Billy -- Burlen, this is my husband, Tom." The seamen nodded, and each gave a sincere finger forehead salute. Billy groaned, he touched the caked blood of the split portion of his forehead.

Tom spoke up, "Glad to meet you, how are your injuries coming along?" They nodded.
Tom added, "How many more shipmates are there?"

Billy spoke up a bit tartly, "Dead or alive?"

Tom slowly said, "I'm sorry, I didn't mean it to come out so cold, or blatant.

Norma intervened, "Tom, they believe the captain is the only one left. They don't know about Larua, but scuttlebutt had it there was a woman on board." She was smiling broadly. Tom asked, "How was your captain to serve under?"

Burlen said, "Seen better, seen worse, but now he's got the rabies!"

Billy spoke out, "He flicked out -- went mad -- but not brutal. He still believes we're sailing to our new port of call for cargo."

Burlen remarked, "Yeah, he ordered us to change the name plates after we were wrecked."

Tom uttered a "Jesus." Billy said, "Yeah, he thinks Number Two is at the wheel."

She asked somberly, "Where is Number Two?"

Billy said, sullenly, "He's dead with Orson and Miller. . . . There are two missing to my count --the cook, Pevee, and Dassat." Billy said subjectively, "I'd like to go back and get some goods out of my locker. It isn't much, but it's all I've got!" Billy stated, "I can't say enough about you folks saving our skins. Have you a bowl of beans or a sandwich. We had an early breakfast, and I need a little in there for my ulcer."

Tom said to Norma, "You want to get those Spanish surprises--left overs? They were wrapped in foil."

She replied, "You bet, Tom."

Tom asked, "What do you men want to drink?"

They looked at each other and smiled, and then Billy spoke up, "Norma knows!" She came back with the quartered sandwiches and placed them in front of their guests. She smiled as they devoured them quickly. She said, "What does Norma know?"

Tom said, "I asked them what they wanted to drink."

They met her smile with their's, and she said, "Okay, guys, one more, then the bar is closed." They congratulated each other with their bright eyes. Tom increased the short distance from the hulk so he could see the ship's main deck rail easier.

Captain Dork commanded, "Have you completed the orders on changing the name plates?"

Albert replied, "Your orders have been carried out."

Captain Dork spoke out, "Have Billy report to Number Two, at six bells, for the wheel. That'll be all."

Albert asked, "What about lunch for the woman passenger?" Captain Dork replied, "Have the cook send two plates to my cabin."

Albert started to leave and heard another order, "Ask Number Two how long will it be until we're out of this storm?" Albert took one last look about and saw nothing that was significant. He was intent on finding Laura. He turned to go, saw a small door. He strolled over and peeked inside. It was a small and empty head. He hurried out with a confused look.

Dassat was behind the door that opened into the drab unlit cubicle. He was observing the movements of Captain Dork, who was hopelessly trying to get into his chair, then back to his desk. Dassat thought about how pathetic it was to watch the obvious, but what a great opportunity. Dassat had witnessed a strange man being accepted as one of the crew.

He thrust himself out of the head and shouted, "Captain Dork, inspection of the changed name can be viewed only on the starboard gang stairs!" Captain Dork gave up trying to capture his illusive feisty chair.

Dassat said, "Follow me, Captain." Dork and his frown went out into the gleaming sun. They approached the steel safety gate. Dassat unhooked it, and pulled it open for him. Captain Dork gave him a faint bow. He briskly stepped through and unpretentiously, let out a crying -- long wail, as his frown was cartwheeling while falling. His black captain's coat made Dassat think of a tumbling penguin. His scream ended as he was instantly killed while slamming into the rocks and seas. He died, as he was born, with his frozen scowl, but a captain's coat covered his thin frame, his scowl remained intact. Dassat rehooked the safety gate and boldly strutted to see the contents of the lower, left hand drawer. He pulled it out. All that was in it, was a small dark bottle, a handful of straws, and two pair of latex gloves. His anger rose and he kicked the drawer shut and screamed, so frustrated, that he had forgotten the shoes, laced around his neck, and he used the wrong foot to take out his frustration. The swollen toe had taken the blunt of the clashing kick. He one legged it, pulling out drawers, frantically tossing papers about indiscriminately, disgusted and hurting. He decided to pee. He went into the captain's head. As he urinated, he grinned at the thought that he should have pissed on the captain's desk. He looked about, there was a low, small cabinet in the corner. He opened up one of its little doors, and saw a crude safe. This cupboard of sorts, was just a guise for his catch. His toe was smarting, but he dropped to his knees, and grasped the old worn metal lever and gave it a husky twist!

Albert had searched frantically, all of the crew's quarters, and ship stores. He raced down the metal stairs to the cargo holes. He had to stop before the second landing because of the miniature sea beneath him. The staircase emitted sufficient light for his clear observation. The surfaces were littered with anything that floats--alive or dead. There was a monumental amount of trash bobbing and moving in the internal sea. Two trapped bobolinks were frantically diving and circling their wings, feeling the strain of extended flight, both searching for a spot to perch. Their spewing chirps responded in echoes, over the splashing, congested waters in the steel coliseum. Albert's eyes scanned the unholy sanctum, and thought what a bastardly evil place this was, entering out of the beautiful sunlight, into this cockeyed, gloom deeps of hell. The large moving shadows caught his straining eyes. He focused on a large bobbing wood crate floating erratically, as its hitch-hiker was frustratingly scrambling from end to the other, its tail swinging in defiance, its pink eyes glassy, back and forth, back and forth. Albert found himself dazzled with the relentless scurrying, recapturing the path, knowing the fight of survival, may or not be acted upon. The frightened bobolinks also were aware of their danger, trapped in a large cage, in reality to them, with just water below. Both creatures of land and air were terrified by their new and unwanted environment.

Albert was crouching, and as he stood up, he meditated that if he had a second wish in his life, he would ask that Laura not be part of this devil's brew. He turned quickly, and ran up the stairs, the echo of steel laced the steps behind him. The bright sunlight blinded him temporarily, and he sucked in the fresh air. The most ridiculous thought rushed into his disrupted mind. He mentally, could not, if his life depended on it, remember what kind of earrings Laura was wearing when she disappeared. Was it some kind of sign?

He went back to the ship stores and grabbed a stack of blankets, and grudgingly covered up the three corpses. He moved to the opposite port, railing where the boatman's chair lay. He waved, and Norma saw his gesture. She shouted, "Tom," pointing. He got out of the helmsman's chair and went out to the aft, and scurried up the ladder to the flying bridge, and hit the button to the dual trumpet horn. Tom's eyes turned to horror when Albert crossed his arms twice--and then gestured with a come to me sign, then pointed down. Tom hit the horn button once.   Norma was on the ladder, her flowing hair just a head above the deck of the flying bridge.

Tom looked down compassionately and said, "I'm afraid Laura isn't on the wreck--Cap signaled--with cross arms--twice, and he's going to lower himself down the side. Give him a hand, Hon. Have Billy help you too. I'm staying here-- I hope to do a better job!" She circled her index finger with a come here, he did, and dropped to his knees. They kissed. She remarked, "you did just fine, blue eyes," She went down the ladder. He heard her call Billy over the din of the grating and shrieking birds. Albert was halfway down, and Tom was still inching the *Magnetic* as close as possible to the wreck's shifting hull. The waves were bashing against the wreck, the current was moving approximately 30° adjacent to the cruiser. He had to feather the twin diesels and rev at two different speeds. The metallic reverberations of gnawing, and scraping continued. The clamoring seemed hideous. He was doing his best, the mind boggling battering was contemptuous, on the subject of Laura. Just what in hell he was going to say when he looked into Albert's eyes, Jesus . . ."

Billy and Norma grabbed the chair, and Albert dropped on the deck, unstrapped, and got out of the safety rig.

He said, with a lackluster voice, "Leave the chair dangling over the side--just in case we might want to board her again, or someone wants. . . ." He dropped to his knees, his hands flew to his face. He completely broke down, the heavy, low frustrated sobs attested to his torture. In between, in desperation, he blurted out words in pieces, "She's gone."

Billy knuckled his eye, and got the hell out of there. Norma went down on her knees in front of Albert, and reached gently, and cradled his head to her breast. She rocked and patted, and shared his grief, her eyes full of tears, she looked up. Tom eased the throttles and walked over to the ladder and looked down. He witnessed his wife's tear stained eyes, looking up to him, for the desperation and pain wrenching scene with the two of them in lock, was an unpretentious release of wrath of a divine chastisement. The sun rays picked up the tears in her eyes, and flashed a garish radiance.

Tom couldn't see the sensual release. Tears had welled up in his eyes. He made a pass and grabbed the mike, and stonily pressed the transmit button and called the Coast Guard. Billy and Burlen, still sitting in the lounge, heard on the speaker, Tom's revealing of the ship wreck, name, time, and there are D.O.A.'s number unknown. They had two survivors on board. Their boat's name, and location. He also added, missing was a Laura Dryfuss, presumed to have been aboard the ship. That Albert Dryfuss was Captain, and his crew of the *Magnetic* would remain at the scene until assistance arrived. . . God, please hurry!"

The two crewmen gestured with their eyes to one another that this was the beginning of the logical fate for their lost shipmates--a decent burial. They were the only ones, to their knowledge, who had escaped death, by disobeying orders to run and hold fast, except the Captain, of course, who gave them the orders. Albert and Norma finally cried themselves out. They mutually helped each other to a standing position. They both handkerchiefed, not wanting to look into each other's naked eyes. She then told Albert, the crewmen had heard through scuttlebutt, that there was a white woman on board, but that they did not see her.

They started to enter the lower helm and Albert stopped. Their reddened eyes met, and he spoke slowly in a sacramental voice, "I felt, and smelled Laura, in one of the small cabins. It was just instinct."

She gave him a needed squeeze, and softly said, "I'm not wired like Tom, but I God damn well know that Laura is still alive, and the four of us will soon be a quadrant." She kissed him on the lips tenderly, to his questionable stare. She said, "Laura told me to kiss you!" They moved toward the lounge.

The two survivors were nervous and shifting their weights as they sat on the settee. Billy spoke out to Albert, "Did you meet Captain Dork?"

Albert replied slowly, "Yes, I did, and you're right, he has lost his mind. He thought I was one of you, and gave me orders to pass on, it was pathetic! I left him in his cabin."

Billy asked, "Did you see any of our shipmates?"

Albert didn't want to talk, but that question had to be answered. He replied, appallingly, "I saw three of your shipmates and covered them up with blankets, except for the Captain. I didn't see anyone else." He turned and hurried in the direction of his cabin. He broke down again, when he said, "Not seeing anyone else." He felt the razor sharp blade, slice in his heart. She lightly grabbed his hand, he gracefully declined by moving his hand and kept on going.

She uttered, "Jesus," and skewed her reddened eyes again. She felt so God damn shitty, about lusts, trashing happiness, with a contemptible slap, and worse.

Tom made his way to the bow port, and released the anchor, that would keep them a safe distance from the groaning wreck. Tom looked at the hulk, the bright sunlight and the scores of sea fowl now perched sedately on the rails, represented it as divine perception. Tom hoped they would send a chopper in first. He cut the engines, joined his wife and two happy seamen in the lounge.

As he came in he heard Norma say, "The bar's open, how about one more for . . . . How about it?

"You betcha" and a "damned right" were the seamen's replies.

She asked Tom, "You could stand one, too?"

Albert wants to be by himself. Tom said, "Please mix one for me, too."

She set about mixing up four strong, straight, neat drinks.

It opened, Dassat's heart was thumping, as his hands were as busy as those of a gopher, pulling and discarding the cluttered contents. His finger pads caught the little yellow ties to a purple small silk bag. His fingers could not keep up with this excited mind. He finally undid the knot, and tapped out a plastic, covered package with two rubber bands, and a fancy laced, female garter around it. The covered pouch was just a little bigger than could be stuffed in one's mouth. He plucked at the rubber band with such anxiety. He ruptured the rubber, and it stung him. He shucked it off the double tied garter, and opened the plastic cover. A partly thick, curled, large rack of one hundred dollar bills filled his eager eyes. He put the roll back in the pretty bag, except for the two rubber bands that broke, and stuffed it in his pocket. His shoes got in the way of his busy arms. He grinned and chastised one of them, mockingly. Lust was beaming in his eyes. He thought, he might as well scavenger the crew's quarters. He thought he was really doing them a favor to keep among friends rather that some son of a bitch of a landlubber. He ravaged their goods, and stuffed them in a pillow case. He felt it was time to debark this wreck. Dassat was as high as a kite, and felt the power of his every move. He thought survival tactics had to be enacted upon -- food. He headed for the galley, his shoes bouncing around his shoulders. He stopped suddenly, cocked his head, and grinned. He felt like God and retraced his steps backwards to the crew's quarters. He kept low on the main deck so as not to be seen from the cruiser. He reopened his locker, and plucked out two packages of brown shoe laces. He jumped on his bunk, then wished he hadn't. The toe of his shoe hit his nose. He took them from around his neck, and untied them, sitting cross-legged, he finger nailed the paper bands that each held 28 inches of made in China laces, and began lengthening the length of his shoe laces. Dassat double knotted for strength, and tested each knot. He tethered it, twice, around his neck, the shoes were dangling at his waist, just right. He depanted another pillow, and clutching the bag, went to the galley. He snatched food, and tossed it into the second pillow case haphazardly. He thought about one of this favorite meats--summer sausage, and flipped the dial clasp latch to walk into the cooler. He pulled on the worn metal handle, it wouldn't budget, it was jammed. He wanted his sausage all the more, and set about to find a pry. A round metal, once chromed bar, encircled the huge commercial stove as a safety decoration. A misused meat chopping block had a rack with three cleavers, randomly placed. He snatched up the largest, heaviest of the butcher's implements. He smashed the cast, chromeless pipe; it popped out and banged on the galley floor. He thrust the bar against the metal handle and pulled grudgingly. The heavy, four hinged, insulated door swung open. He dropped his pry, and stepped through the small opening. He screamed and cursed as he struck a

soft bulky, heavy swinging weight. It struck him again. He cried out, "Son of a bitch."
He stepped back and shoved the huge door wide open. The galley had two dirty portholes. He saw what had attacked him. He looked in awe, cocking his head in wonder. He winced as he captured the ballooning, bugged out eyes of the cook. The non-blinking pupils were staring at the ceiling. His body was slowly turning, as was the half of beef, sharing the same hook. Blood from his gaffed neck was down his full apron, and pools had solidified, forming a crusty blanket below him and the steer. Dassat perceived the weight of hundreds of pounds of the half steer, must have served as a stabilizing deterrent, with the half exposed razor sharp hook. The cook must have been in here just before the ship rammed. That sudden impact had hurled him on the sharp gaff, and gutted his neck. He looked up again, and thought of the odds, and possibilities of that happening.

Dassat silhouetted this hooking with that of Captain Dork, stepping out on an illusion that his gang plank steps were still in place. He looked down at his red toe, that he too had a flirtation with a hook, the sharp barb, that dug through his toe. Then he thought the odds of his toe, finding that hook in that devil storm. He snatched a long skinned sausage and left the cooler, to the meat block. He captured a smaller, razor sharp cleaver, and raised it high, and severed the sausage roll. He put half of it, in his pillow case. As an after thought, he sliced himself a thin piece, and crammed it into his waiting mouth. He chewed with satisfaction. Now. . . it was time to leave this plucked goose. He moved to the main deck, keeping low, and stepping carefully, not to further irritate his swollen toe. Three gulls took flight from the aft as he approached his escape route. He moved to where the sea had claimed and was splashing water and foam over the fantail. He stuck out his good toe, in the water. It was cold. The droning of the clashing metal was less prominent, but the mass of the wreck, still vibrated when the waves bashed against the outside of the hull, and the miniature sea inside, further demonstrated its power. About 60 feet to evade, the slick rock pinnacles that could break his bones, and then be free, and would escape this steel coffin. He tied the throat of his two bags tighter, and stepped into the overwash from the fan tail. Through the frothy waters, he could barely make out the steel rail, and the end of the deck under water. The sun streamed with a bright light, and the sea fowl were circling, gliding and shrieking their long cries. He lunged forward, clearing the mired derelict. His shoes were floating, then quickly succumbed with water as he kicked and stroked coordinately. The thrust forward was initially stronger than the two drowned shoes, their weight like anchors biting deeply around his thin neck. He tried to dig his fingers under the choking shoe strings that were starving his lungs for air. He flayed backwards and surged up was to capture needed air. He managed to slip the fingers of his left hand between his neck and the ties, and switched from overhand swimming to a side stroke. He pulled with his right arm and hand, and kicked like a frog rhythmically. His main propulsion was his legs and feet. He cocked his legs, then drove them intensely to their extended lengths. The water filled shoes would not conform to this interaction of his gliding motion. Having to swim side stroke was actually swimming on your side, limiting your vision in half. His left fingers, still under the ties, protected his throat. His two looted bags he had dropped when fighting for air.

The wave crests were heavier, the closer he was getting toward shore. He wished he hadn't double lassoed the strings around his neck. His deadweight shoes were a bit of a nemesis. The crests did slow him down, but the long swell, was encouraging, it was moving him closer to terra firma. The foaming splashing white water was cold. Dassat recoiled his feet and thrust outward, his bare feet had somehow become entangled in something. His left hand was busy protecting his throat. His right dug into the water and his head popped out, in the long low swell, and he gulped a precious mouthful of air. He doubled up, and dove to see what had snagged his feet. Panic welled up in his brazen eyes. A horrific sight revealed the intrusive battered face of Captain Dork with a smile! Dork's lax, beaten body, was really upside down. The logistics of Dork being upside down had not validated the position of his body. Dassat's intense horror was only perceived in his mind of that incredible satanic smile, a devil omen. He had inadvertently kicked his bare feet during a coiled surge, and his feet entered deeply into the buttoned up captain's coat. He thrashed under the water like a fish on a well set hook. Dassat's ambient body, redundantly, was reduced to meaningless flaying as his shoes whirled around him in the under current and tugged mercilessly against his water swallow throat. Finally gurgling bubbles escaped from his lips. Captain Dork, and Dassat, were for a short time, one entity, locked together, freely moving with the rhythmic rhythm of the undertow.

The emergency channel had been set so all could easily hear the transmissions in any area of the *Magnetic*. The VHF emergency frequency was filled with constant chatter, after the Coast Guard related they were sending a chopper, a cruiser, and a follow-up standby boat with doctors and nurses, if needed. The channel was alive with the possibilities of two ships on the rocks at Trinity Islands. Casualties known, but not how many, survivors unknown. Ship's names as follows, the *Scarab* and the *Baracs*. Details forthcoming. Emergency -- any craft in that area respond -- repeat.

Devlen Dawson heard that startling news on his radio and fell to the floor. He was taken to the emergency room, and it was indicated he had a mild stroke. Reported to be in stable to good condition.

Tom, Norma, Burlen and Billy sat in the lounge. They were nursing their stout drinks silently. Albert was still in his cabin, sitting on the bed, and holding his distraught face. They heard, "Click, Click, Teddy Bear, this is Laura!"